Integrated Resource and Environmental Management

The Human Dimension

Dedication

This book is dedicated to the natural resource manager. For it is this person, above all others, who is faced with the reality of multiple demands, finite limits and a constant need to balance the present with the omnipresence of tradition and the unknown spectre of the future.

INTEGRATED RESOURCE AND ENVIRONMENTAL MANAGEMENT

The Human Dimension

Alan W. Ewert
Indiana University
USA

Douglas C. Baker
Queensland University of Technology
Australia

Glyn C. Bissix
Acadia University
Nova Scotia
Canada

CABI Publishing

CABI Publishing is a division of CAB International

CABI Publishing
CAB International
Wallingford
Oxfordshire OX10 8DE
UK

Tel: +44 (0)1491 832111
Fax: +44 (0)1491 833508
E-mail: cabi@cabi.org
Website: www.cabi-publishing.org

CABI Publishing
875 Massachusetts Avenue
7th Floor
Cambridge, MA 02139
USA

Tel: +1 617 395 4056
Fax: +1 617 354 6875
E-mail: cabi-nao@cabi.org

A catalogue record for this book is available from the British Library, London, UK.

Library of Congress Cataloging-in-Publication Data

Ewert, Alan W., 1949–
Integrated resource and environmental management / by Alan W. Ewert, Douglas C. Baker, Glyn C. Bissix.
p. cm.
Includes bibliographical references and index.
ISBN 0-85199-834-8 (alk. paper)
1. Environmental management. 2. Natural resources--Management. 3. Environmental policy.
I. Baker, Douglas C. II. Bissix, Glyn C. III. Title.

GE300.E94 2004
333.7--dc22 2004004003

ISBN 0 85199 834 8

Typeset by AMA DataSet Ltd, UK.
Printed and bound in the UK by Biddles Ltd, King's Lynn.

Contents

Preface

What constitutes effective natural resource management? From a historical perspective, many of our ancestors believed that the 'best' natural environment was one that provided physical and communal security. Later, much of the public felt that managing our natural resources embodied a sense of husbandry, or protection, where various natural environments are viewed primarily from the 'goods and services' they provided. The goal of this perspective is to conserve these resources for the long term, in order to allow access to the various products and commodities these environments offer, such as timber, water, forage and recreation opportunities.

Because of the economic 'value' associated with each separate resource, various conservation organizations, academic programmes and governmental agencies emerged with the tasks of protecting and facilitating the use of these various natural resource bases or what Samuel Hays (1969) calls the 'Gospel of Efficiency'. For example, the US Department of Agriculture (USDA) Forest Service became the overseer of the country's timber resources, and the US Department of the Interior Fish and Wildlife Service and Park Service became the guardians of wildlife and parks, respectively. The USDA Soil Conservation Service took a leadership role in safeguarding the nation's soil resource, Fisheries and Oceans Canada manages the fisheries of Canada, and so forth. In each case, the resource became both the focus and the main reason for the development of substantial infrastructures consisting of protecting agencies, watchdog groups, concerned citizens, sectors of industry, and groups of people who used that particular resource. Natural resource management, under this perspective, consists of a collection of entities, often in competition with one another, that are concerned primarily with protecting and promoting their own individual natural resource.

Most recently, a perspective has emerged that views natural resources as an integrated whole rather than a collection of individual components. For example, within this perspective, forests become more than reservoirs of timber or wildlife populations, but instead constitute a dynamic and interconnected system of biological and physical entities (soils, plant systems, water holding and transport, nutrient cycles, etc.). In addition, these various systems are increasingly impacted upon by human actions. According to this perspective, effectively managing our natural resources involves considering the physical characteristics of the resource, how humans interact with these resources, and in what ways the management actions imposed on one resource may influence other resources and users.

Integrated Resource and Environmental Management is written with an underlying philosophy and belief that our natural resources should not be viewed as independent and isolated entities managed strictly for the products they provide. Rather, natural resource management should encompass an integrated approach, where our citizens' needs, wants, demands and expectations are considered within the broader framework of *all* of our natural resources. In its most generalized form, we define integrated resource and environmental management (IREM) as both a management *process* and a *philosophy*, that takes into account the many values associated with the natural resources within a particular area. IREM also considers the various linkages between these resources and other systems, and views these resources from a long-term, sustainable perspective.

Toward this end, we have written this book with the intention of it being used as a textbook for students, primarily at the undergraduate level, in the many academic areas pertaining to natural resource management. It may also be a useful resource and/or reference tool for practitioners, researchers and managers currently involved in the field of natural resource management. In addition, because natural resource management transcends political and geographical boundaries, we have attempted to address IREM from a global perspective and not limit our discussion entirely to the USA or North America.

Integrated Resource and Environmental Management consists of three sections. Section 1 presents an overview and history of natural resource management, from both the North American and international perspectives. Section 2 discusses the challenges facing IREM by examining issues such as conflict, property rights, the role of social science in natural resource management, the influence and formation of power in a decision-making context, and the theoretical foundations of IREM. Section 3 addresses the definition and application of IREM from several different contexts, including real-world applications and planning frameworks.

Individual case studies are integrated into the end of each chapter and serve two major purposes. First, the case studies provide an important link between the theoretical concepts presented in the chapter and an actual 'real life' example. As such, each case study has been selected for its ability to highlight the important points produced in each chapter. Secondly, the case study can be a discussion unto itself. Each case study has its own unique set of circumstances and variables for consideration. Whether used as corroborating or support material for the individual chapter or as a discussion point by itself, the case studies serve to illustrate various models associated with IREM that demonstrate the advantages and challenges of using an IREM approach. In addition, discussion questions are included at the end of each chapter to stimulate dialogue underlying many of the issues presented in the text. These questions are designed specifically to provide an example of the complexity and vexing nature of many of the issues facing natural resource management.

The premise underlying this text is simple – we believe that in most situations, employing an integrated resource management perspective is preferable, and ultimately more effective, for the long-term management of our resource base rather than the more traditional perspective of simply considering one or two resources to be of primary importance. We do not believe, however, that IREM is always the answer, or even appropriate in all situations. It is for the readers of this book to make that decision. We advocate, however, the belief that IREM presents a guiding model of both philosophy and practice that is more aligned with most current visions of how resource management will be 'done' in the future. As such, IREM presents the promise of looking ahead rather than simply looking to the past.

Alan W. Ewert
Douglas C. Baker
Glyn C. Bissix

About the Case Study Contributors

Jane Adams is Associate Professor of Anthropology and of History at Southern Illinois University at Carbondale. Her books include *The Transformation of Rural Life: Southern Illinois 1890–1990* (University of North Carolina Press) and *Power and Politics in the Transformation of Rural America* (University of Pennsylvania Press). She is currently working with digital video and web authoring, with research focused on the social transformation of the lower Mississippi Delta.

Anna J.L. Carr is a Research Fellow with the Department of Psychology and Centre for Environmental Strategy, University of Surrey. She has an Honours degree in Communication Studies from Murdoch University, a Master's degree in Economic Development from the University of Waterloo in Canada, and a PhD in Resource and Environmental Studies from the Australian National University. Anna's research has concentrated on both theoretical and applied environmental stewardship in a rural Australian context.

Trevor Davies is a retired production engineer living in Bristol, UK. He received a Higher National Certificate in Mechanical Engineering with an Endorsement in Industrial Administration from the Bristol Technical College and a further Endorsement in the Theory of Structures from Bath Technical College. He is currently a Conservation Volunteer with the Avon Wildlife Trust with its head office in Bristol and works specifically on the conservation of Folly Farm in Somerset, UK.

David F. Duke is Assistant Professor of History at Acadia University, Nova Scotia, Canada. He received his BA in 1991 and PhD in 1999, both from the University of Alberta. His doctoral dissertation is on the subject of the environmental history of the USSR. In addition to his present position at Acadia University, Dr Duke has taught at the University of Alberta and at Minsk State Linguistic University, Minsk, Belarus. His current projects include a comparative study of the environmental policies of the USA and

Soviet Union during the Cold War. He is editor of the forthcoming *Russia and Eurasia Documents Annual*, volumes for 2000 and 2001, published by Academic International Press.

Leslie Duram is an Associate Professor of Geography at Southern Illinois University at Carbondale. Her research interests include: local participation in environmental management, land use change and alternative agriculture.

Ross Firth currently works as a Senior Ranger with the Loch Lomond and the Trossachs National Park in Scotland. He has also worked as a Park Ranger with the Alberta Provincial Park System and the US National Park Service as well as Operations Supervisor with a large government appointed private campground operator. He earned a Bachelor of Arts in Recreation and Physical Education from Acadia University, a masters of Applied Environmental Studies from the University of Waterloo and was qualified as an Emergency Medical Technician – Ambulance (EMT-A) by the Southern Alberta Institute of Technology. His leisure interests include cycling, hill walking and travelling.

Winifred Kessler is the Director of Wildlife, Fisheries, Ecology and Watershed for the USDA Forest Service, Alaska Region. Previous Forest Service assignments included National Wildlife Ecologist (1986–1990) and Principal Rangeland Ecologist (1992–1993) on the Washington, DC, staff. From 1993 until 2000, she chaired the Forestry Program at the University of Northern British Columbia. An active volunteer, she currently chairs the Public Advisory Board for British Columbia's Habitat Conservation Trust Fund, and is serving as the Northwest Section Representative of the Wildlife Society. Dr Kessler's education includes BA and MSc degrees from the University of California at Berkeley, and a PhD from Texas A&M University.

Steven E. Kraft is Professor and Chair of the Department of Agribusiness Economics and Co-Director of the Environmental Resources and Policy PhD program at Southern Illinois University, Carbondale. He teaches courses in farm management, natural resource and environmental economics, and social perspectives on environmental problems. He has been conducting funded research on soil and water policy since the late 1970s and has published numerous articles on the topic.

Christopher Lant is Professor and Chair of the Geography Department at Southern Illinois University at Carbondale and serves as Executive Director of the Universities Council on Water Resources. From 1994 to 2001, he served as editor of the *Journal of the American Water Resources Association*. His research and teaching interests are in water resources and environmental policy analysis.

Tim Loftus is the Director of the Water Quality Laboratory (WQL) at Heidelberg College in Tiffin, Ohio. Through research and extension, the WQL focuses on the impacts of land use and other activities on the aquatic resources of Ohio, the Midwest and Lake Erie.

Gary W. Ness is the Director of the School of Recreation Management and Kinesiology at Acadia University in Nova Scotia, Canada. An exercise physiologist by training, he has long been interested in the promotion of physical fitness and the use of informal areas for recreation and physical activity. For many years, Gary has served on town and school board committees, including several terms as chair of the Wolfville Recreation Commission, the Wolfville Open Space Committee and the Wolfville Community Services Planning Advisory Committee. His passion is photography and railway history; these interests have been combined in the publication of two books on railways in Canada.

Chris Rodri is presently a forest management consultant and Director of the Maya Forest Enterprises. After service with the British Forces, Chris entered academia, earning a BA specializing in Environmental Sciences at the Open University, and gained a Master of Science degree at the University of Edinburgh in Natural Resources Management. He worked for 10 years on environmental issues focusing on natural resource management and biodiversity conservation. Until recently, and for 6 years, Chris managed a tropical research station in Latin America.

J.B. Ruhl is Joseph Story Professor of Law at The Florida State University in Tallahassee, Florida. He teaches and researches in the fields of endangered species protection, ecosystem management and related topics. He is co-author of the recently published law school casebook, *The Law of Biodiversity and Ecosystem Management* (Foundation Press, 2002).

Megan Squires is a recreation and environmental planner. She earned a Bachelor's Degree with honours in Recreation Management from Acadia University, Nova Scotia, Canada, and a Master's Degree in Environmental Studies (Planning) from the University of Waterloo, Ontario, Canada. Much of Megan's research has focused on the recreational impacts in Kananaskis Country and Banff National Park. As a planner, Megan has worked on a number of recreation and environmental planning assignments including projects with the Alberta Parks and Protected Areas Initiative and with Banff National Park.

Lynne M. Westphal, PhD, is a Research Social Scientist with the USDA Forest Service, North Central Research Station, in Illinois. As well as conducting her own research, she manages the Station's Riparian Integrated Research Program. She investigates how trees, open space and other natural resources can improve quality of life in cities and towns. Recent projects include developing a research programme into the simultaneous ecological and economic revitalization of rustbelt landscapes; assessments of empowerment outcomes from urban greening projects; and the public's perceptions, uses and development ideas for the Chicago River.

Acknowledgements

The authors gratefully acknowledge the immense help they received in the writing of this book. In particular, we would like to gratefully thank Dr Alison Voight for her technical reading, editing, organizing and countless suggestions toward the improvement of this text. Alison, please accept our heartfelt thanks. Finally, we would like to acknowledge and thank the authors who have graciously contributed the case studies for use in this text.

1 Introduction

> The nation behaves well if it treats the natural resources as assets which it must turn over to the next generation increased, and not impaired, in value.
>
> (Theodore Roosevelt)

When Theodore Roosevelt made this statement, the nation was immersed in numerous controversies surrounding how the country's natural resources were to be used and managed. These early 20th century controversies, such as what are the most appropriate uses of our natural resources and who should decide how these resources will be used, still perplex society both in North America and elsewhere. Why are these issues so difficult to resolve? Is there something unique about natural resource issues that precludes finding solutions that can be agreed upon by most segments of society?

This chapter discusses why natural resource issues can be so complex and why it is difficult to find widely accepted solutions. It includes a sampling of troublesome issues that resource managers, the public and scientists are currently facing, along with reasons why these topics have become so difficult to resolve. In addition to the issues of complexity, Chapter 1 discusses the various desired futures among the assorted natural resource stakeholders, and the challenge in providing effective and accurate assessment and monitoring systems. Natural resource management concerns, both current and emerging, are varied, numerous, growing in number and becoming particularly burdensome in terms of finding widely accepted solutions. This is due, in part, to the fact that they are often extremely complicated, in both the underlying science and political areas, and, consequently, they present challenges regarding how the various components surrounding an issue can be assessed and/or monitored (e.g. public health or economic versus non-economic outcomes).

Why the Difficulty

A number of factors may explain why natural resource issues often appear insolvable. Controversies revolving around natural resource use often present a unique combination of factors and circumstances rendering any solution especially complex. A sampling of some of these contentious issues is listed in Fig. 1.1.

Several things become apparent from the items listed in Fig. 1.1. First, while admittedly this is only a partial sample, the list continues to grow in number rather than diminishing. Secondly, the wide range of issues reflects the multi-dimensional aspects of resource use and appreciation. Thirdly, these issues often represent a difference of opinion in not only *what* decisions should be made regarding the management of our natural resources, but also in *how* these decisions should be made and *who* should make them. Fourthly, some of these issues are particularly burdensome to resolve for the following reasons:

- The issues cannot always be resolved satisfactorily by turning to science for the answers, i.e. the public may not know, care or believe what the underlying science doctrine says about the issue. (Examples: clear-cutting; reintroduction of certain species, such as wolves or grizzly bears.)
- The issues typically represent long-time neglect and/or abuse. (Examples: depletion of the ozone layer; groundwater depletion.)
- The issues 'cross-cut' political and scientific boundaries. They often involve a number of different countries and political jurisdictions in addition to necessitating a variety of scientific input in order to develop a better understanding of the issue. (Examples: depletion of fisheries; hunting regulations for migrating waterfowl.)
- Any solution to these issues will often be expensive and require a long-term commitment. (Examples: water pollution; elevated atmospheric levels of CO_2.)
- Some damage is irreversible. (Examples: loss of habitat; extinction of species.)

On a broader scale, the rapid advances in technology and information science, the demographic changes of growth, ageing and cultural shifts, the increasing global economy, as well as global environmental changes, have also made resolution difficult to achieve in natural resource management. Moreover,

Cumulative watershed effects
Key indicator species
Biological diversity
Wilderness allocation
Job stability
Atmospheric CO_2 levels
Fisheries depletion
Timber harvesting techniques
Waste containment
Protection or harvesting of old-growth forest
Long-term productivity
Fuels management
Economic development
Road building
Visual resource impacts
Resource-based tourism
Development of open space
Access rights

Fig. 1.1. Sample of natural resource-related contentious issues.

these issues represent factors that contribute to the difficulty in finding long-term and widely acceptable solutions. Specifically, these factors include **complexity**, **desired futures**, and the ability to **assess and monitor** environmental conditions.

Complexity

Natural resource management issues are often difficult to resolve because they involve highly complex systems. These systems contain numerous and interrelated parts, patterns or components which necessitate thinking in terms of: (i) connectedness; (ii) relationships; (iii) context; and (iv) aggregation rather than disaggregation (Rowley *et al.*, 1997). Examples of *environmental systems* include the atmosphere, biosphere, geosphere and hydrosphere. At the same time, a number of *human systems*, such as community, economics, cultures and politics, are interacting with each other as well as with many of the environmental systems, to form a matrix of interacting events, impacts and demands.

These complex systems also involve multiple perspectives and multiple truths. For example, Kimmins (1995) suggests that both experience and scientific knowledge indicate that clear-cutting, as a timber harvesting technique, is neither automatically all good nor all bad. Factors that should be considered in any overall assessment include soil types and fertility, climate, clear-cut type (e.g. location and shape) and, perhaps most importantly, how the harvest site is maintained (e.g. properly constructed roads, reseeding process, etc.).

Other factors which increase complexity are **spatial** and **temporal scales**, and the **magnitude of effect**. From a *spatial* perspective, scales can be global, continental, regional, landscape or local. *Temporal scales* include short-term, long-term and epochal time frames. For example, a resource issue, such as soil loss, can occur in one setting very rapidly, while in another it occurs at a slower but accumulative rate, often over a number of generations. An epochal scale is represented by certain geological scales, such as the Pleistocene and Miocene Epochs.

Magnitude of effect implies cumulative or systematic change. Cumulative changes imply a growth of localized events. Orians (1995) suggests that some events related to cumulative change include loss of biodiversity, deforestation, habitat fragmentation, unsustainable levels of groundwater discharge and introduction of alien or exotic species. Obviously, cumulative threats can be more difficult to ascertain and 'fix' than individual, localized events. Systematic changes occur everywhere, and are even more difficult to control because the impacts are occurring from many different points. Examples of systematic effects are ocean/air pollution, ozone depletion and increases in atmospheric CO_2.

Complexity can also be reflected in unanticipated environmental events. Unanticipated events often occur in a situation where causes of the event are poorly understood, behaviours are unexpected, and action produces unforeseen outcomes (Holling, 1986). Kates and Clark (1996) suggest that unanticipated environmental events can fall into four classifications: (i) rare events with serious consequences (e.g. Bhopal, Three Mile Island, Chernobyl and Mount St Helens); (ii) common events that are poorly understood or elude detection/ prevention (e.g. legionnaires' disease and zebra mussels in the Great Lakes); (iii) unexpected consequences (e.g. CFCs and groundwater pollution); and (iv) expected but mistakenly blamed consequences (e.g. stratospheric ozone depletion and forest damage). Whether or not these events are viewed as serious or insignificant is, in part, dependent on the perspective of an individual's desired future. If an impact, such as increased levels of zebra mussels in the Great Lakes system, is *not* perceived as negatively affecting one's life, this natural resource problem will not be regarded as highly important.

Desired Futures

Traditional natural resource management emphasized input/output models of resource use, where humans were viewed as users of the resource but not necessarily a part of the resource. Moreover, Brown and MacLeod (1996) point out that from an ecological perspective, pristine ecosystems were often viewed as a highly desired end-product of society and, as such, management is often focused on attaining these qualities for a particular setting or location. Thus, officially designated wilderness areas became the end-goal of many management agencies, such as the USDA Forest Service and British Columbia Provincial Ministry of Parks in Canada.

Many desired outputs from natural resource management, however, do not lend themselves to a simple outcome or answer. In some cases, desired outcomes go well beyond localized events and include more far-reaching goals such as the following:

- Long-term sustainability
- Species conservation and maintenance of high levels of biodiversity
- Environmental and human health
- Diminished conflicts over resource use
- Maintenance of high quality of life:
  - healthy economy
  - job security
  - stable communities
  - availability of resources
- Environmental equity with balanced resource use
- Sustainable yields and reduced levels of extraction
- Wise use of resources/conservation
- Preservation of wildlands/access for recreation and aesthetic enjoyment

This list represents a partial collection of desired future outcomes in natural resource management which have appeared in the literature. One reason that natural resource management issues are so difficult to resolve is because the desired goals from resource use are often different and may actually compete with one another. As will be discussed in a later chapter, interference with real or perceived goals is one of the driving forces in resource management conflicts. Given the variability of these desired outcomes, how should they be measured or assessed in order to ascertain whether the management plan has been successful?

Assessment and Monitoring

Another reason that natural resource management issues are troublesome to resolve is that they can be difficult to measure (e.g. What is forest health?) as well as the fact that it is difficult to decide *what* should be measured (e.g. What variables should be measured to determine whether a forest is healthy?). For example, in evaluating the sustainability or health of a particular ecosystem, the following questions could be asked:

- Which attributes or qualities of a system are to be sustained?
- Who will decide this?
- How long should these systems be sustained?
- How will this sustainability be recognized?

In the case of the last question, consider the many parameters that could be identified as potential measures for monitoring ecosystem sustainability. Several of these measuring parameters are listed below:

- Level of productivity
- Rate of decomposition
- Nutrient cycling and losses
- Species diversity and/or richness
- Habitat fragmentation
- Minimum viable populations
- Size of usable habitats
- Resistance and resiliency
- Key indicator species
- Selected outputs (e.g. water quality)

Assessing changes related to natural resources often requires accepting that: (i) there are usually deficiencies in the underlying science; (ii) most ecosystems have catastrophic events as a common change mechanism; (iii) there are varying spatial, temporal and size scales; and (iv) there are often a range of social factors that need to be considered. In addition, assessment and monitoring are more than simply determining change, even if the proper measurements can be utilized. Systematic measurement must also be used in developing a better understanding of the resource issue in question, establishing a baseline for comparison and providing information for decision making. However, science and assessments can only observe, analyse and measure change; other societal components, such as managing agencies or concerned citizens, must make the actual *decisions* regarding acceptable levels of change or use for natural resources.

Thus, even if accurate assessment and monitoring can be accomplished, natural resource issues often require decisions regarding use and management based on factors other than science, such as politics, economics, public perceptions and tradition. Reynolds (2001) points out that determining forest ecosystem sustainability is, in part, dependent on the biophysical integrity of the system, in addition to its economic feasibility and social acceptability. These additional factors add to the perplexity when attempting to resolve natural resource issues.

What is Integrated Resource and Environmental Management?

As a reference point for study and discussion of these chapters, the authors define **integrated resource and environmental management (IREM)** as: *a coordinated management process and philosophy, which takes into account the many resources of an area, its linkages to other systems (i.e. communities, politics, environment) and the consideration of long-term, sustainable use*. IREM is an approach that attempts to integrate both biophysical and cultural elements in the landscape, and thereby increases management objectives beyond single or multiple uses. Ideally, IREM embraces a more holistic approach that combines our greater understanding of ecosystems with a wider range of stakeholder participation.

Not surprisingly, from a global perspective, there are many definitions of IREM and applications for solving complex natural resource problems. It is our intention to explore the dimensions of IREM as a management tool. The chapters that follow build on the previously stated definition of IREM and provide a comprehensive foundation for the understanding and application of IREM principles and practices. For example, this book will address questions such as: 'What are IREM principles and how can these be integrated into natural resource management?' 'What is the difference between IREM and other natural resource management methods?' and 'What are the strengths and weaknesses of using IREM?'

Summary and Conclusions

This chapter has discussed the reasons why it is sometimes extremely difficult to bring a resolution to natural resource issues that are perceived as comprehensive and fair, and would be capable of receiving widespread support across differing segments of society. Each culture and community can strongly disagree about, and compete for, different uses of a natural resource. Businesses and developers want to build shopping malls, apartment complexes and new interstate highway systems. Some members of a community want this type of commercial growth, while others of the same community fight to maintain a quiet, undisturbed rural setting. Knowing that these types of problems are difficult to resolve, however, does not fully disclose where we have been and where we are going in natural resource management. An essential question for the present day is: How will we integrate our management efforts in a cogent manner? Also, perhaps more importantly, should we? In the following chapters, the history of natural resource management and the theoretical foundations of integrated resource and environmental management are discussed. In addition, how IREM fits within an agency framework, and the role that social science plays in making resource-based decisions, are also examined.

This book is divided into **three sections** that outline the development and use of IREM. **Case studies** are included at the end of each chapter to provide examples of how IREM can be applied, and also to evaluate some of the strengths and weaknesses of this particular approach to natural resource management. The **first section** of this book covers the historical as well as current foundations of IREM. Chapters 2 and 3 explore the history of natural resource management and its development in North America. The chapters include the early background of resource exploitation, the development of conservation strategies in North America in the early 20th century, and case studies of the application of IREM. Chapter 4 deals with the complexity of applying IREM in a world where decisions regarding resources are influenced by powerful forces that are often outside the control of the resource manager or community. These driving forces are comprised of demographics, trade and markets, public perception of natural resources, environmental injustice, international conventions, personal choice, and indigenous or native peoples.

The **second section** defines the foundations and causes that have helped establish the context for IREM. Chapter 5 outlines environmental conflict and the changing role of decision making in the public domain. Different perceptions of property rights and changing values, with respect to resource development, have forced decision makers to deal with conflict and include a wide range of stakeholders in decisions regarding resource management decisions. Chapter 6 provides an overview of the increasingly important role that the social sciences play in IREM and current trends that facilitate this relationship. Chapter 7 examines power and decision making in natural resource management. Several theoretical models are reviewed that attempt to explain decision-making power, and the dynamics of the public policy process. Chapter 8 outlines the theoretical basis for natural resource management and provides an assessment of the complex and diverse foundations that have led to the evolution of IREM.

The **final section** of the book defines IREM: What are the components of the process? How should it work and how do we apply IREM? The theoretical foundations, research and conceptual frameworks that define IREM are outlined in Chapter 9. Chapter 10 defines the difficulty in applying IREM in a real-world context and introduces an analytical framework from which to evaluate IREM. Chapter 11 provides a planning framework for IREM by integrating the previous two chapters into a model for the application of IREM. The model provides a broad picture for managers regarding the dimensions of this management process as well as some useful methods for application. Chapter 12 outlines a summary of the book, and examines future challenges for the application of IREM.

Discussion Questions

1. What factors often make decisions more difficult regarding how a resource should be used? For example, deciding whether or not to build a dam: Who will benefit? Who will be harmed? What will the impacts be to ecosystems and human communities?
2. Describe the different components that tend to complicate decision making in natural resource management. Provide some real-life examples of complexity in natural resource decisions.
3. Review the local newspaper and discuss some recent, unexpected environmental occurrences. Why were they unexpected? Where would these occurrences fit in the classification schemes presented in this chapter?
4. In a small group, develop a list of individual desired futures regarding a specific natural resource setting or area (e.g. Grand Canyon). Compare these individualized lists with others in your group. How would you reach agreement about those desired futures, and how might they conflict with those desires held by other members of your group?
5. Select a local natural area, and then construct a list of qualities you would measure for the purpose of assessment and monitoring. Why these instead of others? What characteristics would you look for in determining environmental change? How would you decide whether these changes are acceptable or not?

References

Brown, J.R. and MacLeod, N.D. (1996) Integrating ecology into natural resource management policy. *Environmental Management* 20(3), 289–296.

Holling, C.S. (1986) The resilience of terrestrial ecosystems: local surprise and global change. In: Clark, W.C. and Munn, R.E. (eds) *Sustainable Development of the Biosphere*. Cambridge University Press, Cambridge, UK, pp. 292–317.

Kates, R.W. and Clark, W.C. (1996) Expecting the environmental surprise. *Environment* 38(2), 28–34.

Kimmins, H. (1995) Clear-cutting: a long history to controversial practice. *BC Professional Forester* (2), 21–22.

Orians, G.H. (1995) Cumulative threats to the environment. *Environment* 37(7), 6–36.

Reynolds, K.M. (2001) Using a logic framework to assess forest ecosystem sustainability. *Journal of Forestry* 99(6), 26–30.

Rowley, T., Gallopin, G., Waltner-Toews, D. and Raez-Luna, E. (1997) Development and application of an integrated conceptual framework to tropical agroecosystems based on complex systems theories. *Ecosystem Health* 3(3), 154–161.

2 The Early History of Resource Management

Introduction

The history of natural resource management is closely tied with human beings' tireless pursuit to relieve scarcity for its basic metabolic and cultural needs. When *Homo habilis*, the first tool-using ancestors of *Homo sapiens*, emerged out of the sub-Sahara of Africa some 2 million years ago, there was little evidence of any meaningful management of natural resources. Initially, humans foraged, then learned basic hunting techniques by working within, rather than manipulating, natural systems. At this time, human beings were preoccupied with rudimentary food requirements, and basic shelter and security needs. Later, as *H. sapiens* radiated out from Africa, eventually populating Oceania and the Americas, it was unlikely to have been a sense of adventure that drove humans from their homelands but rather a never-ending search for sustenance and safekeeping.

As humans evolved, they learned first to use and then to keep tools for later use. This simple foresight allowed humans, in their various species incarnations, to gather and hunt more effectively and provide time to mature socially and intellectually. Over time, humans developed their unique capacity to think, to foresee, to accumulate and to pass on intellectual wealth to subsequent generations. With this intergenerational transfer of intellect came an ever-increasing capacity to manage natural resources. In so doing, humans developed beyond mere biological existence to live increasingly as cultural beings and, in time, developed science, more sophisticated technology and various art forms. As humans explored and settled widely, chasing the advantages of retreating ice sheets, evolving landscapes and productive natural resources, so the constantly changing challenges of time and place tended to advantage some settlements but stunted others. Consequently, the history of resource management development is not a story of uniformity and continuity, but is dominated by circumstance, contrast and disruption. The Iron Age, for example, came some 3000 years later to England than to Mesopotamia. Whether a story of ingenuity, foolishness or serendipity, human history and people's capacity to harness natural resources were dominated by place and environment. This is also a story of cultural advancement and fulfilment that varies substantially from one society to the next.

To build a picture of how resource management developed in early societies, we must largely speculate and extrapolate from scant archaeological evidence. Although we can be reasonably confident that it was insufficiency that largely drove humans to venture to new corners of the globe, it was an uneven mixture of happenstance and necessity that led to new technological developments. Even though we lack concrete motivational evidence about prehistoric societies, we know that in more recent history, the Egyptians, for example, harnessed metals to construct irrigation reservoirs to combat seasonal drought. The early Greek literature frequently records resource scarcity, and the Romans continually acted on scarcity, developing many innovative resource utilization techniques such as constructing drainage ditches in areas such as the English Fens to create agricultural land. Throughout the Middle Ages, scarcity in one form or another haunted Western society, leading T.R. Malthus, an English clergyman, in 1798 to stimulate much modern thinking about limited natural resources. Malthus speculated that increasing numbers in humans would eventually lead to overpopulation and decreasing resources. His ideas are often termed 'Malthusian' and are often invoked today in times of resource scarcity.

The following sections provide a brief introduction to the history of natural resources through the ages. Four phases broadly demarcate the significant transitions in natural resource management throughout the prehistoric and historic periods (Simmons, 1991). The first phase encompasses both the early and middle Stone Ages – the *Palaeolithic* and *Mesolithic* periods. The second concerns the *Neolithic*, or New Stone Age period, the third refers to the so-called Historic period, while the fourth, rather broadly defined, examines the Modern period.

Upper Palaeolithic and Mesolithic Periods

This early era represents little more than basic existence and limited social and cultural development. Humans harnessed whatever food they could, with restricted intellect and minimal technology such as stone projectiles. Dominated by natural resource use rather than any sophisticated management regime, this period supported at most about 2–3 people per 100 km^2. It is estimated that about 4 million people existed worldwide in 10,000 BC, a time that coincided with the great retreat of the northern ice sheets and heralded the later Neolithic (Simmons, 1991). In this earlier hunter–gatherer society, it was critical that each able-bodied Hominid produce a surplus to support dependents, and prepare for seasonal and cyclical shortages. Interestingly, there are both striking similarities and contrasts in the development of human culture and natural resource management throughout the world. The Australian aborigines, for example, probably migrated to their continental island about 40,000 years ago. As skilled hunters, they set bush fires to corral game as well as promote new forage growth.

Without external interference, their nomadic life-style remained and on a large scale largely worked in harmony with nature. This way of life successfully endured until western domination in the last 200 years (Australia's Continental Odyssey, 1988).

In North America, the northward retreat of the ice sheets and the subsequent parallel march of the tundra, which in turn was followed by encroachment northwards of the taiga forests, paved the way for man to follow. To reach North America about 25,000 years BC, humans most probably travelled the Siberian Isthmus, enduring immense climatic difficulty before the glaciers fully retreated. The main source of food for these *Palaeo* Indians was probably migrating caribou, whereas their descendants in Atlantic Canada, the *Mi'kmaq*, for which the earliest archaeological evidence dates back to around 10,000–8600 BC, largely depended on the forest for food, shelter and clothing as well as seasonal transportation by dug-out and birch-bark canoe (Johnson, 1986).

Neolithic Period

The New Stone Age, with the evolution of basic farming methods, including the domestication of both plants and animals, provided an energy surplus that in good times could be committed to cultural development rather than solely survival efforts. The cultural advance that came with ever-developing social structure supported greater population density and increasing social order. There are a number of rather well-developed farming communities to be found in the archaeological record. The *Céide* Fields on Ireland's County Mayo coast, for instance, are the largest known European remains of a prehistoric farming settlement. Evidently planned as a communal entity, this farming neighbourhood endured until climatic change buried it in ensuing bog (Zaczek, 1998).

The Polden Hills in the West of England provide intriguing evidence from fossil pollen and commercial peat extraction, concluded from radiocarbon dating, that prehistoric groups around 3200 BC had well-developed coppicing (coppicing is the practice of cutting the trunk of hardwood trees to stimulate branching and growth; the resultant branches are harvested on a sustainable basis to produce charcoal and materials for furniture) and wood-working skills (Glasscock, 1992, p. 20). Not far south and now the site of Dartmoor National Park, there is evidence, first recognized from aerial photograph analysis, that also suggests a well-developed social order sustaining communal farming enterprises during this Palaeolithic era. The Dartmoor Reeves for instance, established around 1300 BC, brought order and sustainability to the commons (Glasscock, 1992).

On the other side of the world, and apparently for thousands of years, the Polynesians took to the sea in square-rigger canoes to find new promised lands. As Kane (1974) indicates, 'Harsh necessity may have forced such departures' (pp. 758–759). On board, they took regular supplies, but also plants, seeds and domesticated animals such as pigs to sustain them in new, previously unexplored lands. These Polynesian societies showed clear foresight about natural resource management. Their exploration of the Easter Islands in the

South Pacific, for example, was only successful because of their resource management savvy and extensive contingency preparations, as these islands offered no indigenous edible plants.

The Historic Period

Moving from pre-history to the Historic Period denotes a significant cultural advancement in the division of labour and in creating sufficient surplus to focus more on intellectual and cultural concerns. In the quest for greater social sustainability, the written evidence suggests that the Greeks were driven most by population pressures – to support their increasingly sophisticated culture. This was aided significantly by their 'considerable metallurgical expertise' (Mannion, 1991, p. 105). The Romans also made a significant impact on the existing cultural landscape throughout Europe. For example, they were keen to colonize Britain for its cereal-growing potential. While pre-Roman community farms survived, larger Roman estates were overlaid on the pre-existing cultural landscape, and in places this pressure on natural resources substantially denuded the soils. They also established substantial iron smelting works requiring extensive charcoal supplies that extensively reduced the intact forests (Mannion, 1991).

In the Americas, the Mayan civilization of the Yucatan Peninsula tells of a flourishing society lasting more than 3000 years. This society in part paralleled but was far removed spatially, culturally and spiritually from Europe's Dark Ages. It prospered without the aid of a written language, and developed a sophisticated living from the tropical rainforests (Garrett, 1989) by building causeways and reservoirs, and establishing farming terraces (Hall, 1975). Centred further south and west in Peru, the Incas built a shorter, but partially overlapping, civilization. The Incas capitalized on horticultural skills, animal husbandry and prolific combative prowess to build an empire of mammoth proportions stretching from central Chile north to Ecuador. Somewhat later and to the north, the Aztec world flourished. They first settled in Chapultepec in the 13th century and continued to dominate the region until 1521 when they surrendered to the Spanish. The Aztec civilization was a society of contrasts and fluctuations supporting an impressive array of artwork and architecture, but also a society that propagated forest degradation and subsequently succumbed to famine (McDowell, 1980).

On the eastern side of the Atlantic Ocean, society seemed much less colourful; however, there were nevertheless interesting resource management developments. The Normans in England in the 11th and 12th centuries, for example, first designated royal hunting forests. William the Conqueror, the first Norman king, was said to have a great love of deer and established the New Forest in southern England, and King Richard of Crusades notoriety later designated the King's wood in Gloucestershire. During the 13th century, a marked reduction in global population suggests a combination of agricultural deficiency, population pressure, disease and climate change as key causes. In Europe, population growth was severely checked by the Great Plague, whereas in China, the Mogul invasion caused population reduction.

Interestingly, during the 13th century, a much less sophisticated Palaeo-Indian society in Mesa Verde, Colorado, provides a poignant lesson in resource conservation. This cliff-dwelling community, dependent on the plateau above for its water, fuel, materials and food, recklessly denuded its life-supporting forests. Over time, this fragile society extinguished its only water and food source, forcing them to relocate hundreds of miles away. Like the Mesa Verde Indians and the Aztecs, the English too denuded their forests, but over a longer period. By the time of the 12th century Doomsday Chronicles, England had only 15% forest cover remaining (Mannion, 1991) which led in the 15th and 16th centuries to an ever-increasing need for tree planting (Blunden and Curry, 1988).

Perhaps the most significant farm management development of the 1500s in England was the enclosure movement. Subsequently reinforced by an Act of Parliament, the Enclosure Act transformed agriculture from a largely communal, socially undifferentiated resource management process to one where workers became virtual slaves on previously common land. Notwithstanding the dire social consequences, as farming efficiency increased from this farm management shift so the resultant surplus labour made the subsequent industrial revolution possible. It was now possible to produce sufficient food to feed those who migrated to the factories (Mannion, 1991).

The Modern Period

Overpopulation and resource scarcity was postponed in much of Europe because of colonization, technological and industrial development, new food-producing techniques, scientific knowledge and modern medicine. The middle of the 1700s brought the industrial revolution to Europe by intensified use of coal, the harnessing of steam, and further reorganization of society from its agrarian traditions. This era brought a rush of technological developments. In 1812, for example, the first canned foods appeared, resulting in more efficient food production, storage and utilization, and in 1834 a rudimentary refrigeration system further revolutionized production processes.

In 1855, Thomas Crapper developed an efficient flush toilet, which unfortunately became a double-edged sword. While improving urban hygiene, it created immeasurable water pollution problems in the countryside. A highly significant scientific development came about when Gregor Mendel unravelled the laws of heredity in 1866. Although there is evidence of genetic hybridization going back some 10,000 years to the upper *Epipalaeolithic* stage (Mannion, 1991), Mendel's discovery accelerated agricultural advancement. It periodically eased Malthusian pessimism and renewed optimism in solving scarcity.

Until the waning of the 19th century, the natural resources of the USA, at least, were seen as inexhaustible. Some natural resources were even considered 'an enemy' of social advancement and development; forests especially were often targeted for destruction. By the turn of, and well into, the 20th century, most of the US resource management laws and supporting institutions were geared to unchecked resource exploitation.

Unbridled forest exploitation did, however, spur on the American Conservation Movement led by so-called 'scientific men' who were well aware that many European countries had long ago denuded their forests. As early as the beginning of the 1870s, the American Association for the Advancement of Science urged that, without conservation, America's forests were doomed. In 1897, the National Academy of Science echoed these concerns, leading to the establishment of national forest reserves. In 1888, Powell's *Lands of the Arid Region* (see Van Hise, 1910) also sensitized both the public and politicians to water management concerns that led to an irrigation division of the US Geological Survey. Not only in forests and waters were warnings sounded; alarm was also raised concerning oil, gas and coal supply, as well as soil destruction (Van Hise, 1910).

An important symbolic milestone for conservation was the appointment by President Roosevelt of the Inland Waterways Commission (IWC) on 14 March 1907. Through the efforts of Gifford Pinchot, W.J. McGee and other key scientists who recognized the connectivity of forests and water as well as the public's interest for navigable waters and irrigation, support coalesced for conservation action. The IWC's first report recognized the 'interlocking character of the problem of natural resources' and pointed to their association with coal, iron and soil conservation (Van Hise, 1910, p. 5). To address these issues, the Commission persuaded the President to hold a White House conference on 13 May 1908. According to Van Hise, never before in the history of the USA had such a far-reaching representation of governors, cabinet members, scientists and Supreme Court justices gathered to discuss any policy question, let alone a concern for natural resources.

This conference addressed wide concerns about minerals, forests, soils and water that so shocked attending politicians that they endorsed resolutions of far-reaching proportions 'covering the entire subject of conservation, pointing to the extravagance and reckless waste of the past, and making clear that upon the conservation of our natural resources depends the foundation of our prosperity' (Van Hise, 1910, p. 7).

Roosevelt's next step was to establish a National Conservation Commission with responsibility for minerals, waters, forests and the soils. However, Roosevelt enjoyed less success with the Sixtieth Congress, failing to persuade them to appropriate funds for the Commission's expenses. In the House of Representatives, the President was struck a more devastating blow; a successful amendment barred the Commission from directing its work to any federal bureau, as done in the past, and stripped it of any effective implementation power.

Interest in conservation at this time was not limited to the USA. The Canadian conservation movement had its birth, for example, in the 1890s when 'voices could be heard' that prevailing government policy was wasteful and destructive (Burton, 1972, p. 29); however, this movement was to have little political resonance until 1906. Then a Canadian Forestry Convention focused on forestry practices but also included water conservation and the uncontrolled exploitation of mineral resources. Political momentum for conservation policy grew during this period and enjoyed a substantial boost from an invitation to join an international delegation of concern at the White House. Official delegations from both Mexico and Newfoundland joined Canada and the USA on

18 February 1909. Perhaps perceptively but overoptimistically, this conference adopted the following declaration:

> We recognize the mutual interests of the Nations which occupy the Continent of North America and the dependence of the welfare of each upon its natural resources. We agree that the conservation of these resources is indispensable for the continued prosperity of each Nation.
>
> (Van Hise, 1910, p. 385)

This declaration implored nations to take immediate and concrete steps to temper unabated resource exploitation.

In Canada, as in the USA, words were cheap, but any conservation advancement was hard fought and invariably overrun by more pervasive development policy. In 1878, for example, the return to office of the MacDonald Government had heralded the doctrine of 'usefulness' for Canada's natural bounty (Brown, cited in Burton, 1972, p. 27). The achievement of legitimate nationhood was to build on natural resource exploitation, and particularly a transcontinental railway was to reap the abundance of the West. Integral to this policy was that the West's resources would remain under federal control for the service of the Dominion. This arrangement prevailed until the three western provinces of Manitoba, Saskatchewan and Alberta were established in 1930 and, along with this, the transfer of sovereign control of natural resources. Consistent with this overriding doctrine of usefulness, it is important to note that the establishment of Banff Springs as a national reserve in 1885, and its subsequent enlargement in 1887, did not result from an outpouring of preservation or conservation concern but was for economic exploitation and to reward the shareholders of the Canadian Pacific Railway.

For about 20% of the world's population, those living in the developed world, increasing natural resource exploitation in the 19th and 20th centuries resulted in a generally increased standard of living. For the remainder, largely but not exclusively living in the undeveloped world, the picture has been far less encouraging. There, the imposition of western development has been far-reaching but not always positive. Motivated by high profits and few moderating environmental controls, natural resource exploitation has been in many ways detrimental to long-term welfare. Foreign exploitation transformed modest subsistence but largely sustainable agriculture to cash crop systems designed exclusively for the benefit of Western cultures.

Many of these schemes have led to extensive soil degradation and increased salinity. Rampant tropical rainforest logging or removal for cash crop agriculture, for example, seriously upsets biogeochemical cycling, reducing once productive biological and cultural systems to economically unsustainable agricultural systems in as little as 5 years. Since the rise of the industrial revolution, both the developed and the developing world have experienced increased mineral exploration with accompanying environmental degradation such as despoiling groundwater and increased air pollution. Here again, technological development has been a double-edged sword often improving worker safety and reducing air emissions and water pollution, but continually extending exploration and exploitation, thus increasing the total environmental degradation.

Although the 20th century, especially in the developed world, has seen various conservation movements, the predominant story has been one of increasing resource exploitation and disregard for environmental and intergenerational consequences. There have been, nevertheless, a number of significant events since the early years of Roosevelt's conservation movement that have drawn attention to, if not put a halt to, rampant and short-sighted exploitation of our natural heritage.

Captivated by the publication of Rachel Carson's *Silent Spring* in 1962, the general public was suddenly awakened to insecticide poisoning in particular, and inappropriately tested resource management technology in general. Within a more professional and academic audience, Hardin's (1968) *Tragedy of the Commons* drew attention to the fallacy of the 'rationally acting man' solely driving resource management decision making. Also, in the late 1960s, a wave of concern in Western culture over environmental degradation led to a rash of remedial efforts to clean up the environment.

These 'end-of-pipe' strategies met with limited success, however, merely transposing an environmental problem from one environmental medium to another. Perhaps the most profound outcome of the early space flights was serendipity; astronauts' report of a 'thin blue line' around the earth drew attention to 'our atmosphere's' vulnerability. In time, space exploration and associated technology development nevertheless have led to more systematic developments in geomatics such as satellite photography and remote sensing that have substantially aided modern resource management.

Summary and Conclusions

For several millennia, human beings largely lived in harmony with the natural environment. Increasingly, however, successive civilizations developed their capacity to exploit natural resources to make life easier. This exploitation of natural resources often paralleled the exploitation of other humans to supply labour whether that was as slaves, virtual slaves as a result of, for example, the Enclosure Act in England, or its modern equivalent, the sweat shops that produce cheap garments and fancy jogging shoes for Western society.

Our modern science and technology has shifted the balance so we as humans are now capable of drastically altering our natural surroundings and, in fact, we now dangerously spoil our life support systems such as clean water and clean air. In the following chapter, we will discuss in more detail the evolution of a conservation movement that has reacted to this despoliation; we will critically examine how so-called development increasingly stresses our environment and natural resources and, in doing so, creates a need for an IREM regime.

In the case study below, you will be introduced to a civilization in the Yucatan Peninsula in Central America that once flourished and dominated the region. While historians hotly dispute the exact reason why this civilization declined, you will see that substantial evidence leads us to believe that inappropriate natural resource management strategies may have contributed substantially.

Case Study
The Mayan Collapse

DAVID DUKE

Preamble. The following case study provides an example of a culture that collapsed due to a variety of reasons relating to internal resource pressures and external environmental factors. The Mayan collapse is germane to this chapter in that it develops specific examples of resource depletion that affected other great cultures.

The main temple, under restoration, at Caracol in the Chiquibul Forest, Belize. This Mayan site is thought to have been occupied from 600 BC to almost AD 1100 and at its peak the city served a population of 115,000–150,000 people. Photo by Glyn Bissix.

Introduction

One of the most extraordinary and sophisticated pre-industrial societies to develop was that of the Maya. The civilization encompassed a region that includes modern-day eastern Mexico and parts of Belize, Guatemala and Honduras (Fig. 2.1). This Meso-American society developed over several thousand years, but its period of greatest activity was from 250 CE to approximately 800 CE, a time span known as the **Classical Period**.

Mayan society was heavily urbanized. However, this urban sector was underpinned and supported by rural activity on a large scale. Urban society was characterized by great cities such as Tikal, with a population of perhaps 100,000 by 800 CE, or Copan at the southern end of the Mayan region, which had a population of at least 20,000 in the same time period. The

A replica Stella at Caracol in the Chiquibul Forest, Belize. The original is reportedly in a private collection in Canada. Photo by Glyn Bissix.

Fig. 2.1. Map of the Mayan region, showing major Mayan cities and modern international borders.

culture of these cities, and others like them, was both sophisticated and complex. For the Maya, religion was a crucial component of daily life, and temples took the form of pyramids, some of which rivalled or even surpassed the Great Pyramids of Egypt. Unlike those structures, however, the Mayan pyramids were built without the aid of draught animals, metal tools or wheeled transportation. The Maya were not intellectually underdeveloped, however, they were in fact highly literate, and their mathematical skill was refined. They possessed an accurate calendar and the ability to predict both solar and lunar eclipses accurately.

Background

There is a great deal of geographical diversity in the region, from broad floodplains in the coastal region along the Gulf of Mexico to highlands in the region now marked by the Honduras–Guatemala border. Between the two, the landforms tend toward gently rolling hills and river valleys, with the hills increasing in height as one moves towards the south. Rainfall is seasonal, with the heaviest amounts occurring in a rainy season that lasts from May until December. Temperatures tend to fall within a narrow range depending on landform location: on the coastal plains, temperatures range from 25 to 30°C, in the upland region in the centre of the Mayan region, temperatures vary from 15 to 25°C, while in the highland South the figure tends to be around 15°C.

Agriculture

The foundation upon which this great society was built was agriculture – and agriculture of only a few crop staples. Three main crops formed the basis of Mayan agriculture: maize, beans and squash. Of these, maize was by far the most important, accounting for perhaps 90% of each adult Mayan's caloric intake. As a consequence of the varied geography and climate, the Maya were able to practice several different agricultural techniques throughout the Classical Period. At the simpler end of the spectrum was maize cultivation through **swidden agriculture** (also known as 'slash-and-burn' agriculture). Here, sections of scrub, grassland or forest would be burned just prior to the onset of the rainy season in the late spring. The action of the rain coupled with human labour would work the vegetable ash into the soil, thus fertilizing it. Several crops of maize could be raised before nutrient depletion necessitated the abandonment of the plot and the repetition of the process elsewhere. In cases of low population density, swidden techniques are sustainable, as natural overgrowth of worked plots will occur over time, providing fresh fuel for a new round of burn-based fertilization in the future. The recovery process is not quick, however (~20 years in tropical and subtropical environments), and this means that too high a population density leads to unsustainability, because insufficient time is allowed for natural vegetative regeneration between burn and cultivation periods.

More sophisticated agriculturally was the Mayan technique of **milpa** cultivation, which could occur in central regions where consistently heavy rainfall was the norm. This was an intensive technique that involved the cultivation of several different crops in rotation interspersed by short but intensively managed fallow periods to allow for soil regeneration (Simmons, 1996).

The most complex and environmentally significant of the Mayan agricultural techniques was the floodplain cultivation of maize on raised islands known as **chinampas**. The fields were artificial islands, constructed by dredging canals in the river's floodplain and piling the nutrient-rich silt behind constructed barriers. Yearly dredgings ensured high productivity thanks to the annual application of dredged silt to the fields, silt that was constantly replenished by river flow. The canals acted as an efficient transportation network throughout the patchworked agricultural zone, and served as an excellent habitat for fish and other marine fauna, the exploitation of which provided a valuable protein component to the Mayan diet (Simmons, 1996).

On uplands, widespread terracing was undertaken (one of the largest contiguous regions of terrace agriculture extended over perhaps 10,000 km^2). It seems that Mayan farmers employed conscious soil conservation and management techniques in their terracing, transferring soil that had been eroded off the hillsides in the course of agricultural activity from the lowest levels where it was deposited by rainfall back uphill to the higher terraces from whence it had come.

Society

The urban population that was supported by this agricultural activity was at least several millions in size, and perhaps much larger than that. Urban society was complex, rigidly hierarchical and highly cultured. Warfare seems to have been widely practised both for political aggrandisement and as a source of captives destined for sacrifice during the elaborate ceremonies around which Mayan religious life revolved. Scholars of the early 20th century believed that Mayan society was theocratic in nature, i.e. ruled by priestly castes; this was a natural assumption given the pre-eminence of religious structures in Mayan cities. More recent evidence has decisively shifted this view, however, demonstrating convincingly that more traditional political leaders governed Mayan society. The lords of each city and its surrounding hinterland acted independently from and in competition with the lords of other cities. It is known that there was a wide gulf between the urban elites and the base of rural and urban commoners upon which they depended.

It is also known that the Classical Period was marked by significant population growth after 600 CE. Monumental construction in most cities increased rapidly, a process that would have required a major increase in labour investment. This labour would have had to have been drawn from each city's rural hinterland. In addition, such major construction required large amounts of wood to aid in the construction process. This again would have placed considerable stress on supplies of wood, since it was also the sole source of fuel for the Mayans, elite and commoner alike.

The Historical Issue

The Mayan civilization discussed here was durable, spanning, as we have seen, several centuries at least. Yet classical Mayan civilization collapsed almost totally in the 800s – and it did so very rapidly indeed, perhaps in as short a time frame as that encompassed by a single generation. Many construction projects were abandoned in mid-phase, with tools being strewn around half-completed columns decorated with incomplete inscriptions. In one case, archaeologists discovered the unburied skeletons of children in the interior of one of the temples at Chichen Itza. Reverence for the dead was very important to the Maya, and a failure to bury corpses must, therefore, be seen as evidence of some rapid catastrophe.

The question is: What was the nature of that catastrophe? Scholars of the mid-20th century theorized that the rich Mayan civilization collapsed because of brutal and increasingly prolonged warfare between rival cities. Other theories now encompass environmental factors. Their proponents argue that unsustainable Mayan agricultural practices led to a general deterioration of the environmental situation that in turn provoked a social collapse. Other theories focus on larger, natural environmental factors such as climate change. It is likely that a combination of all three factors propelled the collapse of Mayan civilization.

Evidence of the Mayan Collapse

The evidence for the eventual Mayan collapse is subdivided into the three following categories: social/political; local environmental; and regional/climate-based environmental.

Social/political

1. Despite the competition between them, Mayan city-states traded heavily with one another for luxury goods for use by the elite within society. It appears that a durable, long-lasting and stable trade balance developed over centuries, within which each city-state controlled part of the trade in a particular commodity, as either producer, controller of trade route or consumer. In the century following 550 CE, the great city of Teotihuacan, not itself Mayan but heavily influential in Mayan culture, trade and society, collapsed. Teotihuacan was located several hundred kilometres beyond the western edge of the Mayan zone, but its economic and military strength was so great that its collapse severely affected Mayan trade and cultural patterns.
2. It is known, from dates on the last ceremonial buildings and other constructions thrown up by the Maya during the Classical Period, that the decline of Mayan civilization began around 790 CE in the west and over the next 30 years spread eastwards throughout the entire Mayan zone. In Copan, for example, at the south-eastern end of the Mayan zone, the last dated construction was abandoned, incomplete, in 822 CE.
3. There is no evidence of massive external invasion, although there is a cultural intermixing between peoples of the Mexican uplands (around modern-day Mexico City) with those of the western Mayan zone in the 9th century CE – exactly around the time of the Mayan collapse.
4. There is evidence of violence in some of the cities. The effects of the violence indicate that it must have been sudden: there is no evidence of siege or defensive preparations in these cases.

Local environmental

1. Studies of skeletons recovered from cities as far apart as Copan in the south and Chichen Itza in the north have found evidence of malnutrition and of the existence of malnutrition-based illness, such as anaemia.
2. Similar studies of skeletons have discovered that there was a sharp increase in deaths among segments of the population that normally survive famines or other natural disasters. Usually it is the elderly or the very young that succumb in such cases, but in the 800s it seems as though the deaths encompassed younger and mature adults as well.
3. Growing populations require larger quantities of fuel for cooking and other daily purposes. Most of the ceremonial buildings constructed in each city were whitewashed with lime plaster that requires large amounts of heat (or 'pyrotechnology') in its manufacture.
4. Studies of sediment cores drawn from lakes in the region have demonstrated a sharp decrease in the levels of tree pollen after the 600s. After approximately 1000 CE, quantities of

pollen, especially of the black mahogany tree, a common species of established forests, return to average levels in the sedimentary cores.

5. In cities such as Copan, modern excavation has discovered that houses located on down slopes on the outskirts of the city and at the bottom of hillsides in the city's hinterland are filled with large quantities of soil and dried mud sediments. The dwellings were clearly abandoned before the general abandonment of the region occurred.

6. It is known that silt loads in the region's rivers increased dramatically.

Regional/climate-based environmental

1. By studying the prevalence of gypsum in lake cores drawn from Lake Chichancanab in the Yucatan Peninsula, scientists have determined that the period 800–1000 CE was unusually warm and dry in the region. Solid gypsum is a precipitate, i.e. when water that contains dissolved gypsum evaporates, the gypsum is left behind, in the same way that salt is left behind if a pan of salt water is boiled away. Precipitated gypsum is found in large quantities in the segment of the core relating to the 800–1000 CE period.

2. There is good evidence that long-term droughts occur every 200 years or so; indeed, such a drought coupled with resource depletion is considered as a prime candidate for the collapse of Teotihuacan that occurred some time in the century after 550 CE.

Conclusion

It is clear that environmental factors can severely stress even a well-established and flourishing civilization. In turn, social factors may exacerbate environmental problems. The lesson may be that, when the imbalance begins, it can accelerate quickly and render the society incapable of absorbing further environmental shocks, possibly not of the population's own making.

Discussion Questions

1. Describe the linkages between cultural advancement and natural resource management in the early development of humans.

2. What do we mean by scarcity? What role does scarcity play in technology advancement in natural resource management?

3. This section describes four major phases of human development, what are they? What are the key advances in natural resource development that led to these transformational changes?

4. In mining minerals, human ingenuity often improves mine site pollution control and miner safety, so why is it that we see continual increases in pollution and worker injuries and deaths as a result of mining?

5. Natural resource development accelerated rapidly throughout the world during the 19th and 20th centuries. Why is that increases in the standard of living in the developing world where resource exploitation occurred have not kept pace with advances in the developed world?

Case Study Discussion Questions

6. Unlike other case studies presented in this text, the outcome of the case outlined in this chapter is already known. Based on the information presented in the case study, attempt to develop a **plausible and specific** theory to account for the collapse of Mayan civilization in the 800s. Employ both the background presented in the main body of the chapter and the archaeological and environmental evidence presented in the case study in the development of your theory. Remember, your theory should incorporate, as far as possible, elements of warfare, human-induced environmental degradation and broader environmental factors which all played a role in the collapse of the Maya.

7. Attempt to develop an IREM-based solution to determine the points at which integrated and cooperative resource management, or other cooperative behaviour among the Maya, could have averted the collapse of their civilization. What does your IREM-based solution suggest to you in terms of the likelihood of social collapse in the face of resource depletion? Is it a likely scenario, or an unlikely one?

8. Consider whether the conditions outlined in this case study are apparent in any of the other case studies presented in this text. Could policies adopted in the modern case studies have been applied to this historical example? Why or why not?

References

Australia's Continental Odyssey (1988, February) *National Geographic* (map supplement).

Blunden, J. and Curry, N. (1988) *A Future for Our Countryside*. Basil Blackwell, Oxford, UK.

Burton, L.T. (1972) *Natural Resources Policy in Canada: Issues and Perspectives*. McClelland and Stewart, Toronto, Canada.

Carson, R.L. (1962) *Silent Spring*. Houghton Mifflin, Boston, Massachusetts.

Garrett, W.E. (1989, October) La Ruta Maya. *National Geographic* 176(4), 424–479.

Glasscock, R. (ed.) (1992) *Historic Landscapes of Britain from the Air*. Cambridge University Press, Cambridge, UK.

Hall, A.J. (1975, December) A traveller's tale of ancient Tikal. *National Geographic* 148(6), 799–811.

Hardin, G. (1968) The tragedy of the Commons. *Science* 162, 1243–1248.

Johnson, R.S. (1986) *Forests of Nova Scotia: a History*. Halifax Department of Lands and Forests/Four East Publications Halifax, Nova Scotia, Canada.

Kane, H.K. (1974, December) The pathfinders. *National Geographic* 146(6), 758–759.

Mannion, A.M. (1991) *Global Environmental Change: a Natural and Cultural Environmental History*. Longman Scientific and Technical, London.

McDowell, B. (1980, December) The Aztecs. *National Geographic* 158(6), 704–751.

Simmons, I.G. (1991) *Earth, Air and Water: Resources and Environment in the Late 20th Century*. Edward Arnold, London.

Simmons, I.G. (1996) *Changing the Face of the Earth: Culture, Environment, History*, 2nd edn. Blackwell, Oxford, UK.

Van Hise, C.R. (1910) *The Conservation of Natural Resources in the United States*. MacMillan, New York.

Zaczek, I. (1998) *Ancient Ireland*. Collins and Brown, London.

3 The Modern History of IREM

Introduction

As the historical overview of resource management development in Chapter 2 suggests, the ancestry of integrated resource management is extensive but not systematic or widespread. The roots of IREM can be traced back to England with isolated examples in the 11th and 12th centuries. The modern IREM era, however, is more firmly rooted in North America in the late 19th century with Powell's 1888 treatise on arid lands, and in the early 20th century with President Theodore Roosevelt's and Gifford Pinchot's Inland Waterways Commission in 1907. It is rather misleading then to suggest that there was a coherent or systematic development of integrated resource management in the 20th century. There are, nevertheless, historic eras, some mentioned in the previous chapter, that offer a useful backdrop or point of reference for examining the modern history of IREM. To examine these modern historical eras and the key issues that emerge from them, this chapter is divided into the following four sections:

1. Modern periods in IREM.
2. The globalization–sustainability tension.
3. Legislative foundations for IREM.
4. Conclusions and future challenges.

The first section, the 'Modern periods in IREM', briefly outlines the major functional periods of the modern era. These overlapping periods are: conservation enlightenment (1880–1930), recession and uncertainty (1920–1939 or 1942), the Second World War (1939 or 1942–1945), national reconstruction (1946–1970) and the contemporary environmental epochs (1970 to the present). The second section, the 'Globalization–sustainability tension', examines a number of resource management applications such as dam construction, farming, forestry and watershed management that have important links to the development of IREM – they highlight key issues that illustrate the challenges for the continued development of IREM. The third section, 'Legislative foundations for IREM', looks briefly at the roots of IREM legislation, while the fourth, 'Conclusions and future challenges', draws conclusions from the preceding discussions and looks briefly to the future and the potential role of IREM in resource and environmental management.

Modern Periods in IREM

Conservation enlightenment: 1880–1930

The era of conservation enlightenment, roughly from 1880 to 1930, defines the initial period of modern IREM development in North America. This period, dominated by dam building and multiple objective resource development, was very much a paradox. On the one hand, the enlightenment's leaders in the USA – President Theodore Roosevelt and Gifford Pinchot – drew widespread attention to the concentrated and unconstrained power of various industrialists and financiers that caused natural resource depletion and serious environmental damage. On the other hand, their natural resource development initiatives played into the industrialists' hands by promoting large publicly funded projects. The net result during this era was that old as well as new industrialists amassed or consolidated large fortunes as these projects often encouraged monopolistic markets with their generous profit margins and resultant political power (Richardson, 1973).

Recession and uncertainty: 1920–1939/1942

The second period of 'recession and uncertainty' overlaps the 'new deal' promises of President Franklin D. Roosevelt in the USA and its legacy. This era marked a substantial departure from *laissez-faire* resource management policy that pre-dated the Great Depression and also denoted a shift of emphasis away from integrated resource management activity. The US government saw its natural resources, especially in the more undeveloped West, as a strong incentive for economic renewal, and such exploitation took place with little regard for conservation or the environment. One possible exception was the massive mobilization of otherwise unemployed youth in a nationwide programme of natural resource development. This took place under the stewardship of the Civilian Conservation Corps (CCC). The CCC worked on a variety of projects; for example, they built numerous log cabins and lodges in America's national parks; they planted millions of trees; and built numerous irrigation projects. This period, prior to the Second World War, first concentrated on rebuilding a shattered economy and then focused its attention on an arms build-up when war seemed inevitable.

The Second World War: 1939/1942–1945

The third era, the Second World War, is really the antithesis of IREM, where concentrated national security interests predominated over concerns for conservation. Most, if not all, natural resources were assessed for their potential contribution to the war effort. The US Department of the Interior was, nevertheless, the scene of a number of interesting political encounters over development versus conservation. For example, one particular battle erupted over the fate of the Olympic National Park's forests; in this instance, sufficient support for conservation was mustered to save its forests from Washington State's woodcutters (Richardson, 1973).

National reconstruction: 1946–1970

Following the war, from 1946 to 1970, 'national reconstruction' dominated the resource management agendas of numerous countries. As will be seen in the following chapter, this era focused on a transition from a war-based economy to an increasingly consumer society that capitalized on a vastly enhanced production capacity. For example, in the UK's farming sector, the expansion in consumer goods production was substantially bolstered by an increasing reliance on government intervention in the marketplace (see Chapter 8, 'Market and State Functions'). This era marked a period of increasing economic optimism, a developing middle class with greater consumer capacity, and growing, but often unacknowledged, environmental degradation. Even the Willamette River in Oregon, for example, a river of the far west often considered rugged and unspoiled, was transformed from a wild, seasonally variable and pristine river to one dominated by large dams constructed for flood control and power generation. It was also highly contaminated by agricultural and industrial wastes.

Modern environmental epochs: 1970 to the present

Following this rebuilding period, the late 1960s and the early 1970s were characterized by the rise of the environmental movement and the development of various overlapping environmental epochs (distinguishable time periods) – each redefined the way we dealt with environmental problems. Rather than replace the dominant environmental management strategies of the preceding epoch, each perspective superimposed its worldview on existing practices. Mazmanian and Kraft (1999) identify three dominant epochs: (i) regulating for environmental protection; (ii) media and strategy integration; and (iii) toward sustainable communities. Each epoch has its own characteristic problem identification and policy objectives, implementation philosophy, points of intervention, and policy approaches and tools of operation. Each epoch had its own information management and data management needs, predominant political/institutional context, and key events and public actions (pp. 10–13).

Regulating for environmental protection

The 'regulating for environmental protection' epoch from 1970 to 1990 was dominated by so-called 'end-of-pipe' strategies that were dependent upon the imposition of command and control regulations for enforcement (see Chapter 8, 'Sustainability and Ecological Modernization').

Media and strategy integration

The second epoch, from 1980 to the 1990s, reflected a swing away from regulation to market-based and collaborative policy mechanisms. Implementation was subjected to, among other strategies, cost-effectiveness tests and incentive-based fees and charges, taxes, and emissions trading. During this period, known as the era of media and strategy integration, the emerging resource management science developed during the preceding epoch addressed its failures. The first era, for example, failed to account for intermedia transfer (i.e. the transfer of one environmental problem such as air pollution to ground and water pollution). The approach of this succeeding epoch focused more directly on solving environmental problems at the source, before they were created, rather than attempting to clean up environmental problems after they were created. Emphasis was placed, for instance, on reducing solid waste through more discriminating consumer behaviour and better packaging rather than dealing with the solid waste problem after it was generated. Automobiles, as one example, were redesigned to aid recycling once they reached the end of their useful life.

Towards sustainable communities

Mazmanian and Kraft (1999) saw the third epoch (from the 1990s to the present) as an emerging conservation re-enlightenment; a period characterized by community, regional, national and international mobilization to reduce natural resource waste, identify and fix environmental problems, and promote a greater sense of social equity. Despite Mazmanian and Kraft's apparent optimism, we argue that there is little evidence to support a notion of more sensitive environmental behaviour on a broad scale. Based on indicators such as CO_2 emissions, ozone depletion and water pollution, for instance, there is much evidence that seems to point to a world of largely unabated natural resource exploitation, increasing environmental degradation and ever-increasing social inequity. While the evidence for a single, unequivocal conclusion is controversial, there is, regardless, growing opposition to global trade liberalization, which is seen as increasingly driving natural resource exploitation and assaulting presently established and much needed environmental regulations (see Chapter 8, 'Globalization', and 'Good Governance and Sustainable Development'). As a consequence of a more pessimistic view, this text's authors prefer to think of this present period as one of a 'globalization–sustainability tension'.

The Globalization–Sustainability Tension

In the general sphere of resource and environmental management, as well as in the more focused field of IREM, there appear to be many more examples of 'what not to do' rather than of exemplary or broadly acceptable resource and environmental management practices. As a consequence, this section examines a number of approaches and/or concerns regarding natural resource and environmental management that challenge any conception of widely accepted natural resource conservation practices. It explores instead the notion of a 'globalization–sustainability tension' that in large measure reflects increasing natural resource exploitation whether examined at the community or broader levels. This trend of increasing natural resource exploitation, we argue, is only partially countered by meaningful conservation measures. To illustrate this largely destructive trend, we examine the following:

- Dams and multiple-objective resource management
- Parks and protected areas as models for IREM
- Integrated coastal zone management
- Forestry: residual use or IREM?
- Sustainable agriculture
- Frontier development
- Ecosystem and watershed management

Dams and multiple-objective resource management

Despite the fact that President Theodore Roosevelt drew considerable attention to conservation needs in North America during the first decades of the 20th century, there are few practical examples that illustrate a conservation or IREM approach. Although a number of large dams were constructed to control flooding, for example in the Miami Conservancy District and the Muskingum District in the 1920s (in response to the Ohio River flooding of 1913), several other resource and environmental management concerns resulted from their construction, such as erosion, loss of top soil and drought. It was necessary to wait until 1928 when the Hoover Dam was constructed for the first large-scale integrated resource management project to be attempted in the USA. This construction was part of a much larger project envisaged in the Boulder Canyon Project Act. Its purpose was to integrate a number of resource development objectives such as to provide irrigation to the Imperial Valley, produce hydroelectric power and provide flood control for the lower Colorado River.

In 1933, a concern for the integration of resource management and social development concerns led to the formation of the Muskingum Watershed Conservancy District. This agency was charged with the 'simultaneous consideration of water and associated land based resources' involving 'multiple purposes, multiple means and multiple participants' (Mitchell, 1986, p. 18). In the same year, the Tennessee Valley Authority (TVA) was instituted. With the Tennessee River extending 1450 km and forming the fifth largest drainage basin in the

USA, this was by any standard an immense integration project aimed at transforming a relatively impoverished agrarian economy into a more diversified and prosperous one. The TVA was established primarily to improve navigation, produce hydropower and provide for flood control, but was also mandated to tackle erosion and create recreation settings as well as attend to public health and address welfare issues. Often touted as a 'project exemplar' for multipurpose natural resource management combining both objectives for physical transformation and social change, it is interesting to note that this project was never substantially imitated elsewhere. In many ways, it had a less than stellar performance that was due in part to its immense costs; its disappointing outcomes were also the result of political infighting among regional and national politicians. Unfortunately, according to Richardson (1973), a focus on the public good never rose above political opportunism and bureaucratic inertia.

Parks and protected areas as models for IREM

While the management of parks and protected areas often seems inconsequential in the broad context of major natural resource sectors such as mining, farming, forestry, fishing and dam building, their consideration is useful to understanding the potential for IREM in the working landscape. Three examples are examined here: (i) the national parks management in the UK; (ii) multiple-use management in Kananaskis Country in Alberta, Canada; and (iii) coastal management in Queensland, Australia.

The UK

Some of the most interesting and successful examples of landscape-scaled, integrated resource conservation management occur within the National Parks of England and Wales. In 1949, the National Parks and Access to the Countryside Act designated several areas for preservation and enhancement of natural beauty as well as to promote enjoyment and attend to the needs of agriculture and forestry (Atkinson, 1992). From their inception in 1949, English National Parks management was oriented to the integration of resource and environmental management as well as multiple use. The Countryside Act of 1968 added a further integrative function, which was the administrative objective of the economic and social interests of rural areas within each park's boundaries.

Although the Dartmoor National Park Authority (DNPA) – one of England's most popular national parks – for example, has formally managed the Dartmoor landscape for about 50 years, the landscape has endured human influences for well over 5000 years (Bissix and Bissix, 1995). There is, for instance, evidence on Dartmoor of communal land management dating back several thousands of years. Dartmoor's upland landscape, with its outstanding historical, biophysical and cultural features, both supports and enriches its human settlement. The present socioeconomic activity of Dartmoor, such as agriculture, light industry and commerce as well as tourism, mineral extraction and recreation, has deep roots that long preceded the park's formal establishment. However, the rapid changes of the last half-century in the English countryside have given the DNPA and its integrated landscape management responsibility increasing significance (Atkinson, 1991). Interestingly, after decades of debate about this approach to protected area management in Scotland, a similar management model has recently been adopted as the Loch Lomond and Trossachs National Park Authority, which is situated just north of Glasgow (Boyack, 2000; see also the case study in Chapter 10).

Canada

Kananaskis Country, in Alberta, Canada, is situated on the eastern slopes of the Rocky Mountains in an outstanding natural landscape (see the case study in Chapter 8). Kananaskis Country was formally established in 1975 (Oltmann, 1997); it is perhaps the most extensively integrated resource and environmental management project of a regionally significant scale in the world. To most Calgarians who live less than an hour's drive from 'K' Country (as it is locally known), this area is their playground. However, K Country is much more than a park or a playground; it is a working landscape where highly developed and high quality outdoor recreation opportunities from Olympic class downhill skiing, hunting and fishing, mountain biking and ATV trails abound and coexist with oil and gas production, forestry, tourism (it was the site of the 2002 G8 Nations Leaders' meeting) and cattle grazing. While the alpine and the foothills of K Country provide some of the most profitable exploration sites for the oil and gas industry in North America, this industry must continually contend with alternative land use pressures from recreationists and cattle grazers. These stakeholders see the oil industry's sour gas wells as a continued threat to human and livestock health as well as to general well-being in K Country.

It is interesting to note that while some recreationists align in fleeting coalitions with cattle grazers to oppose oil and gas developers, other recreationists and preservationists contest highly subsidized cattle grazing in K Country; they see this economic activity as an incursion on wilderness values as well as a threat to environmental quality. Foresters within K Country, on the other hand, see recreationists as an encumbrance to economic efficiency. Often forestry companies must absorb the added costs of accommodating recreationists with little or no financial compensation or recognition. They must, for example, design and construct forest roads to meet the future needs of recreationists once forestry operations are complete (Squires, 1997). Despite the controversy among K Country's user group stakeholders as well as Alberta's taxpayers, K Country remains a leading example of where integrated and multiple-use management can work, even if this occurs with continued growing pains.

Australia

Born out of the concern for 'pollution, overfishing, physical degradation of marine habitats and declining opportunities for access to resources' (Ottesen and Kenchington, 1995, p. 151), Australia has established a number of Marine Protected Areas (MPAs), of which the most celebrated is the Great Barrier Reef Marine Park (GBRMP). The GBRMP lies off the coast of Queensland and was formed by an Act of Parliament in Australia in 1975. The GBRMP is often cited as a classic model for integrated resource management that has, over the years, found effective ways to mediate resource interests to pursue sustainable development. The key feature of the management system is that its boundaries are extensive; they encompass the vast ecosystem stretching over some 2000 km of shoreline and reaching in places over 200 km offshore. The park is managed for multiple uses, with the overriding objectives being for protection, wise use, understanding and enjoyment.

The GBRMP authentically integrates a variety of natural resource values. The Authority holds paramount management power over this natural resource, with the possible exception of the Navigation Act. It enjoys widespread legitimacy. Although zoning is a basic management tool, the Park's activities are guided by a comprehensive strategic plan. Despite the enthusiasm surrounding the Park's organizational accomplishments, major environmental management concerns still abound. The reef itself appears irreversibly threatened by pollution and might be degraded substantially within this century. Queensland's tourism industry is itself a boom story and an ongoing threat to the reef. It is experiencing unprecedented growth which drives the natural resource management agenda onshore that in turn increasingly impacts marine resources and offshore island habitats. Brisbane's (Australia) thirst for water, for example, has destabilized the water table in offshore islands that have been commandeered as urban water supply reservoirs. Quarrying sand – a basic construction commodity – also jeopardizes island ecosystems and indigenous life.

Ottesen and Kenchington (1995) suggest that the debate concerning conservation of marine areas in general rests with two overlapping views on the management of their marine resources. The first focuses on the traditional 'sectioned' approach to marine resource management, while the second is an integrated approach. The traditional approach considers, regulates or stimulates a single natural resource management interest with little concern or understanding for other resource values. An integrated approach, on the other hand, contends that a single sector approach has had little success in marine environments, even when the primary objective is preservation. There are generally seen to be at least three reasons for this.

1. Boundaries have tended to be carved politically (see Chapter 9) rather than reflecting logical ecosystem dimensions. The ocean environment is dominated by a dynamic and powerful connectivity that ebbs and flows over time. These dynamics influence species mix, food sources and habitat that cannot be managed effectively on a smaller or more politically constrained spatial dimension.
2. There is little coordination among sector interests even when there are overlapping resource management and access concerns.
3. There are no mechanisms for competing interests to voice their concerns within the formalized management structure. Consequently, these interests tend to express themselves outside the management system and sometimes as powerful political influences.

Integrated coastal zone management

In response to the concerns outlined above, the integrative management of coastal environments has blossomed throughout the world in recent decades, but with varying effectiveness. In the USA, for example, the Coastal Zone Management Act of 1972, which was substantially amended in 1980 (Goodwin, 1999), set up a mechanism for a collaborative and voluntary Federal–State Coastal Zone Management Program (Humphrey *et al.*, 2000). After 30 years of experience, this Act's reach now extends to over 99% of the US coastline but, as one might expect with such a vast and diverse coastline, there is considerable variability in its implementation effectiveness from urban to rural landscapes and from one state to the next.

The scientific literature on integrated approaches in coastal areas, referred to more formally and universally as integrated coastal zone management (ICZM), sometimes garners criticism similar to single sector approaches. According to Kenchington and Crawford (1993), there are, nevertheless, important lessons to be drawn from ICZM that are relevant to the integrative management of marine areas generally, as well as to terrestrial environments. The more formalized ICZM approach attempts to do the following:

- Reflect an understanding of the natural and human systems and the interrelated links
- Seek an understanding between economic development, protection of the environment and social values
- Provide an over-arching jurisdiction or agreement so that measures may be applied to cover a single ecosystem even though it falls under several jurisdictions
- Involve partnerships between all interested parties, including individuals, non-governmental organizations (NGOs) and all levels of government
- Develop a strategic approach in order to avoid the negative effects of incremental decision making

Forestry: residual use or IREM?

Integrated resource management has for some time been advanced to enhance and conserve the forest landscape. In his inaugural speech upon appointment as the first professor of forestry in an Australian university, J.D. Ovington (1965) regretted the growing rift between foresters and conservationists emanating from the industry's increasing focus on monoculture plantations and single-minded exploitation. He argued that the re-emergence of the concept of multiple use of forestland in the 1960s was a step in the right direction to forge a new alliance between foresters and conservationists. With this approach Ovington suggested:

> Forest areas are required to be managed to provide not only wood and its products but wilderness and scenery, fish and wildlife, water and water catchment protection, as well as livestock and recreational facilities. This shift, from emphasis on a single

resource to consideration of the relative values of a number of resource uses, gives a much broader and sounder basis to forest management.

(pp. 4–5)

While there are numerous isolated examples of multiple-use forestry throughout the world, there are relatively few that go beyond a modest scale. As a case in point, the Colchester/Cumberland Counties Integrated Resource Management Pilot Project in Nova Scotia, Canada, when first proposed in 1996, ostensibly encompassed all Crown lands. The project's overall purpose was to integrate the management of forestry, minerals and the energy industries as well as recreation and wildlife protection. In practice, however, long-term cutting licences, pre-established mineral rights and long established energy easements remain the key determinants that dominate natural resource management decision making rather than *bona fide* attempts at integrating the concerns for conservation and ecosystem health (Bissix and Rees, 2001).

Elsewhere, in the British landscape, for example, market forces have been less sympathetic to the natural landscape than generally found in North America. Conversion from forests to agriculture has long been advanced in Britain (indeed since Roman times), resulting in the lowest forest cover in the whole of Europe. Remaining forests have survived largely as an amenity on large private estates or as speculative investment spurred on by substantial government subsidies. In the 20th century, there were numerous attempts at integrating forestry and agriculture, but few have succeeded economically or as practical conservation models. In response to growing criticism of forestry policy in the 1980s, the Countryside Commission in the UK – a complementary land management agency to the Forestry Commission – made a strong case for integrated forest management policies by arguing for more wide-ranging amenity objectives. Although the Countryside Commission emphasized the continued need for wood fibre in some form, it placed this requirement in the broader context of 'rural employment, recreation opportunities, [and] attractive landscape and wildlife habitats' (Blunden and Curry, 1988, p. 74).

In Spain, where temperate forests of the north give way to semi-arid wooded and treeless *dehasas* (open savanna), integrated forest management has, until recently, been a way of life rather than a target of formal private or public policy. Spanish forestlands as a whole are predominantly managed in small private holdings (~68%), while much of the remainder are in the hands of local government. Most private forests were managed for short-term revenue generation and used extensively for producing construction materials, charcoal and fuelwood. These landholdings were also widely used for producing cork, aromatic and medicinal plants, hunting, grazing and for fruit and nuts. Despite this traditionally broadly based integration of forest values and uses, forest sustainability has been threatened increasingly by shorter term and narrowly framed commercial goals driven by shortsighted government incentives. Since the mid-1970s, for example, the state has encouraged the conversion of the *dehasas* to plantations, which has impoverished biodiversity and increased susceptibility to fire (Palacin, 1992). It is clear that increased resource management intensity has undermined the cultural and biophysical sustainability as low-intensity integrative methods have been replaced by single-purpose and intensive resource management initiatives. Interestingly, a similar state of affairs appears to be prevalent in Portugal in its East Central Region (Rego and Coelho-Silva, 2002).

In North America, forestry has faced the challenge of integrating other values beyond single-use management (see Chapter 6). Forestry in the USA and Canada has evolved around the concepts of multiple use and sustained yield. The challenge of managing public forests for more than mono-crops has, however, stimulated considerable controversy in both countries over the last 25 years. Wondolleck (1988), for example, has documented several examples of the types of conflict that have arisen with respect to the management of public lands by the US Forest Service. In Canada, substantial conflict developed on the west coast of British Columbia as a result of forestry companies cutting old-growth timber on Crown (public) land. Pitched battles between environmentalists, government and timber companies characterized the 1980s as the decade of the 'War in the Woods'. Behan (1990) suggests that an integrated approach forces forest managers to think of forest management beyond simply *supply and demand* to a more holistic view that integrates many other values.

The search for sustainable agriculture

Catchment (watershed) management in Australia has become an important means to address increasing land and water degradation throughout the country. It is interesting to note that environmental problems in Australia have been reported, some the result of aboriginal practices, since Europeans first colonized this continent in 1788. Over centuries – since colonization – intensive agriculture has progressively and systematically undermined water quality and quantity as well as the productivity of its soils. This approach to natural resource and environmental management has threatened both the economic and social fabric viability of the region over the long term. Ranching and cropping, for instance, following clearing and bolstered by irrigation, has continuously degraded the landscape, leading to soil 'salinization' (or, more accurately 'alkalinization'). This problem is widespread in Australia, covering fully 3000 km^2 (Mannion, 1991). Despite its scale, general recognition and acceptance of this problem within Australia has been slow. Finally, forced by impending economic collapse, even in once highly productive farmlands, land managers have begun to adopt some aspects of IREM in areas such as the Murray–Darling basin in the southeast of Australia. The Murray–Darling is the longest river system in Australia, quenching an otherwise parched semi-arid region; it supports one of Australia's must successful wine industries among other important agricultural products such as grazing and cropping. Here, as in other regions of Australia, catchment management combined with sustainable regional development is slowly replacing singly focused agricultural practices.

'Integrated catchment management' (ICM) is one response to the salinity problem throughout Australia. All of the Australian states have formed policies and legislation to deal with the complex problems associated with watershed degradation. There are several variations of catchment management such as whole catchment management (WCM), total catchment management (TCM) and ICM. Each has its own legislative and policy form. As Bellamy (1999, p. 117) notes, 'At present the application of ICM varies widely throughout Australia, with some states such as NSW [New South Wales] having established legislation, while others such as WA [Western Australia] and QLD [Queensland] have no specific legislative powers. In effect there have been six experiments running in parallel but as yet no dominant model has emerged.' The various state approaches have evolved since the mid-1980s, largely driven by salinity management issues. Syme *et al.* (1994) provide an overview and history of the various state policies and legislation related to catchment management that occur across Australia.

ICM has emerged in Australia as a bottom-up approach to integrate water and land management, government agencies and stakeholders. Bellamy (1999, p. 117) defines ICM to be 'formal government policies, programs and laws for the management of land and water resources on a catchment basis'. ICM is a management process that is considered on a long-term horizon not only to change *present* problems and management practices, but to address *future* organizational cultures, local perceptions and values within the catchment area. ICM is a bottom-up approach to solving land and water degradation problems in watersheds, and the key elements of the process consist of voluntary programmes, community-based action and the formulation of partnerships. The challenges to ICM (see Chapter 9) generally consist of resistance of the various levels of government agencies to integration and cooperation (Hooper *et al.*, 1999) and lack of resources to finance and deliver catchment strategies (Syme *et al.*, 1994).

The development of ICM in Australia has had a strong influence in the evolution of IREM. Case studies in catchment management across Australia have contributed considerably to the empirical literature for IREM (see Mitchell and Pigram, 1989; Mitchell and Hollick, 1993; Bellamy, 1999; Bellamy *et al.*, 1999). Lessons from the Australian experience in catchment management are transferable to other regions of the world and will continue to inform best practices in IREM.

Frontier development

A considerable challenge to environmental management, sustainable development and IREM is frontier development. Frontier development refers to large-scale natural resource exploration and exploitation projects that are characterized by 'advanced technologies, high capitalization and a high degree of uncertainty' (Smith, 1993, p. 106). Such projects include hydroelectric development, pipelines, and offshore oil and gas development. They occur in remote areas and typically pit the economic development demands of an external society against the subsistence and cultural needs of indigenous or aboriginal communities. Often driven by exaggerated economic claims, frontier developments rarely live up to their initial billing as social stimulants, and frequently experience unanticipated or purposely concealed environmental impacts. This underperformance invariably produces economic shortfalls, creates environmental problems larger than those they claim to solve, and subsequently fails to meet local requirements or broader expectations (Smith, 1993; Baker *et al.*, 2000).

The greatest challenge in frontier development is to adequately integrate the environmental, economic and cultural needs of the host society with that of the external project sponsor and its supporting society. This challenge was acknowledged most forcefully in the Berger Inquiry in Canada, which reported in 1977. This inquiry grew out of growing opposition to exploiting and transporting Arctic oil from the Beaufort Sea landward to southerly, economically developed destinations (Gamble, 1978; Smith, 1993). As a result of its extensive investigation and deliberations, the Berger Inquiry established the notion of social impact assessment in resource development and provided a normative framework for addressing broad integrative socioeconomic and technical concerns for an ecosystem of a substantial scale. Although a landmark advancement conceptually and a highwater mark for environmental assessment (Mulvihill and Baker, 2001), with few exceptions the Berger framework for frontier development has largely been ignored as standard practice whether in Canada or elsewhere.

In Canada, the James Bay Hydro Development Project stands in stark testimony to the failure of standard impact assessments. Berkes, for example (1988, cited in Smith, 1993, p. 106), using the James Bay Hydro megaproject as his frame of reference, demonstrates 'that the accuracy of most assessment predictions are low. The notion of impact assessment considering all impacts associated with a project is, therefore, a fallacy. Indeed, in the James Bay instance, most impacts were either predicted incorrectly or were not predicted at all.' Smith concludes that although poor science was partially to blame, the lack of knowledge and the 'intrinsic unpredictability' of the impacts of large hydroelectric projects were also instrumental in the failure to foresee such development consequences (p. 107).

One notable exception to ignoring the lessons of the Berger Inquiry was the Great Whale project in Quebec (see Mulvihill and Baker, 2001). This project, a proposed successor megaproject to the James Bay Project by Quebec Hydro, was abandoned by the Quebec Parizeau government in 1993 when it became inevitable that growing environmental opposition and an imminent joint report from the federal and provincial government would lambaste Quebec Hydro's proposal on economic, environmental and cultural grounds. Regardless of any formal environmental and IREM planning process, the Cree and Inuit people, whose livelihood was to be most affected by this project, had managed to bring together substantial worldwide sympathy to make the Quebec government and, perhaps more importantly, its key customer in New York City think again about this project's overall worth.

Ecosystem and watershed management

The International Joint Commission on Great Lakes Water Quality Management also represents a defining moment in conceptualizing environmental management thinking, if not always successful in keeping pace with emerging environmental and social problems. It provides practical ways of integrating, often competing, natural resource management objectives into an acceptable and practical resource management regime. The Great Lakes basin represents a mammoth challenge to sustainable and integrated resource management, linking as it does two of the world's most prosperous nations – Canada and the USA – that collectively impose their worldview through trade and foreign policy on the rest of the globe. The Great Lakes basin encompasses over 480,000 km^2, captures one-quarter of the world's freshwater resources and includes a dense population and industrial concentration. Fully one-quarter of Canada's population and one-ninth of the much larger population of the USA live within this basin (Rabe, 1999).

The predictions of doom for the Great Lakes region a quarter of a century ago have not rung true; many indicators of environmental health have indeed improved. This is in spite of increased population, expanding industry and increased private transportation. The challenges for IREM nevertheless increase in response not only to regional pressures but also to more globally induced stresses such as changing weather patterns and resultant lake levels. Internally, the success in curbing point source pollution, first by command and control measures then by more integrated methods that tackled contamination in the design stage, have seen greater success than addressing the ubiquitous non-point source pollution from pesticides and waste management that are culturally imbedded in agricultural industry and, for example, cosmetic gardening.

Although enjoying some real successes in improving or maintaining environmental quality within the basin, often overlooked in the policies of the International Joint Commission, are the broader environmental costs to 'downstream' regions, whether that be the poisoning of beluga whales in the St Lawrence or the acidification of the Maritime Provinces' lakes. These are never clear-cut issues morally or politically even for the receiving regions, such as, for example, Nova Scotia. Although Nova Scotia (Canada's most easterly mainland province) regularly cries foul about Great Lakes air pollution, which damages its vulnerable inland waterways, it in turn chooses to generate electricity from high-sulphur coal that allegedly acidifies Newfoundland's lakes.

On a smaller scale, watershed management applies the principles of IREM across North America; this is similar to the catchment management experience in Australia. For example, Born and Sonzogni (1995) document a collaborative and integrated approach to managing the Black Earth Creek watershed in south-eastern Wisconsin. The case study in Chapter 9, the Cache River in southern Illinois, illustrates a more detailed example of the challenges facing smaller watersheds in implementing integrated resource and environmental solutions. Margerum (1997) provides an overview of other examples of watershed management across the USA. In Canada, the Fraser Basin Council in British Columbia has established an integrated approach to managing one of the largest watersheds in the province. The Council works in an advisory capacity in the watershed and supports scientific research and facilitation throughout the Fraser River watershed.

Legislative Foundations for IREM

There are a number of legislative foundations that supported the development of IREM processes in the Western world; some of these have been discussed earlier in this chapter, such as the US Inland Waterways Commission in 1909, the International Joint Commission on Great Lakes Water Quality Agreement signed in 1978, which was derived from the Boundary Waters Treaty of 1909 between the USA and Canada, the 1947 National Parks and Access to the Countryside Act in England and Wales, and the enabling legislation that created Kananaskis Country in Alberta, Canada in 1975.

In settling the West, the focus of the US federal government was initially to stimulate development. The US Department of the Interior was established in 1849, and the Homestead Act of 1862 and the Mining Act of 1872 flowed from this initiative, which in turn prompted the divestment of public lands to private hands. It was not until the passage of the Forest Reserve Act in 1891 that a shift in legislative emphasis was seen. This change in federal policy, although welcomed by conservationists, rankled both state governments and private concerns that remained preoccupied with economic development.

Although we must wait until 1969 for the introduction of formalized environmental impact assessment (EIA) in the USA, some environmental protection was afforded by common law under nuisance provisions such as pollution crossing state lines (Sussman *et al.*, 2002). EIA is in itself an affirmation of the basic principles of multiple objective management and IREM. Formalizing the EIA idea began in the USA in 1969 with the enactment of the National Environmental Policy Act (NEPA). This legislation recognized two important principles: (i) the efficacy of science in planning; and (ii) broadening analysis from single resource concerns of technical, financial and legal feasibility to wider integrated interests affecting the broader human condition. The NEPA required a detailed statement on environmental impact, unavoidable environmental effects, project alternatives, local short-term uses and long-term productivity, and irreversible and irretrievable resource commitments.

Initially, in practice, this legislation proved difficult to manage, due in large measure to inadequate biophysical data, insufficient time, scarce social data and problematic weighting systems, but in time the process became increasingly cumbersome. It finally buckled under the weight of litigation and it failed to consider viable alternatives rather than continually offering incremental adjustments to initial proposals. In calling for the rethinking of impact assessment rather than abandoning the process altogether, Smith (1993) appealed for a thorough redesign that more closely integrated the science of environmental analysis with the politics of resource management. As a basic design platform, he called for institutional reorganization to promote impact mitigation. This 'new approach', according to Smith, would rely on an integrative model of natural resource management rather than a process driven or stalled by conventional inter-agency power brokerage (see also Chapter 7, 'Political bargaining models').

In 1966, the Canadian government established a rather ambitious integrated resource management policy by establishing the Department of Indian and Northern Affairs to oversee the cultural and economic development in Canada's north. According to Mitchell and Sewell (1981), however, this department struggled to integrate both the development and preservation of the north's natural resources, as well as seeking economic advancement and cultural protection for its indigenous populations. In 2002, the Canadian government introduced new legislation to replace this Act; it is, however, far too early to see how this revised approach might affect cultural issues and resource and environmental management. Environmental assessment (EA) for the whole of Canada began as a cabinet directive in 1973, was modified and adopted under the Department of Environment Act (1984) as a set of guidelines, and was incorporated in separate legislation in 1992 (Valiante, 2002). The most comprehensive law concerning the environment in Canada is the Canadian Environmental Protection Act (CEPA) of 1988, which was substantially amended in 1999 and proclaimed in 2000 (Valiante, 2002). On the other side of the globe, the Integrated Planning Act of Queensland (1997) and the Resource Management Act (1987) of New Zealand provide examples of legislation that attempts to integrate the development process. While the relative success of these Acts is debatable, they do provide a new genre of legislation that will challenge the foundations and principles of IREM.

Conclusions and Future Challenges

In the preceding pages, we have seen where single-minded solutions, such as flood control, created new concerns such as soil degradation and public health concerns. We saw this in Kananaskis Country where there is a constant need to mediate competing interests and resolve differences fairly. We have also seen that IREM is a continuous process where the need for scientific information is great, but also where managerial and political acumen must scrutinize and mediate science to make informed and practical judgements rather than making purely mathematical or scientific calculations. We also understand that money alone, such as in the TVA project, cannot solve environmental and social development problems unless the political will exists. In addition, we know in frontier development that engineering know-how on its own will not distribute prosperity or social costs equitably. Furthermore, we appreciate how the Joint Commission of the Great Lakes has made enormous strides under complex political conditions to advance IREM principles. Despite these advances, we understand that there remains much work to be done in dealing with, for example, non-point source pollution despoiling the Great Lakes watershed. These more complicated political and hard to pin down environmental problems are likely to challenge Great Lakes' policy architects in the years to come so as to better resolve the interlocking problems of, for example, water quality, navigation, water supply and general ecosystem health.

Even with their limitations and deficiencies, perhaps the most encouraging examples of IREM are those of Alberta's K Country, the national parks in England and Wales, and Australia's Great Barrier Reef Marine Park. It seems no coincidence that they are all in protected areas but, encouragingly, as lessons for other jurisdictions, they are also in working landscapes where residents eke out a reasonable existence in the midst of relatively strict environmental regulations. In all these examples, resource utilization and daily commerce is as important as resource conservation, and vice versa. As such, these IREM examples are exemplars that offer important insights as to how seemingly competing resource management objectives might reasonably coexist.

This brief and rather selective history of IREM attests as much to the intricately woven complexities of the resource management decision-making process as its does to any unifying rational and scientific IREM planning model. Despite a lack of over-arching theoretical and procedural approaches to resource and environmental management, the authors foresee that ecosystem management and its implicit IREM processes are likely to dominate the political and resource management agenda for the anticipated future. This leads to the need to map out in detail in this text the issues that underscore and potentially impede its further development.

The case study at the end of this chapter provides a modern-day history of farm-based, integrated resource and environmental management. This case concerns Folly Farm in the west of England, which began life as a working farm, was held by the landed gentry of a century ago as a working estate – a place where the pleasure of the well-to-do was as important as traditional farm income, and where in more modern times the farm has been worked to enhance the countryside's wildlife as much as it provides more traditional farm income. In many ways it is a living history of bygone farming methods – the farm's managers attempt to revive traditional farm practices, such as charcoal production, that go back thousands of years. Folly Farm integrates farming and forestry and attempts to mimic organic farming methods that were the mainstay of farm production half a century ago. A question that this operation implicitly asks every day is whether it is realistic to think of returning modern-day farms to more integrative farm practices, similar in intent if not in specific practice to those routinely practised in the past. Farmers in the Murray–Darling watershed are contemplating this very same question, and similar decision calculations are being played out in a number of resource management sectors. The goal in farming should be, of course, to make its farming practices more ecologically sensitive and, as a necessary goal to achieve this, make farming more sustainable for the future.

Two hundred years of humans' largely unrestrained and rapidly accelerating natural resource exploitation has increasingly intensified the need for an all-out effort in sustainable development. The history of resource exploitation over the past century and especially the last 10 years plainly shows on a global scale, however, that society's momentum is largely working in the opposite direction. This brief and selective history of resource management shows both IREM's promise and its limitations in addressing the many environmental problems that impede charting a course for a more sustainable society. In the following chapter, we will examine the driving forces of modern-day natural resource management to measure further the stresses on our natural environment and to assess the need for IREM.

Case Study
Folly Farm, Somerset, England

TREVOR DAVIES

Preamble. Although Folly Farm is very successful at IREM by meticulously managing for farm production, commercial forestry management, environmental education and biodiversity management, it poses an interesting question of whether its ongoing management practices, that are made possible by generous public donations, various grants and dedicated voluntary labour, can be used as a useful template for IREM in modern commercial farming. As you will

An overview of the Folly Farm site, Somerset, England. Photo by Glyn Bissix.

The old farmhouse at Folly Farm. Photo by Glyn Bissix.

see, Folly Farm is in essence a museum of traditional farming methods that mimics old farming methods that were practised several decades ago of necessity rather than by design.

Introduction

In response to the food shortages experienced in the Second World War, British agricultural policy, bolstered by generous subsidies, focused on self-sufficiency and developed one of the most intensive agricultural systems in the world. While per acre production was very high, this came at a cost to environmental quality and required massive inputs of fertilizers, pesticides and petroleum products. In the past decade or so, more attention has been paid to restoring the environmental quality of the countryside by reducing certain subsidy forms and introducing others that stimulate more environmentally sensitive resource management methods. Folly Farm is located approximately 10 miles south of Bristol, a city of some half a million people in southwest England. The farm sits in an undulating rural area, close to the Mendip Hills, a limestone upland that also includes Cheddar, the origin of the famous cheese. The site is presently owned by the local Avon Wildlife Trust and is managed by the Trust as a nature reserve which also includes traditional farm operations. It was purchased by the Trust in 1987 from a member of the Strachey family, who were Lords of the local manor – Sutton Court. Prior to acquisition by the Trust, it had been let as a tenanted farm, and for most of the 20th century was used for dairy cattle, sheep and pigs. In the 1780s, it had been landscaped as an ornamental farm (*ferme ornée*) so that the gentry of the manor could enjoy a coach ride and admire the views before dinner.

Farm Description

The farm is set into a curved ridge of land that provides wonderful viewpoints and, perhaps more importantly, has discouraged intensive farming on its steep hillsides (toward the rear of the farm). This has helped maintain a habitat of unimproved natural grassland and has also preserved an area of ancient woodland on its slopes. Significantly, the lower flatter fields that have been ploughed and fertilized repeatedly are less interesting for wildlife. This farmland then has three main components: (i) unimproved grassland; (ii) improved grassland; and (iii) woodland. The unimproved grassland and woodland form the nature reserve, while the improved grassland is given over to sheep and beef cattle grazing. The sheep and cattle are owned and managed by a local herdsman who rents the grazing rights. Interestingly, some areas of improved grassland have recently been planted to extend the woodland, while beef cattle graze the unimproved grassland during the winter months. The 17th-century farmhouse, which is currently vacant, occupies the farm's centre. Presently, the outbuildings are utilized as workshops, tool sheds and rest rooms for the volunteer workforce, and lecture rooms for the various educational courses held there. This building complex is presently being converted into a residential environmental education centre.

Unimproved grassland

Unimproved grasslands constitute 50 acres located on a curved ridge with moderate to steep slopes rising approximately 300 feet, forming a natural amphitheatre. The soils are mainly neutral limestone and clay, which support unimproved natural grassland interspersed with large areas of scrub (mainly blackthorn and bramble), species-rich hedges and mature trees.

The whole of this sector has been designated a 'Site of Special Scientific Interest' (SSSI) which is a central government designation that gives some level of legal protection to help maintain integrity as an unimproved grassland habitat. These species-rich grasses support many important flora such as cowslips, heath spotted orchids and pepper saxifrage, while among the butterflies, marsh fritillary, brown argus and marbled white predominate. Roe deer have been seen browsing, and barn owls hunt over the tussocky grass; in addition, kestrels occasionally hover over the grassland, while buzzards are a common sight soaring at height over these grassy slopes. A key indication of the undisturbed nature of the grassland is the number of large (12–18 inches high) anthills that dot these slopes.

Despite the predominance of natural processes at work in this sector, challenges in maintaining a desired wildlife habitat occur. For example, scrub would overrun the grassland if left uncontrolled. Therefore, a level of control is achieved by cutting back the scrub with both powered brush cutters and hand tools. The debris is then carefully burned to avoid undue soil enrichment. Due to the presence of nesting birds in these areas of blackthorn and bramble, however, control strategies can only be implemented during the October–March period. Poaching, trampling and breaking the soil into wet muddy patches, which is caused by cattle pulverizing the soil with their hooves during wet conditions, is also a problem. This is particularly problematic at gateways and 'narrows' that occur between clumps of scrub, and is especially evident toward the end of the normally wet winter. In addition to degrading wildlife habitat, this also makes the footpaths impassable. As this is an open-access farm with footpaths especially created to encourage visitors, and because this is a SSSI where any form of imported surface material is not permitted, it is important to manage this problem in an environmentally acceptable way. This is normally achieved by widening narrow access points where possible, or by forming alternative routes for walkers.

Improved grassland

The improved grassland areas are fenced mainly for sheep grazing, but are also used, to a lesser extent, for cattle. Grazing results in constantly cropped grass that provides very little habitat for small mammals or a variety of flora. This grazing area was considerably reduced in 1999 by the creation of extensive 'barn owl corridors'. Moving the stock fencing back from the field boundaries by some 30 yards, forming broad continuous runs of rough grassland, created these corridors. This strategy provides an improved habitat for small mammals such as field moles that form the basic diet of barn owls. Nationally, barn owls have been in decline. This is thought to be due mainly to the loss of grass margins that resulted from intensive agricultural development. A recent site survey indicates that moles are already colonizing these corridors, and all owl boxes placed in proximity show signs of roosting. Most importantly, grazing intensity has been adjusted to the reduced grazing area; significantly, organic farming techniques have been introduced recently, and there appear to be few conflicts between the farming activities and nature conservation. Indeed, light grazing of the SSSI is not only acceptable, it is considered necessary to produce a desirable habitat.

Woodland

There are two major woodland areas within the boundaries of Folly Farm. While they have many similarities, Dowlings Wood has been designated an SSSI, while the other, Folly Wood, has not. This difference in designation is due in part to past management practices and in part to location. The more protective restrictions inherent in an SSSI designation impacts the woodlands' use and management. This difference has resulted in Dowlings Wood having a wider variety of flora and fauna species and consequently being of greater natural history interest. In addition to the two main woodlands, there are also three orchards on the farm. One is adjacent to the farmhouse and consists of mature apple trees, while the other two have recently been completely replanted with traditional apple varieties, some of which are becoming quite rare.

Dowlings Wood

Dowlings Wood extends approximately 25 acres over a sloping site that is predominantly fairly wet due to poor drainage of the underlying soil. The trees are mainly ash, pedunculate oak, wych elm and hazel, while the ground cover includes ramsons, cuckooflower, bluebells, dog's mercury, wild garlic, early purple orchids and herb paris. A large badger sett is a prized wildlife asset and is situated on the edge of the wood. Roe deer, grey squirrels and rabbits are also present.

An extensive management programme of coppicing is being carried out to promote greater diversity of the habitat structure and, as a consequence, provides greater interest for wildlife. Coppicing is an ancient forest management technique that involves the selective cutting down of trees to ground level and the nurturing of sprouting of fresh growth from the base. The selected mature trees left standing help provide a range of habitat within the coppiced area. Deer browsing of the tender regrowth is a major problem that has been overcome by surrounding each coop with a 6 ft high chestnut paling fence. While this fencing detracts from the natural appearance of the woodland and sometimes necessitates re-routing of the footpaths, it is considered essential for successful regeneration and sustainable wood production. Bat, bird and dormouse boxes have been sited throughout the wood to encourage population stability as well as help monitor their presence. From an IREM perspective, it is important to note that although a 'way-marked' walk is routed through Dowlings Wood, other, more intensive activity is not encouraged. It is also important to note that the felled timber created from coppicing has been used for making charcoal on-site, firewood and other woodland products.

Folly Wood

Folly Wood covers an area of some 20 acres of mature woodland that has been extended to approximately 60 acres by the recent planting in adjacent improved grassland. As with Dowlings Wood, the mature area has been coppiced at certain locations. Folly Wood is used extensively for educational and recreational purposes, particularly during the summer. It has several footpaths running through it, including one that has been levelled and surfaced with crushed stone to provide an 'access for all' trail. Consequently, these trails are suitable for people using wheelchairs and parents with children in pushchairs. Courses are held throughout the summer term for classes of school children to improve their understanding of, and relationship with, nature and the environment. Much of this activity takes place within Folly Wood.

A deep ravine, through which runs a small stream, dominates the east end of the wood. To one side of the ravine and easily observed from a platform on the 'access to all' trail is a badger sett; two fox earths have also been noted in this area. Among the birds recorded are buzzard, sparrowhawk, great spotted woodpecker, green woodpecker, goldcrest, chaffinch, blackcap, great tit and wood pigeon.

The newly planted areas mentioned earlier were formed on improved pasture previously grazed by sheep that had little biodiversity. The planting was carried out between 1993 and 1997, using mainly one-third ash, one-third oak and one-third a mixture of wild cherry, field maple, crab apple, Scots pine and holly. It is expected that only one in nine of the planted stems will survive to maturity; however, the variety and the unpredictability of which stems and which species survive promises a species-rich and diverse, although even-aged forest structure in the future. Wide rides (areas left for grasses) have been incorporated throughout the planted areas to provide access and to maximize woodland edges to give habitat diversity. In some areas, the saplings were fitted with plastic sleeves to protect against browsing, while other areas were completely surrounded with deer-proof fencing. Experience has shown that protection with tubes has been most beneficial for tree vitality; perhaps because of this shielding, it also provides protection against the elements during the early stages of growth.

Farm Management

The day-to-day cost of running Folly Farm is met by the Avon Wildlife Trust, which is funded by membership subscriptions and occasional grants and legacies. Although Folly Farm is the largest property owned by the Trust, it also manages 30 other nature reserves as well as its head office in Bristol. Central government money, funded through the Countryside Commission, is available for specifically defined tasks such as hedge laying, tree planting and provision of public access. Capital projects may also be partly funded through the Heritage Lottery Fund. One such project has been approved to fund a renovation and conversion of the farmhouse and its outbuildings to make them suitable to accommodate residential courses.

There are no permanent staff currently in residence at the farm, or indeed employed full-time at the site. The general maintenance of the farm buildings and nature reserve is carried out, within their abilities, by a group of volunteers 1 day a week. However, this is an ageing workforce comprised of semi-retired men and women, including this case study's author. This workforce is split into three work parties consisting of six conservation workers, three building maintenance workers and a smaller group who maintain the garden at the front of the farmhouse. The conservation party spends its time clearing scrub, forming and maintaining footpaths, replacing and repairing fences and gates, as well as constructing simple bridges and viewing platforms. Bird boxes have been made and sited, and require annual inspection. This core group's efforts are occasionally augmented by large groups of usually younger people who come in on weekends and carry out the more arduous manual labour such as scrub clearance, an almost continuous task from November to March. Contractors carry out large-scale coppicing, and work such as tree felling and major fencing projects as and when required.

The farm is open to public access at all times, and visitors are encouraged. A car park is provided near the farmhouse from which three 'way-marked' walks emanate; one of which is the 'access for all' trail. A fourth walk is planned and will be available in the future when way-mark signs, stiles and 'kissing gates', etc., are in place. One extremely positive note is that very little vandalism or damage has occurred despite the open and generally unsupervised nature of the site. This may be due to its distance from a population centre; people have to make a special effort to visit the site and in general it seems that visitors are sympathetic to the Avon Wildlife Trust's aims.

Discussion Questions

1. Identify six initiatives regarding the management of natural resources from your community or region that appear to support the liberalization of the global trade agenda and the promotion of a sustainable society agenda (three of each). To what extent is each of these initiatives implemented successfully? To what extent do these examples overlap in their trade liberalization and sustainability objectives?

2. The Muskingum Watershed Conservancy District, which was formed in 1933, was charged with the responsibility of integrating 'multiple purposes, multiple means and multiple participants'. Consider a natural resource management initiative in your home region and either identify how this project involves multiple purposes, multiple means and multiple participants, or think about how it might do so to better serve society.
3. Given the experience of places such as Dartmoor National Park, UK, Kananaskis Country in Alberta, Canada, and the Great Barrier Reef in Queensland, Australia, what do you consider to be the trade-offs for single interest natural resource exploitation industries (e.g. forestry, fishing and mining) that must operate in multiple-use, integrated resource and environmental management situations?
4. The text suggests that there are few, if any, examples of large-scale integrated forest resource management projects throughout the world. What influences forest managers to pursue single purpose – usually forest fibre extraction – objectives rather than practice more integrative, multiple-objective strategies?
5. Australian farmers are increasingly looking toward integrative resource management methods. What factors lead them towards this and what influences seemingly stand in their way? Are there examples in your own region that illustrate similar circumstances?
6. Frontier developments are said to be a substantial challenge to environmental management, sustainable development and integrated resource management, and rarely live up to their initial billing. What factors create the most difficulty in improving the life-style of local inhabitants in these regions? Identify an example of frontier development reasonably close to your own region and discuss the pros and cons of this development.
7. What management challenges do you see as critical in trying to corral a large residential population with a broad range of industrial and recreation interests to follow a coordinated approach to ecosystem-based resource and environmental management?
8. The US National Environmental Policy Act (NEPA) was said to falter because of inadequate biophysical data, insufficient time, scarce social data and problematic weighting systems in developing an environmental impact assessment (EIA). Identify an example of each of these concerns for a real or hypothetical case and explain why each would create difficulty in making a useful impact statement.

Case Study Discussion Questions

9. While it can be said that the primary aim of the Avon Wildlife Trust and its volunteers at Folly Farm is to enhance the area's value as a site of nature conservation, to provide access and to provide interpretation which, hopefully, will improve visitors' educational and recreational opportunities and increase their enjoyment of wildlife, to what extent can it also be seen as a laboratory for sustainable farming and integrated resource and environmental management?
10. To what degree do you believe that the resource management techniques employed at Folly Farm are both useful for enhancing environmental quality and practical in today's farming environment?
11. Who benefits and who loses from farmers adopting more environmentally friendly resource management practices?
12. If a farm near you was to manage its land resources more to enhance biodiversity, to what extent do you believe it is reasonable to subsidize farmers from tax revenue to fund the farmer's extra costs and forgone profits? From what level of government should these subsidies be drawn? What other ways are possible to sustain these environmentally sensitive practices over the long term?

References

Atkinson, N. (1991) *Dartmoor National Park: Second Review 1991*. Dartmoor National Park Authority, Bovey Tracey, UK.

Atkinson, N. (1992) *Dartmoor National Park Local Plan: Including Minerals and Waste Policies – Consultation Draft 1992*. Dartmoor National Park Authority, Bovey Tracey, UK.

Baker, D.C., Young, J. and Arocena, J. (2000) An integrated approach to reservoir management: the Williston Reservoir case study. *Environmental Management* 25(5), 565–578.

Behan, R.W. (1990) Multiresource forest management: a paradigmatic challenge to professional forestry. *Journal of Forestry* 88(4), 12–18.

Bellamy, J. (1999) *Evaluation of Integrated Catchment Management in a Wet Tropical Environment: Collected Papers of Land and Water Research Development Corporation*. (R & D Project CTC7–2) Institutional arrangements for ICM in Queensland. CSIRO Tropical Australia, Brisbane, Queensland, Australia.

Bellamy, J., McDonald, G., Syme, G. and Butterworth, J. (1999) Evaluating integrated resource management. *Society and Natural Resources* 12(4), 337–353.

Bissix, G. and Bissix, S. (1995) Dartmoor (UK) National Park's landscape management: lessons for North America's Eastern seaboard. In: Herman, T., Bondrup-Nielsen, S., Martin, J.H. and Munro, N. (eds). *Ecosystem Monitoring and Protected Areas*. Science and Management of Protected Areas Association, Wolfville, Nova Scotia, Canada, pp. 563–571.

Bissix, G. and Rees, J.A. (2001) Can strategic ecosystem management succeed in multiagency environments? *Ecological Applications: a Journal of the Ecological Society of America* 11(2), 570–583.

Blunden, J. and Curry, N. (1988) *A Future for Our Countryside*. Basil Blackwell, Oxford, UK.

Born, S. and Sonzogni, W. (1995) Integrated environmental management: strengthening the conceptualization. *Environmental Management* 19(2), 167–181.

Boyack, S. (2000) *National Parks for Scotland: Consultation on the National Parks (Scotland) Bill*. Scottish Executive, Edinburgh, UK.

Gamble, D.J. (1978, March) The Berger Inquiry: an impact assessment process. *Science* 19, 946–952.

Goodwin, R.F. (1999, April–September) Redeveloping deteriorated urban waterfronts: the effectiveness of US coastal management programs. *Coastal Management* 27(2–3), 239–269.

Hooper, B., McDonald, G. and Mitchell, B. (1999) Facilitating integrated resource and environmental management: Australian and Canadian perspectives. *Journal of Environmental Planning and Management* 42(5), 747–766.

Humphrey, S., Burbridge, P. and Blatch, C. (2000) US lessons for coastal management in the European Union. *Marine Policy* 24(4), 275–286.

Ketchington, R. and Crawford, D. (1993) On the meaning of integration in coastal zone management. *Ocean and Coastal Management* 21(2), 109–127.

Mannion, A.M. (1991) *Global Environmental Change: a Natural and Cultural Environmental History*. Longman Scientific and Technical, London.

Margerum, R.D. (1997) Integrated approaches to environmental planning and management. *Journal of Planning Literature* 11(4), 459–475.

Mazmanian, D.A. and Kraft, M.E. (1999) The three epochs of the environmental movement. In: Mazmanaian, D.A. and Kraft, M.E. (eds) *Toward Sustainable Communities: Transition and Transformations in Environmental Policy*. MIT Press, Cambridge, Massachusetts, pp. 3–43.

Mitchell, B. (1986) The evolution of integrated resource management. In: Lang, R. (ed.) *Integrated Approaches to Resource Planning and Management*. University of Calgary Press, The Banff Centre School of

Management, Calgary, Alberta, Canada, pp. 13–26.

Mitchell, B. and Hollick, M. (1993) Integrated catchment management in Western Australia: transition from concept to implementation. *Environmental Management* 17(6), 735–743.

Mitchell, B. and Pigram, J. (1989) Integrated resource management and the Hunter Valley Conservation Trust, NSW, Australia. *Applied Geography* 9, 196–211.

Mitchell, B. and Sewell, D. (eds) (1981) *Canadian Resource Policies: Problems and Prospects*. Methuen, Toronto, Ontario, Canada.

Mulvihill, P.R. and Baker, D.C. (2001) Ambitious and restrictive scoping: case studies from Northern Canada. *Environmental Impact Assessment Review* 21, 363–384.

Oltmann, R. (1997) *My Valley: the Kananaskis*. Rock Mountain Books, Calgary, Alberta, Canada.

Ottesen, P. and Kenchington, R. (1995) Marine conservation and protected areas in Australia: what is the future? In: Shackell, N.L. and Willison, J.H.M. (eds) *Marine Protected Areas and Sustainable Fisheries*. Science and Management of Protected Areas Association, Wolfville, Nova Scotia, Canada, pp. 151–164.

Ovington, J.D. (1965) *The Role of Forestry: an Inaugural Lecture*. The Australian National University, Canberra, New South Wales, Australia.

Palacin, P.C. (1992) Spain. In: Wibe, S. and Jones, T. (eds) *Forests – Market and Intervention Failures: Five Case Studies*. Earthscan, London, pp. 165–200.

Rabe, B.G. (1999) Sustainability in a regional context: the case of the Great Lakes Basin. In: Mazmanaian, D.A. and Kraft, M.E. (eds) *Toward Sustainable Communities: Transition and Transformations in Environmental Policy*. MIT Press, Cambridge, Massachusetts, pp. 247–281.

Rego, F.C. and Coelho-Silva, J.L. (2002) Rural change and resource management in Mediterranean mountain areas: the case of Serra da Malcata, Central East of Portugal. In: Ewert, A., Voight, A., McLean, D., Hronek, B. and Beilfuss, G. (eds) *Proceedings of the 9th International Symposium on Society and Resource Management*. Indiana University, Bloomington, Indiana, pp. 217–218.

Richardson, E. (1973) *Dams, Parks and Politics*. University of Kentucky Press, Lexington, Kentucky.

Smith, L.G. (1993) *Impact Assessment and Sustainable Resource Management*. Longman Scientific and Technical–John Wiley & Sons, New York.

Squires, M.T. (1997, October) An investigation of the effectiveness of the multiple-use concept using Kananaskis County, Alberta, as a case study. Unpublished Honours thesis, Acadia University, Acadia, Nova Scotia, Canada.

Sussman, G., Daynes, B.W. and West, J.P. (2002) *American Politics and the Environment*. Addison Wesley Longman, New York.

Syme, S., Butterworth, J. and Namcarrow, B. (1994) *National Whole Catchment Management. A Review and Analysis of Process* (Occasional Paper Series No. 01/94) Land and Water Resources, Research and Development Corporation, Canberra, New South Wales, Australia.

Valiante, M. (2002) Legal foundations of Canadian environmental policy: underlining our values in a shifting landscape. In: Van Nijnatten, D.L. and Boardman, R. (eds) *Canadian Environmental Policy: Context and Cases*, 2nd edn. Oxford University Press, Don Mills, Ontario, pp. 3–24.

Wondolleck, J.M. (1988) *Public Lands Conflict and Resolution: Managing National Forest Disputes*. Plenum Press, New York.

4 The Driving Forces Underlying IREM

Introduction

Throughout history, changing environmental factors, social factors and cultural advances have influenced the human condition by altering the way in which societies have responded to challenges in using their natural resources. It has become apparent, over past centuries, that various societies have employed natural resources in increasingly sophisticated ways in order to meet their ever-expanding needs. This chapter focuses on current forces impacting natural resource management issues and examines the overall need for more sustainable natural resource management practices, such as the use of IREM.

Although natural resources are often shaped by natural processes that are beyond human control (e.g. wildfires in Australia and California, floods, drought, etc.), there is an ever-widening range of human activities that affect the way natural resources are used, managed and consumed. As a consequence, these activities can exert both positive and negative impacts upon the natural environment and society, ranging from individual behaviours such as recycling, to the collective behaviour of communities, industry and other groups (e.g. creating or reducing air pollution). Not surprisingly, both individuals and groups are affected by 'driving forces' which can serve to influence their attitudes, beliefs and subsequent behaviours toward the environment. While we should appreciate that each driving force is important in its own right, it is also necessary to understand that each is often inexorably interlinked. The interconnectedness and interdependence of these driving forces form a complex web of cause and effect relationships, where deliberate remedy or unintended change in one area can cause a serious ripple effect elsewhere. In turn, this 'ripple effect' creates a whole series of unintended or unexpected consequences.

Driving forces are defined as factors and events which influence the types and magnitude of various uses of, and impacts on, natural resources. While there are numerous driving forces affecting natural resources, they generally fall into two categories: *biophysical* and *anthropogenic*. Biophysical driving forces include physical events such as volcanism (Berner, 1990), solar input (Lean and Rind, 1996) and global climate changes (Crowley, 1996). Anthropogenic driving forces include 'change agents' that are caused by, or related to, humans, such as demographic changes and their effects on public values toward natural resources, and public perceptions of natural resources. The chapter will focus on a number of driving forces currently impacting natural resource use and management. These driving forces include: (i) demographic changes; (ii) individual choice and markets; (iii) environmental inequity and justice; (iv) policy tools; (v) environmental philosophy; (vi) public opinion and the media; (vii) technology; (viii) aboriginal concerns; (ix) international conventions and conferences; and (x) why these factors serve to create the need for a more integrated approach to resource management. Finally, while there are numerous other driving forces (e.g. land fragmentation, urbanization, environmental agencies, etc.), the underlying discussion of any one of these issues tends to emphasize the complexity and interconnectedness of both the problems and forces impacting the situation.

Demographic Changes

Management of our natural resources can no longer be thought of as periodic interventions, separate from the broader realities of an individual's life or that of society. Not only do many aspects of an individual's life influence and use natural resources, but many environmental issues, such as air and water quality, also have direct impacts on individuals and, collectively, society (Portney, 1992). Moreover, demographic variables such as age, gender and residence can play important roles in how individuals rely on natural resources and how these natural resources are generally perceived by an individual (Virden and Walker, 1999; Cordell and Tarrant, 2002).

For example, it has long been understood that differences exist between those visitors to public lands who come from an urban environment, and those visitors coming from a more rural environment. In many cases, the rural-based visitor is often more interested in a more primitive experience, while the urban-based visitor may be attracted to developed sites with more amenities and facilities (Knopp, 1972). Part of the explanation for this difference is due to the variation in the available opportunities and levels of stress of the urban resident. Recent research, however, has suggested that this rural–urban dichotomy is breaking down, if indeed it ever really existed (Spencer *et al*., 1992). In its place there is a growing tendency toward a 'homogenization', or blending of cultures, regardless of a person's place of residence. Indeed, one of the most pressing environmental issues for some locations is the current increased migration of people from urban to rural locations, thus creating rural subdivisions and 'wildland–urban interface' lands. Moreover, many of these wildland interface lands also represent 'gateway' or 'portal' communities which attract large numbers of transient visitors who are just passing through but bring their own unique set of impacts to the natural resource and local culture. Moreover, these interface lands increasingly present a challenge to the resource manager in terms of local communities, impact on the natural resources, watershed management and wildfire management (Bradley and Bare, 1993; Chase, 1993; Ewert *et al.*, 1993; Ewert, 1996a; Wolosoff and Endreny, 2002).

Thus, it is not surprising that a number of demographic factors, such as place of residence, can have profound impacts upon many of the attitudes and consequent behaviours of people. In this chapter, the demographic characteristics examined are population growth rate, population age, growth of minority groups, population distribution, diversity of household types and disparity of socioeconomic resources. In addition, underlying this discussion is the assumption that an individual's predisposition toward natural resources is often not a simple set of attitudes and behaviours, but rather is dependent upon a number of complex variables such as desired expectations, intentions to behave, persuasion and beliefs (O'Keefe, 1990; Petty *et al.*, 1992).

Population Characteristics

There are four major patterns regarding population characteristics which persuasively affect behaviours and beliefs surrounding natural resources in significant ways. These patterns include: (i) decreasing rates of population growth, particularly in North America and Western Europe; (ii) an ageing population; (iii) an increasing number and proportion of minority citizens; and (iv) a continuing redistribution of the population in urban and suburban areas.

Taken as a whole, these four patterns represent an overall situation that is fluid and dynamic. Generally, these patterns represent a public that is *increasingly* concerned about natural resources, but is *less* familiar and knowledgeable about those resources.

Population growth rate

Three processes serve to determine population growth: fertility, mortality and migration (Murdock and Ellis, 1991). Population growth is projected to continue to grow in the future, with anticipated maximum growth rates being reached between the years 2040 and 2050, and declining thereafter. As a result, this population growth will influence natural resource management in a number of ways. For example, as the population increases, there will be a concomitant increase in the demand for particular natural resources, such as wood fibre. This increased demand will place more 'pressure' on managers of specific natural resources (such as timber resources) to provide more of that product. The competition for various natural resources will probably increase over time. This is particularly true in locations faced with heavy competition between different segments of society, all demanding access and use of the same resource base.

Ageing of the population

By the year 2050, it is estimated that nearly 23% of the US population will be 65 years of age or older (Spencer, 1989). Heinrichs (1991) suggests that not only is the overall population becoming older, but individuals are retiring at an earlier age. Siegel and Taeuber (1986) report that older populations will require more health-related products and a higher level of service. Regarding natural resource use, two criteria need to be considered: (i) health of the person; and (ii) degree of engagement. Doka (1992) points out that older adults, who are healthy and seeking out new and stimulating environments, are the more likely to appreciate activities in natural environments such as parks and other protected lands. For example, there will be an increased demand for more facilities to be built such as resorts and marinas, with a corresponding decrease in the demand for remote, backcountry opportunities. Thus, from the perspective of providing public accommodations in natural landscapes, management will need to provide for opportunities that are physically active and amenity-rich, while at the same time allowing for the capabilities and desires of the older visitor.

Growth of ethnic diversity

Concurrent with the issue of ageing is the increase in minority populations. In North America, this is evident by the fact that Anglo/White populations tend to be older than minority populations. Minority groups represent faster growth rates within the total population when compared with Anglo/White populations. For example, minority members will grow from 25% of the overall population in the USA in 1990, to more than 40% by 2050.

Differences also emerge when examining perceptions of natural resource use. For example, within a recreation and leisure framework, Anglo visitors have often engaged in activities such as backcountry hiking and hunting. With the increasing presence of other cultures and ethnic groups, other activities may become more prevalent. In addition, recent research has suggested that culture and ethnic background have some influence on the resource users' motivations, appeal and perceived problems concerning specific public and protected lands areas (Ewert and Pfister, 1991; Floyd, 1999).

Ethnic, racial and gender differences among the public also appear to lead to variations in how natural environments and resources are perceived and preferred among different groups (Virden and Walker, 1999). Whether or not there are differences, these findings are in line with the suggestion that the way individuals view natural environments and, by extension, natural resources can be strongly influenced by their sociocultural background (Saegert and Winkel, 1990; Williams and Patterson, 1996; Cordell and Overdevest, 2001).

On a similar note, Allison (1993) points out that simply providing opportunities is not a sufficient management strategy. Rather, specific actions such as taking into account the cultural aspects and expectations of a particular racial or ethnic group need to be considered when developing resource management plans. Based on work by Hollingshead (1992) and Allison (1993), this 'strategy building' needs to consider the following components.

- Look for multi-forms and dimensions within a cultural group. There is no single set of characteristics or attributes that define a particular cultural group. The needs and expectations of group members are varied and not monolithic.
- Beware of using standards that are based on White or dominant group characteristics. These characteristics are often varied, change over time or have become irrelevant to the situation.
- Recognize the need to educate and communicate with users, and between the management agency and segments of the public. Certain groups favour different forms of communication for information. For example, some groups may rely on television to receive information regarding natural resources, while others will be more likely to use word-of-mouth, or leaders of the community.

Whatever the case, the public is composed of many groups, each of which has specific sets of perceptions, communication patterns and belief systems. How the public views, perceives and uses natural resources can have profound implications for future management strategies. For example, natural resource managers will need to be sensitive to the varying needs and expectations expressed by minority visitors. Language is one obvious example in which agencies should anticipate an increased need to have multilingual specialists on staff. In addition, differences have been noted in the research literature between Hispanics and Anglos, including: (i) time orientation; (ii) interpersonal relationships; (iii) 'power distance' or the level of conforming to authority and regulation; and (iv) personal space (Marin and Marin, 1991).

Distribution of the population

Population distribution refers to how the residents are distributed within a given area. For example, an individual could live in a large city in the western USA, a small village in England or a remote outback location in Australia. In general, most developed countries such as those in North America, Australia and Europe have a continuing set of increasing population concentration and migration patterns based on climate and economic factors. In the USA, there has been an overall migration and increased growth of population in the southern and western regions as compared with the northeastern and midwestern regions. In addition, by 1990, 77.5% of the population lived in metropolitan areas, compared with 22.5% who lived in non-metropolitan or more rural environments (Long, 1988). Murdock and Ellis (1991) suggest that this trend will continue, but decrease in magnitude.

Lessinger (1987) believes that one important aspect of population distribution will be the movement of populations to wildland–urban interface lands, or those lands located near large urban environments (Ewert *et al.*, 1993), thus increasing the fragmentation of these resources. According to Garreau (1991), this 'micro-level migration pattern' can already be seen in the development of 'edge cities'. If true, this micro-level migration pattern will put intense pressure on many recreation and natural resource systems by creating adjacent development and increased impacts on remote areas.

Taken together, from a management perspective, two factors become apparent when considering the distribution of the population. First, where the majority of our citizens live will have a great influence on how they will interact with the natural resource base (Kelly, 1989). Pressure with regard to the use of these resources will increase on non-developed, open spaces, if for no other reason than increasing numbers of people will need open space and natural areas as locations to escape from 'everyday' life and experience rejuvenation and catharsis (Kaplan and Kaplan, 1989; Hammitt, 2000).

Second, migration patterns will also have a tremendous, often negative, impact on natural resource use such as water use, air pollution, development of agricultural lands, and so forth. In some locations, both in North America and globally, specific natural resources such as groundwater or arable land will be overwhelmed by these migrations.

Finally, understanding change in selected demographic variables implies developing a heightened awareness of emerging management needs. For example, Struglia and Winter (2002) suggest that population and demographic projections can aid in addressing four environmental management challenges: (i) anticipated patterns of change in how the public uses and identifies with a particular natural resource or setting; (ii) the development of more effective visitor information systems, by managers and researchers, which are responsive to new cultural and/or ethnic groups; (iii) increasing growth in user numbers will probably precipitate greater levels of impacts, pollution and encroachment into wildland areas; and (iv) the need for collaborative partnerships that can be anticipated and developed in specific areas and for predicted challenges.

The following section discusses additional forces that will serve to impact the management of natural resources.

Additional Driving Forces Related to Demographic Issues

In addition to the four major population characteristics, just discussed, there are several other factors that will affect public beliefs and behaviours toward natural resources. These factors include: (i) the continuing diversity of household types; and (ii) a continuing disparity in socioeconomic resources.

Diversity of household types

Because of the ageing population base, the rate of household formation will slow down. Social and economic forces, however, will continue to create highly diverse households, often characterized by one-parent families or diverse forms of union (Bumpass and Sweet, 1989). Of even greater interest are the associated changes in life-style. The research is consistent in suggesting that different life-styles and developmental periods of life (e.g. childhood, late adulthood) often lead to different expectations and demands (Kelly, 1989). Overall, the census data predict a reduction in 'traditional family' units, a growth in units other than two-parent unions, a growth in multi-income family units with a corresponding decrease in available discretionary time common to other family members, and smaller sized consuming units including smaller families (Fosler *et al.*, 1990; Marcin, 1993). This fact has implications as to how natural resources will be used and what types of values will be placed on these resources.

For example, the 'traditional' 2-week vacation is rapidly becoming unavailable to much of our population. It will be replaced by visits to natural areas that are shorter in duration, usually taking place over the weekends, and will involve less planning and travel time, thus placing a higher demand on natural environments that are closer to large urban settings and more accessible, in terms of both range of allowable activities and costs.

In addition, Place (2000) suggests that early childhood experiences are critical in formulating beliefs about natural resources. Without any exposure to natural resources, much of the public may have a belief system about natural resources that is inaccurate or non-existent. As a result, issues and behaviours such as littering, cutting vegetation, noise pollution, over-consumption or the role that natural areas play in providing services and 'goods to society' may not be appreciated or understood.

Disparity in socioeconomic resources

Disparities or differences in socioeconomic resources for different populations have led to unequal opportunities regarding income, occupational mobility and educational attainment (Farley and Allen, 1987). Census data suggest that while overall income levels are relatively high, poverty rates also remain high and are growing for some minority and other ethnic groups. From a resource management perspective, these socioeconomic disparities can impact the values an individual may place on a particular natural resource and patterns of behaviour. For some, economic considerations and needs will serve to offset concerns for ecosystem sustainability.

Taken together, demographic variables can often serve as influencing agents regarding how people use and perceive natural resources. Moreover, anticipated demographic changes will affect how people use and value natural resources. In many cases, these changes will be dynamic and profound, exceeding those demographic changes experienced in the recent past (Cordell and Overdevest, 2001). For example, Table 4.1 summarizes the effects of demographic changes in natural resource management and, specifically, public and protected land use.

Table 4.1. Demographic change and implications for natural resource management.

Demographic changes	Implications
Ageing of population	Need to accommodate older visitor. Greater demand for amenities. Demand for activities that are less physically demanding. Greater demand for 'front-country', less demand for backcountry opportunities.
Growth of minority groups	Potential change in activity mix. Need for managers to have more comprehensive and wide-ranging communication training.
Population distribution	Movement to wildland–urban interface lands. Greater percentage of 'urbanized' visitors.
Less available leisure time	More reliance on technology. Visits that are shorter, closer to home and cheaper. Demand by visitors for more activity choice.

Adapted from the Proceedings of the Second Canada/US Workshop on Visitor Management in Parks, Forests and Protected Areas, Madison, Wisconsin, 13–16 May 1992.

Public Perceptions of Natural Resources

Concomitant with the effect of demographic changes experienced by society are the perceptions held by the public concerning the use and valuation of natural resources, i.e. in what ways does the public value natural resources, and how do these 'values' translate into public policy regarding resource use and protection? When considered in aggregate, these perceptions have been linked to environmental behaviours, values and beliefs, normative behaviours, participation in environmentally related movements, and resource use. This section will examine some of the factors that influence the formation of perceptions and how these perceptions can in turn influence behaviours.

Forming perceptions

The literature is relatively consistent about the importance of developing processes to facilitate participation in the natural resource decision making, legitimacy in developing policy, and setting goals (Frissell and Bayles, 1996; Slocombe, 1998). Other studies have examined how individuals can be classified taxonomically depending on how they view the environment from a utilitarian or preservation perspective. For example, Kolb *et al.* (1994) describe individuals as having either utilitarian or ecosystem perspectives. Within this framework, those with a utilitarian perspective tend to view forests as primarily a source of timber products. Conversely, individuals possessing an ecosystem perspective see the same forest and envisage a range of opportunities including forest health, recreation and water quality.

Related to the ecosystem perspective, Bennett (1976) suggests that how we use our natural resources is, in part, an artefact of cultural factors, perceived or real needs, and opportunities. Thus, according to Bennett, society's view of the health of an ecosystem, or particular natural resource such as a forest, is contingent on how well we can use it. Controversy arises when different uses of the resource conflict with one another. For example, timber harvesting is one use that can often preclude or interfere with another use, such as recreation. Thus, while one group of users may see the forest as being healthy and capable of sustaining use (timber harvesting), another group (e.g. outdoor recreationists) may see the very same resource as being diminished or degraded, based on a previous use (timber harvesting).

Within the previously described context, one way perceptions are formed about natural resource uses is through the perceived benefits that can be accrued through these resources. For example, Fine Jenkins (1997) points out that many of the controversies involved in natural resources are manifestations of differing perceptions of values and anticipated needs.

Another example of the dynamic nature of perceptions of natural resources is the growing concern over the relationship between ecosystem and human health. For example, McMichael (1997) describes how issues related to global environmental change can influence human perceptions of threats to health and acceptable uses of natural resources. These concerns include: (i) an increase in various vector-borne diseases such as malaria; (ii) rises in sea-level; (iii) rises in mortality and morbidity rates due to increases in thermal stress; (iv) changes in food production patterns; and (v) increases in rates of various cancers. In addition, a number of countries now face 'demographic entrapment' in that the projected or current population numbers exceed the carrying capacity of the environment. King *et al.* (1995) suggest that demographic entrapment will often lead to destabilization and economic vulnerability. Thus, perceptions about how natural resources should be utilized are also dependent on what perceived or real threats are being felt by the public.

The development of perceptions related to natural resources can also be traced to the influencing role that various institutions play in our society. Force and Machlis (1997) define social institutions as collective solutions to universal social challenges or needs which include law, religion, agriculture, education, commerce, leisure and natural resource management. For example, a number of authors suggest that religion can exert an important influence on the perceptions and beliefs that an individual has towards natural resources (White, 1967; Schultz *et al.*, 2000). In a similar fashion, other studies have shown that level of education (Van Liere and Dunlap, 1980), type of educational discipline (Ewert and Baker, 2001; Hodgkinson and Innes, 2001), participation in outdoor recreation (Nord *et al.*, 1998) and socioeconomic status (SES) (Kline and Armstrong, 2001) also play important roles in the formation of perceptions of natural resources and the management of those resources.

Process and decision making

A more recent collective solution that has emerged in the decision-making process concerned with natural resources is participatory involvement in which various stakeholders and other concerned groups have some input into the process (Horn *et al.*, 1993; Bettinger and Boston, 2001; Parkes and Panelli, 2001). Within this decision framework, proposed models have assumed that the group has a complete set of information from which to make decisions and/or recommendations (see also Chapter 7). This 'rational' model assumes that this complete knowledge base is necessary to reach effective solutions. In the second model, termed the semi-rational or procedural rational model, data are incomplete and often necessitate selecting an alternative that is 'good enough'. In a third model, labelled by Bettinger and Boston (2001) as the 'garbage can' model, both data and goals are incomplete and the participatory members are often transitory. As a result, the decisions are often poorly derived and even conflicting.

To sum up to this point, the literature is becoming relatively consistent in suggesting that the overall values being assigned to various natural resources is moving from a strict utilitarian perspective to one that is more multidimensional, involving a wide variety of values and expectations. For example, a forested area not only presents an opportunity for timber harvesting, but also increasingly represents an area more amenable to recreation, water quality, wildlife management, etc.

Whatever the process or method used in participatory decision making, Cortner (1996) argues that utilizing two-way communication, informal methods, and early and consistent participation will aid in the overall effectiveness. In addition, future questions regarding participatory models will need to address the fundamental issues of the proper roles of the various parties involved in the issue (e.g. special interest groups, managing agencies, etc.) and the various values that they have (Wondolleck, 1992; Cortner and Shannon, 1993; Cortner, 1996).

Theoretical frameworks

Not surprisingly, a number of theoretical frameworks have emerged that seek to explain how environmental perceptions, values and behaviours are formed. The broad spectrum of these theories includes the following: (i) an individual view of the environment is a matter of one's worldview (Dunlap and Van Liere, 1978); (ii) people make rational choices about the environment (Ajzen and Fishbein, 1980); (iii) behaviours are linked to altruistic values (Herberlein, 1972); or (iv) people make decisions based on a combination of knowledge, moral beliefs and feelings of personal responsibility (Kaiser *et al.*, 1999). One of the more recent models to emerge is that proposed by Stern *et al.*, 1999). Termed the VBN or values–belief–norm theory, Stern and his colleagues have attempted to combine components of a number of previous theories, as previously mentioned, suggesting the following connection: **values→beliefs→ personal norms→behaviours**.

The research by Stern *et al.* (1999), using this theoretical framework, lends support to the belief that an individual's predisposition to pro-environmental action is strongly influenced by his or her perceptual base and personal moral norms. Stern *et al.* also suggest that environmental behaviours are closely linked to four causal variables, two of which are connected to the previous discussion of demographic changes and perceptions. These causal variables include: (i) attitudinal factors such as perceptions and beliefs; (ii) personal capabilities, such as those associated with the demographics of age, SES, etc.; (iii) habit or routine; and (iv) contextual forces. In the latter case, contextual forces include issues such as personal choice, perceptions of environmental equity, the influence of the media and international influences. The following section discusses a number of these contextual forces in greater detail, including: individual choice and markets, environmental inequity and justice, policy tools and environmental philosophy.

Individual choice and markets

Based on circumstance, each individual goes through life with a set of opportunities that provide him or her with various choices in the consumption of natural resources. These opportunities and choices join together to form a collective arrangement of natural resource consumption and an overall pattern of environmental stewardship. In more complex societies, individual roles have become increasingly specialized and, in order to meet rising quality of life expectations, individuals develop associations with others to produce certain goods and services for trade. The result is a market that creates a supply and demand for a range of natural resources in various quantities, and in an assortment of resource flows and distribution patterns. While individual consumption is often many steps removed from a decision to develop a resource (e.g. a nickel mine in Voisey Bay, Labrador, or a diamond mine in Sierra Leone), each individual's consumption pattern stimulates demand that can lead to pressure to 'produce' more of the resource.

According to classical economic theory, each 'individual' has specific preferences and acts rationally to optimize personal welfare, while everyone else tries to do the same. In the context of natural resource decision making, a rational choice is seen as the course of action that most efficiently allows an individual to attain his/her desired outcomes. This means that individuals will act in their own self-interest and, when necessary, will act competitively against the interests of others to maximize personal benefit. In the context of individual nations, such as the USA, the UK and Japan, these 'individual' countries have developed complex consumer markets and highly sophisticated industries that must rely on an almost insatiable appetite for natural resources. These countries have largely transformed their economies into consumer societies where the demand for goods and services and, as a consequence, demand for natural resources throughout the world, is no longer driven by what could reasonably be considered rational demand but is one that is largely driven by a rather irrational craving for consumer goods. This heightened consumer demand is a largely self-driven phenomenon and can have profound consequences for natural resources.

What can be discerned from the previous section is that individuals in the developed world consume considerably more natural resources throughout their lifetime than do those born in developing countries. It behoves those who live in developed countries to rethink their consumption patterns in order to make them more equitable and just. The idea of a sustainable society is one such initiative and, as will be seen throughout the rest of this chapter and the remainder of this book, such an initiative is much easier to talk about than to actually put into practice. IREM processes that take into account the diversity of resource needs and wants can be effective in developing new resource consumption patterns of use.

Environmental inequity and justice

The colour of your skin or the ethnic group to which you belong would not seem to be obvious influencing factors in explaining resource and environmental management behaviour. However, on a community or regional scale, there is growing evidence that in the USA, toxic or other unhealthful management projects are situated increasingly in areas where, for example, Blacks, Hispanics and other lower socioeconomic segments of America live. Similar patterns of injustice are found in other countries such as in the UK, Australia and Canada. There are three complementary explanations. The first suggests that less economically advantaged groups, regardless of colour and ethnicity, migrate to where work is likely to be in greatest demand and where housing is most affordable. Such conditions can be found near industrial, environmental remediation sites (e.g. landfills) and mining sites. A second explanation suggests that disagreeable projects including landfills and polluting industries such as mining operations are more likely to find planning approval where the politically weakest groups live. A third possible explanation suggests that the more affluent in society are able to migrate away from environmentally risky neighbourhoods, just as the middle class have moved from the inner city to the suburbs.

Lester *et al.* (2001) suggest three phases of the environmental injustice movement. The first phase, in the 1970s, focused on civil rights in the inner cities; the second, in the 1980s, concentrated more directly on environmental racism; and the third phase, emerging in the late 1980s, focused more squarely on environmental justice or lack thereof. In legitimizing the case for environmental 'injustice', researchers must show a clear relationship between environmental risk and an identifiable marginalized group, such as the poor or Blacks or Hispanics in, for example, a White-dominated society. Research by a number of scientists, including Lester *et al.*, suggests that there is a significant relationship throughout the USA between the increased proportion of Blacks or Hispanics in a county's (local government) population and that county's exposure to environmental risk and pollution-related factors such as total toxic releases, toxic stack air releases and toxic gas releases.

In theory, environmental injustice argues that the communities associated with class, ethnic group and race are directly linked to the level of environmental risk, and that privileged groups in society are able to organize themselves politically in a more effective way, in order to keep environmentally risky developments away from their communities. This latter phenomenon is known as the 'not-in-my-back-yard' or NIMBY trend. NIMBY continually works to the advantage of predominantly White middle and upper class communities and redirects unpleasant resource management developments to where political mobilization is weakest. Regardless of the specific cause and effect relationship for environmental injustice, empirical research seeks to explain the relationships between race, ethnicity and socioeconomic standing regarding vicinity and environmental risk. From a neo-Marxist perspective, developed countries adopt stronger environmental regulations, thus creating the tendency for capital and environmental risk to be exported to developing countries where resource management regulations and political mobilization are weaker. This is sometimes referred to as 'neo-colonialism' (Mol, 2001, p. 63).

Policy tools

The way in which government responds to public policy needs, public opinion and media treatment may have profound impacts on resource and environmental management policy. The methods by which various policy tools are applied are critically important to how resource and environmental problems are dealt with in society.

In assessing the appropriate role for government, it is important to recognize that, prior to the Second World War, most Western governments had a relatively limited role to play in people's everyday lives. Beginning with the Second World War, it was learned that the nation state (e.g. the USA or the UK) could accomplish important national objectives by using a variety of policy tools such as regulations, propaganda and incentives. Following successes in public policy during and immediately following the war, the post-war era brought a period of national optimism where the public developed a belief that government could solve difficult social problems including environmental and resource management concerns through various forms of governmental intervention.

Interestingly, over 20 years ago, O'Riordan (1981) described a policy 'pecking order' that he referred to as a 'hierarchy of national goals' (p. 20). This hierarchy, as he described it, habitually placed environmental policy towards the bottom of the stack of public policy priorities. This action regularly left environmental problem solving to the lowest levels of government or to the weakest governmental departments where technical expertise, enforcement ability and remedial capacity were the least effective. Since O'Riordan provided this perspective, significant improvements, as well as noteworthy declines, have been made in various environmental policy arenas. More recently, there has been what Howlett calls 'a hollowing out' process in government (see Howlett, 2002, p. 25). Government functions that in relatively modern times (perhaps the last 40 years) were the responsibility of senior government (e.g. national and state/province) are increasingly transferred to regional and municipal governments, to quasi-governmental organizations or to the private sector. Such shifts in responsibility have often relied on voluntary compliance or self-reporting mechanisms rather than direct regulation, with sometimes devastating results, such as in Walkerton, Ontario, Canada. In Walkerton, a transference of responsibility from professional experts in water quality testing at the provincial level to poorly trained custodians at the municipal level resulted in disaster, causing several deaths and serious illness for hundreds more (O'Connor, 2002).

Knowing what level of government is most effective and efficient to handle various resource and environmental management problems is a difficult task, and making the appropriate choice can drive needed change or impede it. For example, at the federal level in the USA, some issues (e.g. such as requiring the use of unleaded gasoline in cars and the installation of catalytic converters on automobile exhaust systems) have been handled quite effectively. These enforcements have often provided significant environmental quality benefits. Interestingly, however, the automobile industry and consumers alike sidestepped the Corporate Average Fuel Economy (CAFÉ) regulations introduced in the USA in 1975, which was part of the Energy Policy and Conservation Act. CAFÉ regulations were designed progressively to improve the gasoline consumption of private motor vehicles by increasingly adopting stricter regulations for the whole production fleet of each automobile manufacturing corporation. Given a fixed number of automobiles, this would have reduced the total amount of pollution. Instead, consumers turned to gas-guzzling sport utility vehicles (SUVs) and similarly less energy-efficient mini-vans and pick-up trucks. These vehicles, which at the time of legislative introduction accounted for a relatively small proportion of privately owned and operated vehicles, were exempted from this legislation, largely because of their relatively small aggregate contribution to pollution at the time of legislative enactment. Unfortunately, a massive assemblage of SUV, mini-van and pick-up truck owners have now combined to create a substantial pollution problem. Moreover, from a political perspective, these owners have formed a powerful self-interest lobby to persuade federal legislators to keep these vehicles out of CAFÉ regulations. Interestingly, at least as far as road safety is concerned, the perception of greater safety with SUVs is not supported by scientific evidence (Yun, 2002).

Environmental philosophy

Rather than a driving force for resource and environmental management, De-Shalit (2000) argues that environmental philosophers and their thinking have had surprisingly little influence over policy making and are separated from mainstream policy making by too great a concentration on obscure theoretical questions with little direct relevance to society. This detachment, he argues, stems from environmental philosophers' preoccupation with a distinctly narrow view on environmental ethics rather than a more pluralistic and embracing view on how to involve ordinary citizens with environmental and resource management issues in their everyday lives. To draw philosophers and environmental philosophy closer to the mainstream of societal behaviour, De-Shalit argues that society at large must be given a greater profile in shaping philosophical questions and theories.

We have seen from various discussions in this chapter that the general public, on the one hand, supports policy through various public opinion polls to increase environmental protection, but on the other, through their actual behaviour, circumvent policy designed to protect it. They do this, for example, by living in large homes far from their workplaces and by overwhelmingly choosing to travel in large pick-up trucks, mini-vans and SUVs over more environmentally friendly alternatives. How might philosophers cut through such contradictory actions to provide useful ways of viewing this problem, and argue for some sort of policy alternative that appropriately claims the moral high ground? To be relevant and to be listened to, philosophers must combine moral, factual, desired and practical considerations that inform policy makers of both the ethical and practical choices inherent in various policy alternatives. In the case of SUVs, for example, it is too simplistic merely to consider gasoline consumption necessary for larger vehicles and their associated increased air pollution in order to determine what is morally right and wrong. Policy makers must also consider perceived safety, consumer preference, vehicle production capacity, gross national product (GNP), and so on, in their decision-making process.

Would a coherent and well-thought-out philosophical analysis make a difference in the eventual policy decisions made with respect to natural resource use? Might the emergence of a coherent environmental philosophy in the future change these decisions or how they are made? From a strictly bioethics perspective, any reflection concerning the future health of the biosphere that largely ignores the social dynamics of the present would probably suggest that automobile travel is detrimental to the environment and consequently should be curtailed. A less biocentric and more human-centred model might endorse various forms of environmentally friendly public transportation, whereas a view reflecting the present social patterns of the USA, that is highly reliant on private automobile use, may prefer a perspective recognizing that private automobile transportation will prevail for the foreseeable future. A debatable topic is which environmental philosophy is more likely to lead to civilization's salvation and sustainability, and which one is more likely to influence public and corporate policy in the shorter term.

Although IREM is an approach to resource management that is substantially more influenced by concerns for the environment than the dominant natural resource management regime of the developed world, it tends to be considerably more pragmatic than most bioethical prescriptions. By its increased adoption, the authors of this text believe that resource management will take on a decidedly more 'environmentally friendly' approach to harnessing both renewable and non-renewable natural resources. In some ways, IREM can be considered a radical approach to natural resource management that goes against the grain of the dominant ideology inherent in free market economics. It implores resource managers to think in the long term, act more in the interest of community and protect the environmental commons. IREM processes support the idea that wise conservation and concern for the environment can also be economically viable and attractive (Daily and Ellison, 2002). Nevertheless, IREM also has a substantial anthropocentric orientation which reflects a philosophy where society is of parallel concern and importance to nature.

To admit its limitations, this book is most probably constrained by a philosophy of what De-Shalit calls the 'Western fix' (De-Shalit, 2000, p. 214) view of solving environmental problems – a conformist driving force that aims to preserve cherished institutions such as democracy, free markets and our standard of living, that is not necessarily in the best interests of global environmental health. We are, however, unashamedly supporters of the dualism philosophy that sees no sustainability or stability in environmental management when this 'sustainability' occurs at the expense of social injustice on a global scale. This is especially relevant where the wealthy in society increase their share of prosperity while the poor are increasingly forgotten (Hutton, 2002).

Public Opinion and the Media

Until a few decades ago, most natural resource and environmental managers were rarely concerned about, or affected by, public opinion. Decisions were made largely using technical data developed by technical experts, and projects were implemented with little thought to consulting the public. Since the 1960s, the general public has been much more concerned about, and has demanded a far higher level of involvement in resource and environmental management policy (Tarrant and Cordell, 2002). This growth in public involvement has, to a large extent, resulted from a symbiotic relationship between the expression of public opinion and the media. As environmental issues increasingly gained public attention, the various media, such as television, periodicals and newspapers, responded with increasing coverage. In some circumstances, this increased coverage was in direct response to public concern, and in other situations it was the result of investigative reporting that increased public awareness, concern and involvement.

Just how far public opinion and media treatment of resource and environmental management affairs impact or should impact government, corporate decision making and civil behaviour is open to debate. However, it is clear that in some circumstances, public opinion is far ahead of government and corporate action, and in other situations it lags far behind. For example, in a Gallup poll taken in the USA during March 2001, slightly over half of the respondents thought that protecting the environment should have preference over energy consumption, although only a quarter were highly troubled about environmental conditions in general (Dunlap and Saad, 2001; Sanger and Alvarez, 2001). In the UK, there is a similar relationship between public opinion and government policy formulation. On the one hand, many citizens unthinkingly put pressure on Britain's road network, opting to drive their cars to work every day rather than use public transportation or ride a bicycle. Until recently, British politicians have responded rather predictably by supporting budget appropriations for road improvements to meet this escalating demand. However, again, public opinion polls in the UK suggest that politicians lag behind public opinion and should not necessarily continually approve funding for road improvements. As a case in point, 55% of the British public say that environmental concerns will dictate their voting behaviour, 61% want greater investment in alternative transportation rather than more road building, and 68% believe they have greater concern for the environment than the major political parties' leaders (Drake, 2001).

Despite this close association of public opinion and media behaviour, it is important for resource managers to understand that public opinion and media coverage is just one of several considerations in resource and environmental policy making (see Chapter 7). It is also key to recognize that pollsters measure public opinion with varying degrees of scientific precision, and that the media report on news but may offer analysis with varying degrees of editorial and ideological bias. This requires natural resource managers to be extremely cautious when attempting to merge public opinion and technical expertise to make wise decisions; they need to develop their decisions using considerable political savvy and weigh their decisions with what is possible and what best meets their identified objectives.

Technology

As was emphasized in Chapter 2 and again in Chapter 3, *Homo sapiens*' advancement was closely connected to their ability to transform previously 'worthless' biophysical matter into a means of providing food, shelter, energy, transportation and comfort. Advances depended upon humans' ability to accumulate and pass knowledge from one generation to another, and continually to invent and refine ways of using biophysical matter for human welfare. Technology has always been, and continues to be, a major driving force in natural resource management and is a major determinant of a society's overall well-being.

While there are numerous avenues of technological innovation, five broad groups of technologies continue to revolutionize natural resource management and create new ethical dilemmas regarding their application in society. Those highlighted in this chapter are 'Physics and chemistry', 'Electronics and information technology', 'Biotechnology', 'Geomatics' and 'Geoengineering'.

Physics and chemistry

Advances in the basic sciences of physics and chemistry continue to revolutionize natural resource management, regularly providing solutions to old problems and inadvertently creating new ones. Developments in our knowledge of physical chemistry, for example, have allowed us to design a vast array of synthetic materials such as plastics, to work on behalf of humankind. However, each new development brings with it an enormous unknown concerning its effects on the environment (e.g. PVCs). In many cases, those materials found to be the most useful to humankind find their way into the environment in large quantities as pollutants, and pose difficult challenges for environmental management. For example, ozone depletion in the upper atmosphere is one case where technological use of human-made chlorine chemicals known as chlorofluorocarbons (CFCs), previously used widely in aerosol propellants and as cooling agents in refrigerators and air conditioners, created immense environmental problems that are not easily reversible. Ozone protects the biosphere from harmful ultraviolet light, and scientific evidence suggests that depletion is increasingly and adversely affecting the viability of fauna and flora throughout the globe. As a second example, a pesticide such as dichloro-diphenyl-trichloroethane (DDT), although banned in the USA since 1972 because of its environmental side effects, remains the key combatant to the spread of malaria in many tropical developing countries. DDT and its derivatives migrate and bioaccumulate in the Arctic where few direct applications have been documented. Unfortunately, this accumulation has led to widespread health problems in the Arctic and poses serious concerns for the future viability of predator species such as polar bears.

Electronics and information technology

Miniaturization and increased capacity in electronics, especially microcomputers, brings to natural resource management a vast array of powerful tools that were impossible to envisage just a few short years ago. Nowadays, resource managers, even those in remote locations, can have practically unlimited access through the Internet to published natural resource management research using electronic sources (see, for example, http://www3.interscience.wiley.com/journal finder.html). In addition, public agencies, universities and private corporations increasingly make technical publications available in a timely manner (see, for example, http://www.r5.fs.fed.us/ecoregions/) to Internet users by means of readily available, downloadable and free software. Such technological innovation puts massive amounts of information at the resource manager's fingertips but usually does not offer practical ways to synthesize all this information to help make wise decisions. To fill this gap, academics and consultants have stepped in to provide various decision support systems (DSSs) that require varying levels of technology support. They all have the common denominator of cutting through this great quantity of data to focus on what is absolutely essential for the resource manager to make good decisions rather than get bogged down with masses of information that is merely nice to know (see El-Swaify and Yakowitz, 1997).

Biotechnology

While the field of biotechnology is very broad, our discussion here refers to the transfer of desirable genetic traits from one organism to another, taking the genetic material from one organism carrying those traits and attaching this genetic material to the genes of another organism. Commercially viable products are available when sufficient quantities of this new biological material are produced to meet a corresponding market demand. As with synthetic materials, the true environmental impacts of using genetically altered material cannot be fully known until it is put to use in society. This involves using the environment as a research laboratory where the consequences of some unpredicted outcome might be catastrophic or simply trivial. For example, proponents of biotechnology argue that improved and more resistant cereal crops will substantially reduce the need for pesticides, thus reducing the overall environmental risks and costs. Despite assurances from the biotechnology industry, the issue of biotechnology in Europe has reached fever pitch, with public awareness campaigns to boycott biotechnology foods. In North America, however, arguably the world centre of biotechnology research and development, the public seems much less informed and nowhere near as concerned.

Geomatics

Geomatics is the broad term that denotes the application of earth science technology such as remote sensing (RS), geographical information systems (GIS) and ground surveying using geographical positioning systems (GPS). Together they have revolutionized natural resource assessment and management. Until recently, the management of catastrophic weather events, for instance, seemed beyond the capability of resource managers. Typically, resource managers reacted after the fact, to 'mop up' flood damage. In the wake of the disastrous Red River flood of 1997 in North Dakota and in southern Manitoba, considerable work has been done on both sides of the USA–Canada border to minimize the future impacts of flooding. This work includes accurate land and water surveying using GPS, which feeds practically seamless geocoded data into GIS mapping processes that model the Red River geomorphology and hydrology. These computer models are then used to provide more accurate flood predictions. Such data provide key information before the fact on the best sites on which to build dykes and apply other flood damage reduction strategies. Fields, for example, can be prepared, crops planted, and livestock protected in elevated enclosures or barns, with greater assurance of the risks involved. During flooding events, these models inform flood relief agencies, farmers and local residents (using home or office computers) of impending flood conditions so that various damage mitigation resources, including simple technologies such as sandbagging, can be applied at the most opportune times and locations.

RS has not only revolutionized resource protection as in the Red River Valley, it is used extensively in natural resource exploration, planning and remediation. For example, geologists now rely on relatively inexpensive shock wave instruments, known as seismic reflection surveying, to identify structural traps that may contain oil and gas beneath the ocean floor rather than initially using extremely expensive test-hole procedures, as was routinely necessary in the past. RS also allows geologists to identify magnetic fields that can provide basic geological information. RS can be used on bedrock structures, such as fault lines that are hidden below the earth's surface; and in some situations RS can more efficiently lead field geologists to bedrock that has economic potential. Furthermore, RS assists companies such as Chevron to monitor its field operations in order to assess its compliance with increasingly stricter environmental regulations. It also helps Chevron make wise and timely decisions in mop-up operations, as in the case of oil spills (Pfeil and Ellis, 1995).

Geoengineering

As indicated in Chapters 2 and 3, human ingenuity has had considerable impact on natural resource management, from building aqueducts in Egypt, throughout the Roman Empire, and in the Mayan Peninsula in Central America, to irrigate farmland. In addition, simple but ingenious technology has been used to build dykelands in The Netherlands and Nova Scotia in order to create land from the sea for farming. Considerably more ambitious technology has also been utilized to change the course and scope of waterways, as in James Bay Hydro Development Project in Canada (see Chapter 3). This massive hydroelectric-generating project provides power not only to Quebec, but also extensively in the USA. In the Aral Sea in Kazakhstan (near Russia), where excessive water diversion of natural inflows for irrigation has reduced this once mighty inland sea by 40%, this project has brought increased prosperity to some regions and reduced other areas to abject poverty. It has also severely impacted the region's biodiversity, with substantial long-term implications.

Summary

There is no doubt from these five broad groups of technologies and many other examples that humankind has considerable ability through its geoengineering prowess to change land and seascapes for its own indulgence. Schneider (1996) even reports that a 'National Research Council report proposes using 16-inch naval guns to fire aerosol shells into the stratosphere in hopes of offsetting 'the radioactive effects of increasing carbon dioxide'.' Given the many unintended knock-on effects of past projects and having the capacity to greatly transform our natural resources for both good and bad, which creates both winners and losers, this raises serious ethical questions as to whether any large- or small-scale geoengineering project should be attempted at all (see the case study in Chapter 7).

In the last few centuries, technology advancement has turned out to be both a blessing and a burden. Technological breakthroughs have provided, until relatively recently, the necessary conditions to support larger global populations with higher overall levels of social welfare. In this context, it is important to understand that the industrial revolution with its increased capacity to harness natural resources brought with it substantial social costs as well as environmental damage. This pattern of inequity and environmental damage has been repeated until the present. It is essential to note that the continued accumulation of unwanted side effects has increased the need for vigilance to ensure that the reapplication of old and the use of new resource management technologies do not create net costs to society or an unreasonable or uneven distribution of benefits. It is for this reason that the need for new management approaches such as IREM has arisen.

Aboriginal Concerns

In the 17th, 18th and 19th centuries, a strong wave of colonization came from Europe to the New World. Often funded by investors such as the Merchant Venturers from Bristol in England, various explorers focusing on building their sponsors' wealth, and later others, such as settlers, showed little compassion for the native people they encountered or the land and seas they exploited. Some indigenous groups were, nevertheless, quite welcoming, but no matter what the reception, the objective of Euro-colonialists was always clear – to gain access to and to exploit natural resources. When defiant native populations had to be subdued or pacified, colonial powers in the safe havens of European capitals were often reluctant to commit sufficient resources to either eliminate resistance or fairly negotiate land and natural resource rights. Consciously or not, colonialists used a variety of strategies, including dishonest treaty building, coercive land purchases or rents, and other forms of duplicity to attain their objectives. Whether any of these strategies were approached in good faith or not, few were raised to national consciousness until the latter part of the 20th century, when native groups such as American Indians in the USA and Maori tribes in New Zealand began first to discover and then assert their previously negotiated rights.

A number of texts have begun to deal with issues pertaining to native affairs on natural resource management in the USA. A text by Sussman *et al.* (2002) for instance, entitled *American Politics and the Environment*, provides reference to the influence of the US Bureau of Indian Affairs and gives some attention to Native American resource management issues. Interestingly, a monograph by Cordell and Overdevest (2001) entitled *Footprints on the Land* provides only cursory discussion of the impact of Native Americans on resource management despite lengthy treatments of present and potential influences of African, Anglo, Asian and Hispanic Americans. Importantly, a text by Mazmanian and Kraft (1999) recognizes the historical relationship of indigenous peoples to the land and acknowledges contemporary American Indian contributions to local environmental problem solving, and Booth and Kessler (1996) examine native worldviews and how these impact native beliefs regarding the management of natural resources.

A paper by Flanders (1998) provides a useful historical perspective of US land ownership, natural resource management and Native American issues. He defines Native Americans as consisting of American Indians and the northern indigenous groups in Alaska known as the Yupiit, I-upiat and Aleut. He points out that the official policy of the USA from 1887 to 1933 was to take tribal land, provide 160 acres (65 ha) to every Native American adult, and sell the remaining 'surplus' lands to non-natives. Under the Indian Reorganization Act of 1934, this policy was reversed, with Congress restoring, and in some cases increasing, tribal lands; provison was made for the formation of constitutional governments, and educational programmes that assumed the continuation of Native American culture and society were instituted. The question Flanders ponders is why did

Congress take such a position when it largely carried out its earlier policy, which was part of a broader goal to assimilate Native Americans into mainstream American life, without question. He concludes that this revised policy was part of a far-reaching but complex strategy to manage common property rights, especially water rights west of the Mississippi where irrigation was often indispensable for agricultural productivity.

Using this strategy, the federal government was able to streamline the management of water rights by first asserting indigenous resource rights, and then, upholding its custodial role over Indian affairs, it was able to control the allocation of these resource rights. Managing these common properties with so many private and diverse interests asserting claims, Flanders argues, would have been impossible under state and/or private control. Had land been apportioned in 65-ha lots, as was common east of the Mississippi during colonization, the chances were that many farmsteads would have been unsustainable. Such apportions were insufficient to sustain the average family unit in these semi-arid lands; many units would have been without adequate water supply as the doctrine of 'prior rights' would have prevailed that appropriated all available water, sometimes to land owners far from the riparian.

Interestingly, the precedent for federal intervention over native resource rights was made in 1908 when the federal government asserted a treaty right for water for the Fort Belknap reservation in Montana. This precedent claimed similar rights for American Indians settled on reservations and was directed against outside interests, mainly non-natives, who would otherwise claim a prior right to water flowing through reservation land (Shurts, 2000). This policy of asserting indigenous rights over the years helped the federal government assert its pre-eminence in natural resource management in the West and allowed it to take a leading role in natural resource management and allocation. To a large extent, the states, private interests and Native Americans acquiesced to Congress in natural resource management.

In some instances, this strategy allowed the federal government to execute far-ranging resource reallocation projects such as the construction of the Hoover Dam on the Colorado River. This single construction project has had far-ranging impact on natural resource allocation and management in the US Southwest. On the one hand, the Hoover Dam provides water to large cities such as Phoenix and Los Angeles and transforms deserts into farms and golf courses throughout much of Arizona. On the other hand, it restricted resource development in the Upper Colorado and denied Mexico its international treaty rights to the Colorado's waters (Raven *et al.*, 1993).

While the assertion of water rights in the West has in many respects deflected much of the direct conflict among private interests and state government in the USA, a visit to various Internet sites suggests a rather unsettled picture concerning Native Americans and natural resource management issues throughout the USA. The full extent of these impacts is yet to be played out in natural resource management. For example, in 1998, the Oneida Indian Nation used the courts to convince the federal government to expand a previously unsettled lawsuit dating back to 1974. Now the revised lawsuit includes land claims against the

County of Oneida in the state of New York for 250,000 acres as well as back-rent and damages. Some other land and water rights settlements are much larger. Although a temporary stay by a federal judge for the State of New York to pay nearly US$248 million to the Cayuga Tribe for the illegal purchase of 64,000 acres of land some 200 years ago puts this particular claim on hold (Reeves, 2002), it clearly illustrates the turmoil that may be festering below the surface concerning land tenure throughout the USA (Cronin-Fisk, 2001). At issue here is a federal law passed in 1790 that states emphatically that Congress must approve the purchase of Indians' resource rights and, at least in this particular case, no such approval was given.

Besides the failure to follow its own laws, some points of argument have resulted from mainstream government gradually attacking the traditional rights of indigenous populations. For example, in the state of Wisconsin, the state government has repeatedly tried to restrict 'on-reservation' hunting and fishing practices of the Anishinaabe Nation (Silvern, 2000). While there have been a number of significant settlements in favour of Native Americans, not all settlements have gone their way. For instance, again in Wisconsin, the Menominee asserted a right to fish and hunt off-reservation, based on Article Six of its Treaty of 1831, but the US Supreme Court extinguished this claim. In fact, the 'Indian question', as it is often referred to in the USA, has plagued successive presidencies, with little sense of resolution. By the mid-1870s, the beginning of the disappearance of the buffalo signalled a greater need for American Indians to explore other forms of subsistence including agriculture. Whereas non-native settlers could and did move on from areas to which agriculture was unsuited, American Indians were largely restricted to reservation land that was frequently unsuitable for farming. In addition, they were often discouraged from pooling capital to buy technology that may have solved many of their problems, and they were also discouraged from building ethnic community networks, which proved so successful among many European settlers, including the Czechs and Germans (Bateman, 1996).

In the end, poverty ensued among many Indian nations and, despite a succession of presidential decrees to address this problem, it is persistent. For example, President Johnson inaugurated the era of self-determination, and this was followed by President Nixon's special message on the 'American Indian' in July of 1970. The Indian Self-Determination and Education Assistance Act of 1975 followed. A similar message came from President Ronald Reagan, who largely echoed Nixon in reaffirming a government-to-government relationship and the goal to self-determination for American Indians (Stull, 1990).

The focus of the 1970s, 1980s and 1990s on increasing natural resource revenues on Indian lands was largely unsuccessful, and the era of self-determination, for the most part, resulted in greater economic dependence of American Indians on the federal government. The question that looms strongly on the horizon, then, is to what extent can Native Americans' rights to natural resources break the poverty cycle in the future and to what extent will Native Americans' reasserted claims to land and resource rights change the complexion of natural resource management in the USA and elsewhere.

Over the last decade or so, native peoples' land and natural resource management claims have swept several nations, including Canada, where in some regions quite extensive shifts in land management from provincial and territorial governments to native and co-management arrangements (as in British Columbia) have occurred. Some shifts have led to seemingly impossible parallel management regimes where First Nations and provincial governments have imposed separate and often conflicting management prescriptions on both land and water resources. Other approaches have included the 'co-management' of various areas where specific lands are managed by a coalition of First Nations, provincial and federal groups. For example, Bradford Morse, a University of Ottawa law professor, predicts that First Nations (as North American Indians prefer to be called in Canada) will soon have sovereign jurisdiction over 5–10% of land in Canada's ten provinces. First Nations have won landmark decisions in the East that acknowledged aboriginal rights to natural resources that go well beyond previously recognized limits for indigenous natural resource custodial care and consumption (Beltrame, 1999).

In Zimbabwe (formerly the British colony of Rhodesia), the violent reclaiming of lands by native populations bears witness to the historic injustice of colonial powers and the complexity of modern resource management interests, especially when repatriation is accompanied by plummeting agricultural production and a predicted overall increase in poverty (Maslund and Newton, 2002a). In April 2002, the re-elected president of Zimbabwe called for international aid to combat a famine. The cause of this famine is, however, a focus of much heated debate between past colonial masters and the Zimbabwe government. Many believe it is a direct result of the Zimbabwe government's land tenure policy. In South Africa, after the displacement of the white minority regime and despite the National African Congress' balanced budget – whose fiscal policies are the envy of many established Northern hemispheric governments – the indigenous population became increasingly impatient with the slow pace of land reform and measurable gains in economic welfare (Mabry, 2002). As in Canada, Australia was slow to embrace the inherent and historical rights of Aborigines. Australia only recognized the Aborigine's right to vote in 1962 and only acknowledged them in the national census beginning in 1967. Since that time, Australia has, as a nation, accelerated its recognition of Aborigine rights to the present point, where some 15% of the land base is either owned or controlled by Aborigines (Beck, 2000).

International Conventions and Conferences

Although it is difficult to pin down the impact of any particular international convention or global conference on shaping the internal policies of nation states and that of large and small corporations, there are events that clearly became important milestones in the development of resource and environmental thinking and policy. Whether causes or effects, these events influenced resource and environmental planners, they mobilized the public interest, energized Environmental

Non-Government Organizations (ENGOs), and increasingly sensitized politicians, corporate officers and the electorate to the growing urgency of sustainable resource and environmental management. In October 1982 in Bali, Indonesia, for instance, the International Union for the Conservation of Nature and Natural Resources convention resulted in guidelines for establishing a comprehensive network of protected areas. This initiative was eventually linked by the World Commission on Environment and Development to a much broader strategy for sustainable development in 1987. This United Nations commission, chaired by former Prime Minister Brundtland of Norway, entitled, 'Our Common Future', committed signatories to 'save species and their ecosystems', while the Rio Earth Summit in 1992 in Brazil, which included the Convention on Biological Diversity under the United Nations Environment Programme, promoted both the importance of protected areas and environmentally sensitive management of the working environment (Bissix *et al.*, 2002).

In Canada, the impetus of the 1987 Brundtland Commission led to the National Task Force on the Environment and Economy that, in turn, helped the formation of provincial-level Round Tables on the Environment and the Economy. It was in these forums, occupied by politicians, corporate and civic leaders as well as environmentalists, that influential decision makers were exposed to the urgency and value of sustainable development strategies. More recently, the Kyoto Accord of 1997 exposed many developed countries' reluctance to put at risk their short-term economic prosperity and potentially the loss of jobs as well as political support. This Accord imposes strict air quality conditions on world nations, where the developed world must make greater contributions than the developing world. On average, emissions are to be reduced by over 5% of 1990 emission levels. While the European Union (EU) ratified the accord in May 2002 and has pledged to meet those targets, to date the USA has not ratified the Accord and has withdrawn the USA from this agreement.

The Need for Integrated Resource and Environmental Management

The complexity of managing natural resources in the context of the many driving forces that impact daily decisions requires integration at a variety of levels. As this chapter has revealed, complex problems are often a result of interconnected causes that require a more integrated response to deal with the complexity. Margerum (1997, p. 459) notes that 'the movement towards integration has emerged from changes in scientific information, the recognition of a wider array of issues and stakeholders, and the increasing complexity of environmental issues'. Management activities have to transcend traditional single management objectives and assimilate a wider range of political, organizational and natural boundaries.

What does integration mean? If *integration* is an important part of a growing trend in natural resource management, how is it applied? Hooper *et al.* (1999) provide an overview of integration in their paper and suggest that various measures are available to achieve integration and that they fall along a continuum with an increasing amount of intervention. Thus, at one end of the continuum, we have a minimal approach to integration, with voluntary actions such as goodwill, trust, respect and willingness to cooperate as the means of integrating interests. In the middle, integration consists of cooperative action where agencies and individuals follow prescribed goals and specified planning processes. Then, at the other end of the continuum, coercive action is taken in order to get individuals and agencies to cooperate. A new lead agency is formed to prescribe integration procedures and ensure cooperation amongst stakeholders.

More recently, Harriman and Baker (2003) suggest that integration has to be formed around a common interest. In other words, people tend not to integrate their interests without a reason. They propose that integration be arranged around a 'substantive' common interest and different parties can align their particular views and values to a shared theme. This is particularly important in field operations where time and budgets constrain a comprehensive approach to resource management. Integration of stakeholders' interests and values needs to be strategic and focused in order to manage for the widest range of resource values within a given landscape.

Integration is a management action that selects both the process of *how* we integrate and on *what* issues. It is a deliberate scoping action that requires intervention. Integration is not always necessary, and some problems can be managed without integrating a wide range of interests or stakeholders. It can be a costly process that may not bring the appropriate benefits for the solution of a problem. An integrated approach is best used where there are complex problems and a need can be established amongst stakeholders that there is value in coordinating interests. The complexity of driving forces and their impacts on natural resource management provide an increasing need for integration in resource management decision making.

Summary and Conclusions

Although there are still thought to be small pockets of Stone Age civilization yet to be uncovered in the more remote parts of Polynesia, and there remain other societies that make very modest demands on the world's natural resources, most societies, even in the developing world, make substantial demands on our natural resources. As we have seen in this chapter, the driving forces impacting resource and environmental consumption are increasingly interwoven and powerful and, when considered in total, continuously stretch the globe's capacity to meet these escalating demands. Ironically, as developing nations transform into more complex societies relying increasingly on greater quantities and more sophisticated services from natural resources, and they increasingly join the ranks of the so-called 'developed' world, they progressively damage the earth's life support systems and reduce the potential for a more sustainable society.

As various societies increase their standard of living, each family's desire for a large number of children decreases. However, this decrease in the rate of population growth does not necessarily provide environmental benefits. We have seen that increased development which reduces family size is a double-edged sword, as each surviving child makes proportionally greater demands on the globe's natural resources. A child born in the USA or Canada, for instance, will consume approximately 20 times the natural resources of a child born in a more modest developing country such as Bangladesh. Given this analysis, it should be clear that overpopulation in the developing world is only a partial cause of the world's environmental health problem. By far the greatest threat leading to environmental catastrophe comes from the collective behaviour of developed nations and their consumption patterns. These patterns produce excess greenhouse gases, destroy the atmospheric ozone layer and produce dangerous amounts of toxic residues.

After reading this chapter, it should now be obvious that the driving forces influencing natural resource and environmental management are multifaceted and complexly interwoven. They combine to threaten the vital life support resources of the globe, such as clean air, drinking water and acceptable levels of greenhouse gases. While technological development allows society increasingly to access once inaccessible natural resources and employ them with increasing efficiency, technology on its own has not provided any adequate solution to the problem of overexploitation and pollution. Our dominant cultural, market and political characteristics have combined to increase both individual and collective consumption in the face of increasing efficiency rather than to stabilize or reduce it. In the end, even though international conventions and signed agreements have targeted substantial reductions in consumption over the past 10 years, the reality has been acceleration in consumption, an overall increase in environmental pollution and an increasing escalation of the problem.

What can be done about this? While many environmentalists call for radical solutions to address the problem of environmental degradation and non-sustainable exploitation of natural resources that require drastic changes to our consumer society, IREM offers one approach to make resource and environmental management more sustainable.

As was seen in this chapter, the problem of sustainable resource and environmental management is a multifaceted one that does not lend itself to simple, single-dimensional solutions. Often there are competing interests, differing perceptions or other factors such as demographic characteristics which serve as 'driving forces' that shape the individual's and society's perspective of natural resource use.

On the surface, IREM appears as a rather conservative adjustment to present-day resource management systems that is hardly likely to make much of dent in present-day practices. If applied carefully and sensitively, however, it has the potential to transform renewable resource management to sustainable levels and has the possibility of substantially reducing the pollution effects of non-renewable resource exploitation. As will be seen in the following chapters, the human dimensions of resource management are highly complex and extremely challenging. Just how IREM will be applied in some sectors and situations of high complexity remains to be seen.

Case Study
Socioeconomic Turmoil in Resource-dependent Communities of Southeast Alaska

WINIFRED KESSLER

Preamble. The following case study provides a real-life example of how various 'driving forces' have forced change upon a rural Alaskan community. These driving forces include changes in demography (e.g. migration into the area) and public perceptions of acceptable land use, changing economic markets and opportunities, and public policy. In this case, communities in southeastern Alaska have increasingly moved from resource extraction (e.g. timber harvesting) to tourism. As we shall see, this move has not been without a variety of emerging issues, related both to the local communities and to the natural environment.

Introduction

Southeast Alaska is a wild and rugged area stretching along the 'panhandle' of Alaska. The area consists of a mountainous strip of coastal mainland and thousands of islands of all sizes. The lowlands and slopes are cloaked in a temperate rainforest dominated by Sitka spruce and western hemlock, interspersed with bogs called 'muskegs'. The mountains rise steeply, giving way to rock, ice and alpine scrub at the higher elevations. Saltwater fiords, bays and inlets dissect the terrain. Small, isolated communities occur at sheltered points along the coast. Travel between communities is by boat or aircraft, as few roads exist in this rugged land.

Most of the area is contained within the Tongass National Forest. At 17 million acres, the Tongass is by far the largest national forest in America. The year 2002 marked the centennial of the Tongass, which was established as a forest reserve during Theodore Roosevelt's presidency. For countless generations, the forest's bounty of wildlife, fish and plant resources has supported a rich heritage of Native peoples, such as the Haida and the Tlingit, whose language had no word meaning 'starvation'. Although some places have been logged extensively, about 90% of the forest remains roadless and in its original, old-growth condition.

Today, the forest is managed under the comprehensive Tongass Land Management Plan. A major revision of this plan was completed in 1997 following more than 10 years of hard work and contentious debate, at a cost of around US$13 million. Ideally, the plan would have resolved many of the thorny land-use issues, providing agreement and certainty as a basis for integrated resources management. This has not turned out to be the case, for reasons rooted in the past and influenced by rapid change in the present.

Background

Alaska has an enduring image as the 'Last Frontier.' For much of the past century, people were drawn here because it was one of the few remaining, truly wild places where living an independent subsistence life-style was still an option. Adventurous newcomers joined the

Native peoples who had subsisted on Alaska's natural bounty for countless generations. Reliance on natural resources has always been central to the culture and economy of Alaska.

Starting in the early 1900s, the abundant fish and timber resources of the Tongass National Forest were seen as the obvious foundations on which to plan economic development in southeast Alaska. However, the ruggedness and remoteness of the area severely limited the economic efficiency of the timber industry. Based on national planning during the Depression, a scheme was launched in the 1950s to entice larger companies to locate in southeast Alaska, establish mills and provide employment for local people. Companies were offered long-term contracts for a guaranteed supply of cheap timber in return for establishing mills in southeast Alaska. One such contract went to the Ketchikan Pulp Company and another to the Alaska Lumber and Pulp Company. In return for building and operating mills in Ketchikan and Sitka, the companies were guaranteed access to 13 billion board feet of timber over 50 years. The plan was to harvest most of the Tongass old-growth forest systematically to maximize pulp production on a long-term basis, thereby providing for economic growth and stability in the region. The plan succeeded in attracting people and infrastructure development to support what became, along with fisheries, the economic backbone of the region. The isolated communities that took root owed their very existence to the harvest of Tongass timber.

The long-term sales were modified in 1980 with the passage of the Alaska National Interest Lands Conservation Act, which established large wilderness areas that were off-limits to logging. To compensate for the diminished timber base, the forest companies were guaranteed a government subsidy of up to US$70 million annually. In addition, a harvest level was mandated from the areas not set aside as wilderness.

While well intentioned from a socioeconomic standpoint, the timber programme on the Tongass became the focus of intense controversy during the 1980s. By that time, ecological studies in the rainforest were revealing significant negative impacts associated with the dominant harvest method of clear-cutting. Scientists found that critical habitats of important wildlife species, such as Sitka black-tailed deer and bald eagles, were the same high-volume forest stands located along the beach that loggers targeted for harvest. Fisheries became a major concern as road building and clear-cutting were found to increase siltation and to otherwise degrade spawning and rearing habitats required by salmon. As the natural disturbance ecology of the rainforest became better understood, clear-cutting was viewed as an inappropriate practice that detracted from the ecological integrity and the visual quality of southeast Alaskan landscapes. The scale of forest harvest also fuelled the controversy; enormous clear-cuts were clearly visible to all who flew over the Tongass or travelled through it by boat.

By the late 1980s, the Tongass National Forest became a political battlefield and the subject of intense national scrutiny. High-profile environmental campaigns were successful in effecting change. Passage of the Tongass Timber Reform Act did away with the timber subsidy and made significant changes in the contracts to better protect the rainforest. Within a few years, the pulp mills closed and the long-term contracts were terminated. As the timber industry shrank to a shadow of its former self, a new economic engine was firing up in southeast Alaska. The new industry, large-scale tourism, would bring income and jobs to the area, as well as new conflicts and controversies for land and resource management.

The Current Issue

Today, tourism in southeast Alaska largely revolves around the cruise ship industry. What began as a modest enterprise in the 1980s has become a major force in the economic fabric of the region. In 2002, the larger cruise ships brought up to 718,551 passengers and 311,511 crew members to southeast Alaska, totalling over 1 million visitors during the 4-month summer season.[1] This number exceeds the entire Alaskan population of around 600,000 residents. In Juneau, the state capital and a key stop along cruise ship routes, the cruise industry generates nearly US$100 million in direct local spending and more than US$8 million in local government fees and taxes. In smaller communities such as Ketchikan and Skagway, it dwarfs the extractive resource industries on which commerce used to depend.

The presence of the cruise industry is felt throughout southeast Alaska. On a typical July day in Juneau, 10,000 passengers and crew swarm ashore to increase the city's population by more than a third. A large cruise ship dock now dominates the Ketchikan waterfront, including all manner of souvenir shops, fast food stands and tourist concessions where before there were none. Water and air pollution, noise, crowding and fundamental clashes in values are some of the problems wrought by this imposing economic force.

Noise associated with the tourism industry is a source of major conflict in Juneau. Here the most popular activities for cruise ship visitors include 'flight-seeing' over the surrounding wilderness and helicopter landings on the Juneau ice field above the city. Throughout the day, flocks of helicopters pass over areas where Juneau residents live, work and try to relax. During the 2001 tourist season, Juneau residents were subjected to the noise of 17,783 helicopter landings on the ice field. The agency responsible for issuing the permits, the USDA Forest Service, has authorized up to 19,039 landings per season through 2004, with a 5% growth allowed in the 3 years following. The landings are allowed 7 days a week, from 8.30 a.m. until 8.00 p.m.

Unable to tolerate the noise any longer, some Juneau residents formed the Peace and Quiet Coalition and began a vigorous campaign to contain the problem. They succeeded in getting a referendum on the 2001 municipal ballot, which featured flight-free Saturdays and limitations on the number of permitted landings. Although the measure failed, it clearly signalled the seriousness of the problem. Hoping to resolve the issue, the Forest Service initiated a dispute resolution process using professional mediators. The effort was declared a failure after many months of lively meetings. According to the mediators, the parties lacked sufficient common ground to proceed.

The noise dispute in Juneau is symptomatic of the angst that southeast Alaskan communities are experiencing in the midst of significant changes over which they have no control. In Ketchikan, an arena for the Great Alaska Logging Show was built on the harbour site formerly occupied by the timber sort yard. Tourists pay US$29 each to 'witness the excitement as Alaska's frontier woodsmen do battle!' While a great success financially, the show rankled the locals because it brought in professional entertainers from outside rather than employing 'real loggers' from the surrounding islands.

So far, the large cruise ships do not stop at the small town of Wrangell. A majority of Wrangell residents would prefer to keep it that way, according to events that took place in 2000. The town had been offered a family's bequest of US$6 million for the construction and long-term maintenance of a museum and cultural centre to serve Wrangell residents and visitors alike. Architectural plans were drawn up, a site was selected in the heart of the harbour, and a concrete foundation was prepared for the facility. However, a rift developed within the community between those who supported the project and others who opposed it. The opponents feared that it might attract the attention of the cruise industry, and bring change over which the locals would have little control. The debates intensified so much that a special referendum was called. Despite the fact that over US$1 million had been invested thus far, the nays won and the project was terminated.

[1] Visitor and revenue figures obtained from the 2002 issue of *Cruise News* (volume 3, number 1), a publication of the North West Cruiseship Association (www.alaskacruises.org).

The impacts of tourism in southeast Alaska reach beyond the seaports and into the surrounding wildlands. A key attraction of any Alaskan cruise is to view the scenic wilderness for which the state is famous. Motorized boat tours and flight-seeing trips make this possible by fitting within the tight time constraints of most cruises. However, is it possible to have a wilderness experience with boats and planes buzzing around?

This question came to a head in Misty Fiords National Monument, a spectacularly scenic portion of the Tongass National Forest. Misty Fiords was once a favourite haunt of kayakers and other backcountry users drawn by the area's solitude, unspoiled character and spectacular landscapes. Today, those users have largely abandoned Misty Fiords in favour of more remote areas. The popularity of the area for tourism has brought new problems for resource managers. One example is a seemingly innocuous, 15 ft × 15 ft floating dock that appeared in Rudyerd Bay in 1992. Over the years, the little dock grew into a 100 ft × 75 ft, U-shaped structure that is pivotal in the operations of one fly–cruise business. Since 1998, that company has been taking 90 passengers on a 3 h cruise from Ketchikan into Rudyerd Bay, ending at the floating dock where a fleet of floatplanes awaits them for the return trip to Ketchikan. Disembarking from the planes are another 90 people who board the boats for the return cruise to Ketchikan.

The Forest Service began receiving complaints about this operation, and looked into the permit status from the responsible state agency. It turns out that no permit had been issued. The company had managed to avoid the prohibition against permanent structures in the wilderness area by moving the dock every couple of weeks!

The impacts of tourism go beyond annoyance for many rural residents, who perceive that their traditional livelihoods are threatened. In a land where roads and grocery stores are few, many families rely on wild animal and plant resources to maintain a nutritious and affordable diet. In the case of Native peoples, dependence on wild foods reflects an ancient history of tradition, culture and sense of community. Subsistence use of wild foods is so important that it is provided for in both state and federal law. The Alaska National Interest Lands Conservation Act gives priority to subsistence uses by rural people over sport and commercial uses of fish and game.

The job of providing for subsistence uses is not too difficult as long as the resources and access opportunities are plentiful, relative to the people wishing to use them. However, increasingly, conflicts are arising between rural subsistence users and those with other interests such as sport fishing and wildlife viewing. The conflicts are partly biological, requiring decisions about the capacity of the resource to sustain harvest, and how that harvest needs to be allocated. However, social factors are also important, as people who have long hunted, fished and gathered in their traditional areas are feeling 'invaded' by non-resident hunters, anglers, ecotourists and others.

The cruise industry is projected to grow even bigger in the years ahead. We can expect the issues to intensify as Alaskans' 'love–hate' relationship with tourism plays out in communities and backcountry areas of southeast Alaska.

Conclusion

For better or worse, the economic future of southeast Alaska is increasingly bound to outsiders who visit but do not remain. The industry brings important income and jobs to a resource-dependent region where the extractive industries have been greatly diminished. However, local people and communities are experiencing change that they did not plan for, and do not feel in control of. These feelings, along with commercial tourism's impacts on the environment and psyche of residents, give rise to angst and discord within the communities of southeast Alaska.

Resource managers are challenged to balance the needs and desires of visitors and operators with protection of the outstanding lands and resources that make the region so attractive for tourism. Also, because the social issues are so great, public involvement and conflict resolution will be essential elements of integrated resource management. The real challenge ahead is to anticipate change, prepare for it and engage the public in managing it to achieve desirable or acceptable outcomes.

Discussion Questions

1. Read your local newspaper for a local resource management issue and discuss how the public's understanding of that issue has been affected by this media coverage.
2. Discuss and describe the relationship of natural resource use and management and the following variables
- socioeconomic status
- demographic change
- ethnicity/race

3. Using the information provided in the chapter, discuss your views on how natural resources should be used and managed.
4. One of the basic rules of the IREM approach is to manage the interconnections between the human and biophysical aspects of natural resources. Select a natural resource management issue and identify five key variables or aspects that are important to that situation, and explain the impact that these variables have on each other. For example, timber harvesting and stream siltation are often directly related to each other.
5. Identify five personal choices you have made within the last week and describe the impacts these choices have had on the use and management of natural resources.

Case Study Discussion Questions

7. What ideas do you have for solving the helicopter noise problem in Juneau?
8. Can you envisage ways in which local communities can get better control over the impacts brought by the growth of tourism?
9. What can managers do to address heavy use in remote scenic areas such as Misty Fiords National Monument?
10. Is it possible to have a large-scale tourism operation and sustain and protect the local culture at the same time?

What Happened: the Adopted Solutions to the Case Study Questions

I. Helicopter noise: the Forest Service is working with the city of Juneau to locate a new helicopter staging area away from town, which would allow helicopters to reach the ice field without flying over residential or business areas. Also, the helicopter companies are testing and acquiring new equipment to reduce noise transmission. Although the first attempt at mediation failed, it is likely that public processes will play an important role in future resolution of this issue.

II. Community control over impacts: these needs are being addressed through taxes and fees paid by cruise lines to cover state and local governments' cruise-related expenses. Cruise lines now pay between US$65 and US$80 per passenger to cover such costs as port maintenance and emergency services. Some communities have levied an additional 'head tax' on cruise ship passengers to ameliorate local impacts. Juneau, which charges a US$5 per passenger fee, has invested in improved power and sanitary facilities at its port, thereby reducing effluent discharge by cruise ships. The fees are also used to improve trails and parks, parking facilities, and city services that benefit local residents as well as visitors. By ensuring that benefits (and not just impacts) flow to local people, the cruise industry can improve its standing in the communities on which it depends.

III. Misty Fiords: the USDA Forest Service entered into a contract with the University of Arizona to complete a study of visitor distributions and flow patterns in Misty Fiords. The study products include a GIS-based model that managers can use to identify problem areas and emerging trends. With the information produced by this work, managers can use the permitting system to make needed adjustments in intensity, distribution and movement patterns of visitor use. Also, they can direct backcountry rangers and other resources to the locations where they are most needed.

References

Ajzen, I. and Fishbein, M. (1980) *Understanding Attitudes and Predicting Social Behavior*. Prentice-Hall, Englewood Cliffs, New Jersey.

Allison, M.T. (1993) Access and boundary maintenance: serving culturally diverse populations. In: Ewert, A., Chavez, D. and Magill, A. (eds) *Culture, Conflict, and Communication in the Wildland Urban Interface*. Westview Press, Boulder, Colorado, pp. 99–107.

Baird-Olson, K. (2000) Recovery and resistance: the renewal of traditional spirituality among American Indian women. *American Indian Culture and Research Journal* 24(4), 1–35.

Bateman, R.B. (1996) Talking with the plow: agricultural policy and Indian farming in the Canadian and U.S. Prairies. *Canadian Journal of Native Studies* 16(2), 211–228.

Beck, B. (2000, 9 September) Survey – Australia: a sorry tale. *Economist* 356, 8187, p. S12.

Beltrame, J. (1999, 6 December) Land claims by Canadian tribes gain court ruling and a treaty increase the pressure: non-natives are nervous. *Wall Street Journal*, p. A27.

Bennett, J.W. (1976) *The Ecological Transition*. Pergamon Press, New York.

Berner, R.A. (1990) Atmospheric carbon dioxide over Phanerozoic time. *Science* 249, 1382–1386.

Bettinger, P. and Boston, K. (2001) A conceptual model for describing decision-making situations in integrated natural resource planning and modeling projects. *Environmental Management* 28(1), 1–7.

Bissix, G., Levac, L. and Horvath, P. (2002) The political economy of the wilderness designation in Nova Scotia. *Proceedings of the 2001 Northeastern Recreation Research Symposium* (Northeastern research station – general technical report NE-289). US Department of Agriculture, Washington, DC, pp. 377–382.

Booth, A.L. and Kessler, W.B. (1996) Understanding *linkages* of people, natural resources, and ecosystem health. In: Ewert, A. (ed.) *Natural Resource Management: the Human Dimension*. Westview Press, Boulder, Colorado, pp. 231–248.

Bradley, G.A. and Bare, B.B. (1993) Issues and opportunities on the urban forest interface. In: Ewert, A., Chavez, D. and Magill, A.

(eds) *Culture, Conflict, and Communication in the Wildland Urban Interface*. Westview Press, Boulder, Colorado, pp. 17–31.

Bumpass, L. and Sweet, J. (1989) National estimates of cohabitation. *Demography* 26, 615–625.

Chase, R.A. (1993) Protecting people and resources from wildfire: conflict in the interface. In: Ewert, A., Chavez, D. and Magill, A. (eds) *Culture, Conflict, and Communication in the Wildland Urban Interface*. Westview Press, Boulder, Colorado, pp. 349–356.

Cordell, H.K. and Overdevest, C. (2001) *Footprints on the Land: an Assessment of Demographic Trends and the Future of Natural Lands in the United States*. Sagamore, Champaign, Illinois.

Cordell, H.K. and Tarrant, M.A. (2002) Changing demographics, values, and attitudes. *Journal of Forestry* 100(7), 28–33.

Cortner, H.J. (1996) Public involvement and interaction. In: Ewert, A. (ed.) *Natural Resource Management: the Human Dimension*. Westview Press, Boulder, Colorado, pp. 167–179.

Cortner, H.J. and Shannon, M.A. (1993) Embedding public participation in its political context. *Journal of Forestry* 91(7), 14–16.

Cronin-Fisk, M. (2001, 22 October) 200-year-old land dispute nets $247.9M. *National Law Journal* 24(9), p. A6.

Crowley, T.J. (1996) Remembrance of things past: greenhouse lessons from the Geologic Record. *Consequences* 2(1), 3–12

Daily, G.C. and Ellison, K. (2002) *The New Economy of Nature: the Quest to Make Conservation Profitable*. Island Press, Washington, DC.

De-Shalit, A. (2000) *The Environment: Between Theory and Practice*. Oxford University Press, New York.

Doka, K. (1992) When gray is golden: business in an aging America. *Futurist* 26(4), 16–20.

Drake, W. (2001, July/August) Green Britain. *Environment* 43(6), 7.

Dunlap, R.E. and Saad, L. (2001) Only one in four Americans are anxious about the environment. *Gallup Poll Monthly* 427, 6.

Dunlap, R.E. and Van Liere, K. (1978) The 'new environmental paradigm': a proposed instrument and preliminary results. *Journal of Environmental Education* 9, 10–19.

El-Swaify, S.A. and Yakowitz, D.S. (eds) (1997) Multiple objective decision-making for land, water, and environmental management. *Proceedings of the First International Conference on Multiple Objective Decision Support Systems (MODSS) for Land, Water, and Environmental Management: Concepts, Approaches, and Applications*. Honolulu, Hawaii.

Ewert, A. (1996a) Gateways to adventure tourism: the economic impacts of mountaineering on one portal community. *Tourism Analysis* 1, 59–63.

Ewert, A. and Baker, D. (2001) Standing for where you sit: an exploratory analysis of the relationships between academic major and environmental beliefs. *Environment and Behavior* 33(5), 687–707.

Ewert, A. and Pfister, R. (1991) Cross-cultural land ethics: motivations, appealing attributes and problems. *Transactions of the North American and National Resource Conference* 56, 146–151.

Ewert, A., Chavez, D. and Magill, A. (1993) *Culture, Conflict, and Communication in the Wildland Urban Interface*. Westview Press, Boulder, Colorado.

Farley, R. and Allen, W. (1987) *The Color Line and the Quality of Life in America*. Russell Sage Foundation, New York.

Fine Jenkins, A. (1997) Forest health: a crisis of human proportions. *Journal of Forestry* 95(9), 11–14.

Flanders, N.E. (1998) Native American sovereignty and natural resource management. *Human Ecology* 26(3), 425–449.

Floyd, M. (1999) Race, ethnicity and use of the National Park System. *Social Science Research Review* 1(2), 1–24.

Force, J.E. and Machlis, G.E. (1997) The human ecosystem. Part II: social indicators in ecosystem management. *Society and Natural Resources* 10, 369–382.

Fosler, R., Alonso, W., Meyer, J. and Kern, R. (1990) *Demographic Change and the American Future*. University of Pittsburgh Press, Pittsburgh, Pennsylvania.

Frissell, C.A. and Bayles, D. (1996) Ecosystem management and the conservation of aquatic biodiversity and ecological integrity. *Water Resources Bulletin* 32(2), 229–240.

Garreau, J. (1991) *Edge City: Life on the New Frontier*. Doubleday, New York.

Hammit, W.E. (2000) The relation between being away and privacy in urban forest recreation environments. *Environment and Behavior* 32(4), 521–540.

Harriman, J. and Baker, D. (2003) Applying integrated resource and environmental management to transmission right of way maintenance. *Journal of Environmental Planning and Management* 46(2), 199–217.

Heberlein, T.A. (1972) The land ethic realized: some social psychological explanations for changing environmental attitudes. *Journal of Social Issues* 28, 79–87.

Heinrichs, J. (1991, March/April) The future of fun. *American Forests* 21–24, 73–74.

Hodgkinson, S.P. and Innes, J.M. (2001) The attitudinal influence of career orientation in 1st-year university students: environmental attitudes as a function of degree choice. *Journal of Environmental Education* 32(3), 37–40.

Hollingshead, D. (1992) ‘White’ gaze, ‘red’ people – shadow visions: the dis-identification of ‘Indians’ in cultural tourism. *Leisure Studies* 11, 43–64.

Hooper, B., McDonald, G. and Mitchell, B. (1999) Facilitating integrated resource and environmental management: Australian and Canadian perspectives. *Journal of Environmental Planning and Management* 42(5), 747–766.

Horn, B., Agpaoa, L., Bailey, J., Chambers, V., Kissinger, J., McMenus, K., Morris, G., Smith, R. and Zwang, C. (1993) *Strengthening Public Involvement: a National Model for Building Long-term Relationships with the Public*. USDA Forest Service, Washington, DC.

Howlett, M. (2002) Policy instruments and implementation styles: the evolution of instrument choice in Canadian environmental policy. In: Van Nijnatten, D.L. and Boardman, R. (eds) *Canadian Environmental Policy: Context and Cases*, 2nd edn. Oxford University Press, Don Mills, Ontario, Canada, pp. 25–45

Hutton, W. (2002, 28 April) Log cabin to White House? Not any more. *Observer*. Retrieved 28 April 2002, from http://www.observer.co.uk/comment/story/0,6903,706484,00.html

Kaiser, F.G., Ranney, M., Hartig, T. and Bowler, P.A. (1999) Ecological behavior, environmental attitude, and feelings of responsibility for the environment. *European Psychologist* 4(2), 59–74.

Kaplan, R. and Kaplan, S. (1989) *The Experience of Nature: a Psychological Perspective*. Cambridge University Press, New York.

Kelly, J. (1989) Leisure behaviors and styles: social, economic, and cultural factors. In: Jackson, E. and Burton, T. (eds) *Understanding Leisure and Recreation: Mapping the Past, Charting the Future*. Venture, State College, Pennsylvania, pp. 89–112.

King, M., Elliott, C., Hellberg, H., Lilford, R., Martin, J., Rock, E. and Mwenda, J. (1995) Does demographic entrapment challenge the two-child paradigm? *Health Policy and Planning* 10, 376–383.

Kline, J.D. and Armstrong, C. (2001) Autopsy of a forestry ballot initiative. *Journal of Forestry* 99(5), 20–27.

Knopp, T. (1972) Environmental determinants of recreation behavior. *Journal of Leisure Research* 4, 129–138.

Kolb, W.E., Wagner, M.R. and Covington, W.W. (1994) Concepts of forest health: utilitarian and ecosystem perspectives. *Journal of Forestry* 91(9), 32–37.

Lean, J. and Rind, D. (1996) The sun and climate. *Consequences* 2(1), 27–36.

Lessinger, J. (1987) The emerging region of opportunity. *American Demographics* 9(6), 33–37, 66–68.

Lester, J.P., Allen, D.W. and Hill, K.M. (2001) *Environmental Injustice in the United States: Myths and Realities*. Westview Press, Boulder, Colorado.

Long, L. (1988) *Migration and Residential Mobility in the United States*. Russell Sage Foundation, New York.

Mabry, M. (2002, 11 March) South Africa is not Zimbabwe: the two countries have similar pasts, but crucial differences. *Newsweek* p. 13.

Marcin, T. (1993) Demographic change: implications for forest management. *Journal of Forestry* 91(11), 39–45.

Margerum, R.D. (1997) Integrated approaches to environmental planning and management. *Journal of Planning Literature* 11(4), 459–475.

Marin, G. and Marin, B. (1991) *Research with Hispanic Populations*. Sage, Newbury Park, California.

Maslund, T. and Newton, K. (2002a, 11 March) The grievance of all grievances. *Newsweek* 139(10), p. 36.

Mazmanian, D.A. and Kraft, M.E. (1999) *Towards Sustainable Communities: Transition and Transformations in Environmental Policy*. MIT Press, Cambridge, Massachusetts.

McMichael, A.J. (1997) Global environmental change and human health: impact assessment, population vulnerability, and research priorities. *Ecosystem Health* 3(4), 200–210.

Mol, P.J. (2001) *Globalization and Environmental Reform: the Ecological Modernization of the Global Economy*. MIT Press, Cambridge, Massachusetts.

Murdock, S. and Ellis, D. (1991) *Applied Demography: an Introduction to Basic Concepts, Methods, and Data*. Westview Press, Boulder, Colorado.

Nord, M., Luloff, A.E. and Bridger, J.C. (1998) The association of forest recreation with environmentalism. *Environment and Behavior* 30(2), 235–246.

O'Connor, D.R. (2002) *Report of the Walkerton Inquiry: the Events of May 2000 and Related Issues*. Publications Ontario, Toronto, Ontario, Canada.

O'Keefe, D. (1990) *Persuasion: Theory and Research*. Sage, Newbury Park, California.

O'Riordan, T. (1981) *Environmentalism*, 2nd edn. Pion, London.

Parkes, M. and Panelli, R. (2001) Integrating catchment ecosystems and community health: the value of participatory action research. *Ecosystem Health* 7(2), 85–106.

Petty, R., McMichael, S. and Brannon, L. (1992) The elaboration likelihood model of persuasion: applications in recreation and tourism. In: Mandredo, M. (ed.) *Influencing Human Behavior*. Sagamore, Champaign, Illinois, pp. 77–102.

Pfeil, R.W. and Ellis, J.W. (1995, 30 April) Evaluating GIS for establishing and monitoring environmental conditions of oil fields. *American Association of Petroleum Geologists Bulletin* 79(4), 595.

Place, G.S. (2000) The impact of early life outdoor experiences on an individual's environmental attitudes. Unpublished doctoral dissertation, Indiana University, Bloomington, Indiana.

Portney, K.E. (1992) *Controversial Issues in Environmental Policy: Science vs. Economics vs. Politics*. Sage, Newbury Park, California.

Raven, P.H., Berg, L.R. and Johnson, G.B. (1993) *Environment*. Saunders College, New York.

Reeves, H. (2002, 13 March) Metro briefing New York: Syracuse: Indian payment stopped for now. *New York Times* p. B8.

Saegert, S. and Winkel, G.H. (1990) Environmental psychology. *Annual Review of Psychology* 41, 441–477.

Sanger, D.E. and Alvarez, L. (2001, 29 June) Conservation-mindful Bush turns to energy research. *New York Times* p. A18.

Schneider, S.H. (1996, Summer) Engineering change in global climate. *Forum for Applied Research and Public Policy* 11(2), 92–97.

Schultz, P.W., Zelezny, L. and Dalrymple, N.J. (2000) A multinational perspective on the relation between Judeo-Christian religious beliefs and attitudes of environmental concern. *Environment and Behavior* 32(4), 576–591.

Shurts, J. (2000) *Indian Reserved Water Rights*. University of Oklahoma Press, Norman, Oklahoma.

Siegel, J. and Taeuber, C. (1986) Demographic perspectives on the long-lived society. *Daedalus* 115, 77–177.

Silvern, S.E. (2000) Reclaiming the reservation: the geopolitics of Wisconsin Anishinaabe resource rights. *American Indian Culture and Research Journal* 24(3), 131.

Slocombe, D.S. (1998) Defining goals and criteria for ecosystem-based management. *Environmental Management* 22(4), 483–493.

Spencer, G. (1989) *Projections of the Population of the United States, by Age, Sex, and Race: 1983 to 2080* (Current Population Reports-Series P-25–1018). US Bureau of the Census, US Government Printing Office, Washington, DC.

Spencer, R., Kelly, J. and Van Es, J. (1992) Residence and orientations toward solitude. *Leisure Sciences* 14(1), 69–78.

Stern, P.C., Dietz, T., Abel, T., Guagnano, G.A. and Kalof, L. (1999) A value–belief–norm theory of support for social movements: the case of environmentalism. *Human Ecology Review* 6(2), 81–97.

Struglia, R. and Winter, P.L. (2002) The role of population projections in environmental management. *Environmental Management* 30(1), 13–23.

Stull, D.D. (1990) Reservation economic development in the era of self-determination. *American Anthropologist* 92(1), 206–211.

Sussman, G., Daynes, B.W. and West, J.P. (2002) *American Politics and the Environment*. Addison Wesley Longman, New York.

Tarrant, M.A. and Cordell, H.K. (2002) Amenity values of public and private forests: examining the value–attitude relationship. *Environmental Management* 30(5), 692–703.

Van Liere, K. and Dunlap, R.E. (1980) The social bases of environmental concern: a review of hypotheses, explanations, and empirical evidence. *Public Opinion Quarterly* 44(1), 181–197.

Virden, R.J. and Walker, G.J. (1999) Ethnic/racial and gender variations among meanings given to, and preferences for, the natural environment. *Leisure Sciences* 21(3), 219–239.

White, L., Jr (1967) The historic roots of our ecological crisis. *Science* 55, 1203–1207.

Williams, D.R. and Patterson, M.E. (1996) Environmental meaning and ecosystem management: perspectives from environmental psychology and human geography. *Society and Natural Resources* 9, 507–521.

Wolosoff, S.E. and Endreny, T.A. (2002) Scientist and policy-maker response types and times in suburban watersheds. *Environmental Management* 29(6), 729–735.

Wondolleck, J.M. (1992) Resource management in the 1990s. *Forest Perspectives* 2(2), 19–21.

Yun, J.M. (2002, April) Offsetting behavioral effects of the corporate average fuel economy standards. *Economic Inquiry* 40(2), 260–270.

5 Environmental Conflict and Property Rights

Introduction

Driving forces act as agents of change, and, during this change, conflict in natural resource management is often generated. Over the past four decades, conflict in the preservation and development of natural resources has been a common theme worldwide. Changes in knowledge about the environment and increasing public involvement in decision making have altered how traditional resource agencies and companies make decisions. Blockades, international protests, civil disobedience and boycotting of natural resource products have characterized some of the conflicts over natural resource management issues. IREM, as a process, has to recognize conflict as an agent of change and deal with conflicting views on the environment and associated management strategies.

This chapter will examine the nature of conflict and its role in the management of natural resources. An overview of property rights, and values associated with property rights, provides a means of analysis for environmental conflict. Finally, we include a section on environmental dispute resolution as a means of dealing with certain types of environmental conflict.

Dukes (1996) has recently characterized five developments, or themes, in the USA, that have given rise to the level of public disputes: (i) the dominance of bureaucracy in state governance; (ii) a diffusion of power away from central authority and an increase in the role of public involvement and competing interest groups; (iii) changes in legislation and public policy, and an increasingly activist judiciary; (iv) increasing uncertainty and scientific and technical complexity of public problems; and (v) recognition of the inability of existing institutions to meet the new demands.

Environmental conflict is often a result of a combination of these five themes. Indeed, it is the interplay of these themes that has given rise to IREM, and the need to manage natural resources with different methods that recognize diverse stakeholder requirements and the complexity of natural systems.

Resource-based conflict is often generated over issues such as irreversible environmental damage that include a wide range of participants and stakeholders. This can occur at a variety of scales from the local level, such as the siting of a neighbourhood incineration unit, to international-level disputes where governments, corporations and NGOs are involved. The causes of conflict can also be highly varied and may deal with a wide range of issues such as land use, forest clear-cutting, water quality, disposal of toxic substances or mining. As Tillet (1991) suggests, underlying most environmental conflict is value conflict. Values may differ over views of human rights versus the rights of the environment, or issues of life-style and quality of life. The elements that comprise different environmental conflicts are often case-specific and very complex.

What *is* common to all environmental conflict, however, is property rights; both the access to resources and the decisions as to how resources are allocated are founded in property rights. 'Environmental conflict stems from divergent views about how to allocate and utilize land, air, water and living resources' (Glavovic *et al.*, 1997, p. 270). How should we plan our land uses and allocate resources? Any answer to this question is based in a normative theory of property rights. Thus, answers may vary according to different theories of property rights: they should be distributed efficiently; everyone should have a share; they ought to go to the highest bidder; or they should be sustainable for future generations. People may, or may not, be aware to which theory or combination of theories they subscribe; however, in most cases they have a system of beliefs that justify their answer. In many situations, the different answers to this question cause conflict. Competing notions of ownership and property rights hold differing implications for the management of resources and its effect on the physical and cultural landscape.

The public perception of property rights and ownership has changed over the last two decades and, as a result, conflict has been generated over a wide range of environmental issues. Much of the environmental conflict that we have experienced in North America stems from competing rights of ownership. In some cases, the issue of rights is defined in terms of individual private rights opposing a perceived set of social rights, and in other cases there are conflicting social stands on how resources should be used. For example, much of the forest-related conflict in North America involves the private timber rights of logging companies versus a larger social justification of rights as articulated by environmental, native and local citizen groups. The conflict that characterized the James Bay project was centred around different justifications of the social costs and benefits of a large-scale hydroelectric project in northern Quebec, Canada.

In conjunction with an increasing public awareness of environmental issues, the rights of land ownership have also changed. Limited supplies of natural resources, environmental degradation and the need to procure a sustainable future are common social issues of today. Changes in notions of land ownership have accompanied this increasing environmental awareness. Environmental legislation, growth management policies and provincial/state measures to protect agricultural land represent some of the regulations that have placed restrictions on private ownership. The change in the structure of property rights has resulted from an alteration of social values. However, this change has not come without conflict.

How we justify our right to ownership ultimately affects the way we design our rights to property. In turn, the means by which we design and plan our property rights through laws and legislation affects the distribution of costs and benefits within the natural and cultural landscape. IREM is based in the articulation and development of property rights for diverse stakeholders.

Conflict

Conflict can be defined as a 'situation in which one actor (either an individual, a group or a nation) is engaged in opposition (violent or non-violent) to another actor(s), who is pursuing what are, or appear to be, incompatible goals' (Eldridge, 1979, p. 1). Coser (1956, p. 8) interprets social conflict as a struggle over values and 'claims to scarce status, power, and resources'. In addition, Himes (1980, p. 61) adds that social conflict structure is composed of 'working parts' that include: (i) two social roles – each of which may include one or more actors; (ii) two social positions – placing the roles in confrontation; (iii) a set of scarce values that are the goals of the struggle (Himes uses 'values' to mean 'power, status or resources') and additional values that condition the conflict in various ways; (iv) a set of norms that embody the social values and systemize the conflict relationship; and (v) a relationship characterized by a series of struggles between the actors – more or less patterned by traditional and emergent norms.

Conflict within society has been viewed by conflict theorists as either a natural and healthy part of human relationships (Coser, 1956; McEnery, 1985; Laue, 1987), or as an unusual occurrence or disequilibrium within society that needs to be corrected (Wehr, 1979; Schellenberg, 1982). Laue suggests that conflict is an integral part of society:

> Conflict is never solved; we talk of conflict resolution, not conflict solution . . . Whether family or international disputes or anything in between, conflict incidents may be solved, but conflict *per se* is never solved. Each solution creates, in a Helgelian sense, a new plateau or a new synthesis against which the next conflict scenario is played. Society never 'solves' conflict totally. Conflict incidents or episodes are solved and then re-solved and re-solved.
>
> (p. 18)

Although there is a difference of opinion among theorists as to the nature of conflict within society, there tends to be a consensus that conflict has its limits in society, beyond which it is disruptive to that society. For example, Wehr (1979, p. 15) believes that conflict can be categorized along dimension poles that are 'completely non-institutionalized, normless, violent, no holds-barred conflict and highly ritualized, routinized legitimized disputes'. Filley (1975, p. 2) also distinguishes between two types of conflict, 'some for example follow definite rules and are not typically associated with angry feelings on the part of the parties, while others involve irrational behavior and the use of violent or disruptive acts by the parties'.

There is a role for conflict in society, but it is not always clearly defined. Himes (1980, p. 18) defines the division of conflict in society as 'legitimate and non-legitimate' which is marked by a 'crooked and twisted line'. Conflict is defined as legitimate when it reflects the norms of a society. Terms such as 'struggle, protest and non-violence' characterize legitimate conflict. Terms such as 'violence, rebellion and aggressive war' encapsulate non-legitimate conflict, which exceeds the limits of societal consensus. Himes stresses that what is perceived as legitimate and non-legitimate conflict varies between actors in society and changes in time and place throughout societies.

Conflict, as a positive force in society, prevents stagnation and provides a medium for problems to be aired and solutions to be attempted (Deutsch, 1987). It helps to revitalize norms and values, and also contributes to the emergence of new norms. However, according to Coser (1956, p. 157), 'conflict tends to be dysfunctional for a social structure in which there is no or insufficient toleration and institutionalization of conflict'. Although conflict may have a beneficial role to play within society, certain safety valves have to be institutionalized to accommodate this role. As Deutsch (p. 38) notes:

> Conflict can neither be eliminated nor even suppressed for long. The social and scientific issue is not how to eliminate or prevent conflict, but rather how to have lively controversy instead of deadly quarrels.

Classical theories of conflict and power reveal conflict as a means of revitalizing existing norms and values to create a 'new framework within which the contenders can struggle' (Coser, 1956, p. 157). Weber, specifically referring to law, points out that the clash of interests leads to the modification and creation of law. Simmel (1955) expands theories of social change beyond the law to contend that conflict acts as a stimulus to establish new rules and norms.

Issues involved in conflict can be generally categorized as (Dorcey, 1986; Brown and Marriot, 1993):

- cognitive or issues of fact – perceptions by each party of the correctness of the information;
- interest-based – dealing with the costs and benefits of resource distribution: who gets what and how much;
- behavioural – the history and resulting relationship that have determined how the parties react to each other; and
- attitudes and values – different perspectives of fairness, culture, concepts of justice and morality.

Conflict in Environmental Management

Conflict often involves issues of resource allocation. Dorcey and Riek (1989) identify environmental conflict resulting from both substantive and procedural issues. In the management of natural resources, substantive disputes arise surrounding four sets of issues: (i) project development and resource use effects; (ii) multiple use of resources and areas; (iii) regulations, policies and legislation; and (iv) resource ownership and jurisdiction. Procedural issues are defined by access to decision making and how decisions are made in public and private management of resources.

Both the substantive and procedural issues associated with environmental conflict deal with resource allocation and thereby involve a distribution of property rights. As Glavovic *et al.* (1997, p. 270) state, 'At its deepest level, environmental conflict is the division that arises over competing demands for individual and collective rights . . .'. Any attempt to resolve substantive or procedural disputes that arise as a result of property rights must address a normative theory of property rights. When individuals share similar values and the conflict is a result of cognitive, interest or behavioural differences, an understanding of theoretical differences may not be important in settling the dispute. However, where major differences occur regarding how resources should be allocated, an understanding of the disputing person's value differences is essential.

Property Rights

Resource use conflicts are often generated when people hold different values as to how resources 'ought' to be planned and allocated. Substantive and procedural disputes in environmental management are frequently based in conflicting value differences involving the distribution of property rights. People's perceptions of property and their rights to property are historically embedded in defined cultural landscapes. However, this is a constantly changing landscape that is defined in new ways and by changing values. Different claims to resources are often based in different philosophical justifications of the rights to property. These different claims hold implications for how we plan and structure our institutional frameworks for managing the landscape, because IREM is not a value-neutral activity. The planning process forms a crucial link between normative theories regarding how land should be allocated and the institutional structures by which it is allocated.

How we define our right to a resource will often affect the structure of property rights that is put in place. Thus, the way in which a right is justified can affect the way in which it is articulated and defended as a means to accessing resources. For example, within the context of a cultural theory approach, the justification of property rights is carried to a level where individuals devise institutions, not only to control behaviour within their own social context, but to 'create political environments which inhibit the exercise of other social contexts' (Buck, 1989, p. 8). Wildavsky (1987, p. 8) adds that individuals exert control over each other by institutionalizing moral judgements that 'justify relationships that can be acted upon and accounted for'. When these normative values clash, often there is conflict regarding how the institutional structure for resource allocation should be formulated. Land use conflicts in the planning of resources are frequently based in different values that affect the use and distribution of resources.

Property rights are commonly identified as a right to own or possess something, such as land or an automobile, and to be able to dispose of it as one chooses. However, this is only one aspect of property rights that focuses on the exclusive right to ownership. To have a right to property also implies an enforceable claim to the use or benefit of something. The concept of a property right distinguishes between momentary use or possession of something, and a claim to the thing which will be enforced by society or the state. For example, the claim can be in the form of a licence or lease to common property that gives the individual exclusive rights to secure a portion of that property, such as a fishing licence. So, property rights distinguish not only exclusive ownership of private property, but also rights shared and observed with others in common property. As Randall (1987, p. 157) notes, 'property rights specify both the proper relationships among people with respect to the use of things and the penalties for violating those proper relationships'.

Property rights are often referred to in the literature (Scott, 1983; Harrison, 1987; Randall, 1987) as a bundle of rights, in which ownership is distributed in a variety of ways. The bundles may define rights of exclusion, transferability or enforcement. Within our society, the courts recognize 'bundles of rights' in different ways, with different means of tenure such as freehold, leasehold, easements or rights-of-way. Additional property rights can also be placed over these bundles in the form of riparian rights or licences. Several parties may have different rights to a single parcel, with varying degrees of ownership; for example, a single piece of Crown land may contain a hydro easement, mineral rights, grazing leases and a tree farm licence.

Schmid (1978, p. 6) suggests that bundles of rights confer certain 'opportunity sets' for the individual. The opportunity set defines the various lines of action open to the holder of the bundle. Also, the relative capacity of the individual to make use of the rights is important in defining his/her opportunity set. The available resources, technology and knowledge determine the extent to which a person can exercise his/her property rights.

Both property rights and the opportunity provided by those rights are conditional to a time and place. Every society describes a unique relationship with its people to the available resource base, and thereby formulates a system of property rules. As Usher (1984, p. 391) suggests, 'Systems of property rights are a cultural artifact.' Systems of rights may be as diverse as the Songlines of the Aboriginals of Australia, which define territory according to ancestral songs that defined a stretch of country according to verses in the song. A man's verses were his title deeds to territory (Chatwin, 1987). The meaning of property is not constant (Macpherson, 1978) and it changes over time, across societies and within societies.

Property rights require recognition by others of one's claim to resources through relationships of power, kinship or convention. Levels of recognition within a society may range from a formal declaration recognized by legislation to an informal custom. Property rights form a complex set of social relationships that require recognition and enforcement by the collectivity.

Bromley (1989) has defined property rights as a triadic relationship that depends upon three sets of variables: (i) the nature and kinds of rights that are exercised, and their correlative duties and obligations; (ii) the individuals or groups in whom these rights and duties are vested; and (iii) the objects of social value over which these property relations pertain.

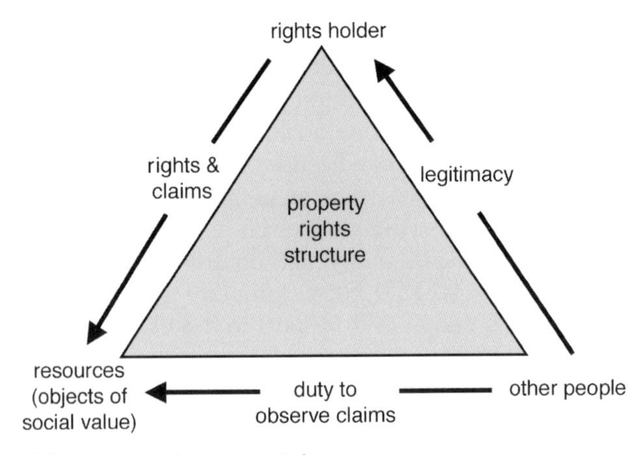

Fig. 5.1. Social structure of property rights.

Thus, the property rights structure that is put in place by a society is dependent on the interaction of the rights holder and that society. These components are outlined in Fig. 5.1. Other people must see the rights holder's claim to a resource as legitimate and, in turn, people have a duty to observe the rights holder's claim. The property rights structure determines who has the legal ability to impose costs on others, the right to the stream of benefits from the resource and who controls access to the resource. Ely (1914, p. 108) has observed, 'The essence of property is in the relations among men [sic] arising out of their relations to things'.

In countries such as Australia, Canada and the UK, the evolution of common law provides precedents for the recognition of property rights. Judges have made decisions based on historically established rights and principles, and have gradually changed the bundle of rights associated with property. Demsetz (1967) proposes the theory that property rights evolve as resources become more scarce and valuable. Systems of rights become more sophisticated as the demand and competition for resources increases. Emel and Brooks (1988) observed this with the formulation of groundwater rights on the High Plains in the USA: as the resource became scarcer, property rights became more defined and enforced. Scott and Johnson (1983) provide a good example of the historical development of common law property rights by examining the evolution of mineral rights in western Canada. The changing character of property rights in minerals evolved in response to the special characteristics of the users of alluvial and hard-rock minerals, and of the economic and technical problems (changing opportunity sets) unique to that industry.

Statute law provides a basis for the legal establishment of 'new principles' governing our recognition of changing property rights. This takes time. Although statutory law may appear to have formalized certain property rights, an enforcement of those rights is required to realize the change. For example, even with the establishment of a new law, it may not provide a binding application until it is proclaimed and contains specified regulations. Legislation or policies that are unpopular with a powerful segment of society, or strong lobby groups, may pay only 'lip service' to certain rights, with government doing little to administer the regulations. Estrin and Swaigen (1978) suggest that, in practice, statutes and regulations are no more binding than guidelines and policies; most of the significant legislation at any level of government is discretionary. The dynamics of changing property rights are based within competing interests and property claims throughout society. The outcomes of these power struggles often determine the eventual recognition of certain claims to property.

Competing and changing claims for property rights are often the basis for conflict and change within society. Different justifications for claims to resources legitimize the access to resources. These justifications are value-based and often form the basis for environmental conflict. Within the context of environmental management, the questions can be raised: 'How do we define our justifications to property?' and 'How does this relate to the structure of rights that are placed on the landscape?'. Certainly, the way in which individuals justify their rights to property affects the structures of rights that are put in place. In addition, the value assumptions in their justifications for access to resources affects how different actors will or will not resolve their conflict.

Environmental Justifications of Property Rights

No single set of theories characterizes an environmental justification of property rights. Rather, the spectrum of theories ranges from principles of deep ecology (Devall and Sessions, 1985), to guidelines for the environmental professional (Spitler, 1988). O'Riordan (1989) observes the following:

> Environmentalism is an uneasy amalgam of beliefs and prejudices which are political in the sense that they shape values and allegiances, and tilt slightly the prevailing distribution of power, away from capital and the established interests of manipulators...The challenge for modern environmentalism is, therefore, to overcome a paradox. On the one hand, environmentalism is becoming subsumed within the political struggle for green votes: in that narrow sense it is succeeding. On the other hand, environmentalism as a mosaic of contested positions could be splintered into competing segments so that its more powerful underlying social critique is lost.
>
> (p. 81)

Environmental ethics and the concept of environmental property rights are important to understand because much of the resource use conflict that has 'sprung up' over the last decades involves an environmental stand by the public and lobby groups. The environmental perceptions vary from local concepts of the home and hearth to larger issues such as global climate change.

Copp (1986) provides a general understanding of the positions in environmental ethics, using the following three categories: instrumental views, environmental values and environmental morality. Each of these perspectives of the environment will be briefly examined.

Instrumental views of the environment

This range of views presents nature as instrumental: as stockpile and sewer or sanctuary and library. The environment is valued in so far as it is useful to humankind and helps us to achieve other things we value. These views are clearly based in an anthropocentric perspective, and a spectrum of traditional justifications for the rights to property resides within this view of the environment. The instrumental view of labour and property is based essentially on the utility of both labour and property to the individual. Labour is a cost incurred by people who want to consume the goods made available by the earth. It is necessary because nature will not satisfy human needs without the expending of labour. Similarly, property rights are only justified using the same type of instrumental rationalization. Ryan (1984) observes that the instrumental view of property is one of the 'commonest views to be discerned in recent industrial sociology' (p. 9). Within the context of the instrumental view of the environment, several theoretical justifications can be used, including first occupancy, labour theory, utilitarian theory and political liberty. A comprehensive overview of these divisions of the justification to property rights is provided by Becker (1977) and Ryan (1984, 1987).

Environmental (intrinsic) values

Within this perspective, the environment is valued intrinsically, as something valuable in itself. As Copp (1986) notes:

> If nature is intrinsically *valued* by us, then we have a kind of goal or purpose regarding nature. We then are motivated to treat nature in ways appropriate to that purpose. And there will be ways of treating nature that contribute to our achieving that purpose as well as ways that do not contribute. However, the idea that nature is intrinsically *valuable* does not directly imply anything about our attitudes towards nature.
>
> (p. 184)

Nash (1989) adds:

> Nature has intrinsic value and consequently possesses at least the right to exist. This position is sometimes called 'biocentrism,' 'ecological egalitarianism' or 'deep ecology,' and it accords nature ethical status at least equal to that of humans. The antipode is 'anthropocentrism', according to which humans are the measure of all value... Human beings are the moral agents who have the responsibility to articulate and defend the rights of the other occupants of the planet. Such a conception of rights means that humans have duties or obligations toward nature. Environmental ethics involves people extending ethics to the environment by the exercise of self-restraint.
>
> (pp. 9–10)

Leopold (1949), in *A Sand County Almanac*, demonstrates the intrinsic value of nature when he writes (p. 239), 'The land ethic simply enlarges the boundaries of the community to include soils, water, plants, and animals, or collectively: the land. A land use decision is right when it tends to preserve the integrity, stability, and beauty of the biotic community. It is wrong when it tends otherwise.' Many resource managers and environmentalists share this view of the natural environment.

Environmental morality

This perspective(s) of nature incorporates the view that we owe certain duties towards nature and the idea that 'some aspects of the non-human world have rights against mankind' (Copp, 1986, p. 185). This school of thought is distinctly different from environmental values because a person who intrinsically values nature does not necessarily believe that duties are owed to nature or that the environment has rights.

The different philosophical justifications that provide a basis for access to resources are not clear-cut. However, there is a distinct division between instrumental views of the environment and intrinsic or environmental morality ideals. Essentially, the instrumental justification for access to resources contributes to the 'dominant social paradigm', which is formed by a 'constellation of values, attitudes, and beliefs through which individuals or collectively, a society, interpret the meaning of the external world' (Pirages and Ehrlich, 1974, p. 43). Many resource uses are based in this value system where single resource values are perceived in terms of their economic importance or use to humans.

Many of the intrinsic or environmental morality values are based in the 'new environmental paradigm' (Dunlap and Van Liere, 1978; Albrecht *et al.*, 1982; Geller and Lasley, 1985), which provides an alternative means to understanding property rights. The environmental movement has posed a challenge to the dominant social paradigm for land use and resource management practices, and has provided a basis for much of the resource use conflict in North America over the last four decades. How property rights are justified in natural resource management decisions is critical to the restrictions and concessions that are placed on the structure of the rules by which we make decisions on the landscape. The decisions are value-based and contain underlying assumptions as to how we view our right to access and use natural resources. Inherent in IREM practice are different justifications to the rights to the environment; and often different views cannot be easily reconciled.

Environmental Conflict Resolution

Environmental conflict resolution has grown as a professional practice over the last four decades as a means of dealing with the diverse range of conflict in resource-based disputes. Different methods of dispute resolution evolved as alternatives to traditional court-based approaches, which were often costly and prolonged. Traditional legal and administrative mechanisms have proved to be poorly structured to deal with environmental conflict because of restricted access for interested stakeholders, and a tendency to exacerbate polarization and foster

Table 5.1. A continuum of alternative dispute resolution processes.

Cooperative decision making	Third-party assistance with negotiations or problem solving		Third-party decision making		
Parties are unassisted	Relationship-building assistance	Procedural assistance	Substantive assistance	Advisory non-binding assistance	Binding assistance
Conciliation	Counselling/ therapy	Coaching/ process consultation	Mini-trial arbitration	Non-binding arbitration	Binding
Information exchange meetings	Conciliation	Training	Technical advisory boards	Summary jury trial	Mediation–arbitration
Cooperative/ collaborative problem solving	Team building	Facilitation			Dispute panels (binding)
Negotiations	Informal social activities	Mediation			Private courts/ judging

Adapted from Moore and Priscoli (1989).

adversarial relationships (Glavovic *et al.*, 1997). Methods for resolving conflict using mediation and negotiation are frequently referred to as alternative dispute resolution (ADR). Bingham (1986) defines ADR as an embracing term that refers to a diverse range of (usually) voluntary approaches which allow disputing parties to meet face to face in an effort to reach mutually acceptable solutions. It can be used where conflict has already occurred or, where there is a potential for conflict, as a management tool. Practitioners sometimes emphasize the proactive approach in using ADR as a means to *manage* conflict rather than resolve it (Brown and Marriot, 1993; Glasbergen, 1995).

The range of dispute resolution processes varies from informal, consensual agreement to formal, well-established practices in law. Moore and Priscoli (1989) provide a continuum of ADR processes that range from cooperative to third-party decision making (Table 5.1). A brief overview of the different methods will be discussed under: *conciliation*, *negotiation*, *mediation*, *mini-trial*, *arbitration* and *public inquiries/panels*. This list is not inclusive of all the forms of ADR but provides an overview of different methods that have been implemented successfully over the last few decades.

Conciliation

Conciliation is often an initial and informal process that can be conducted with or without a third party. The purpose is to reduce tension by exploring common ground with respect to technical issues, interpretation, improving communication and exploring opportunities for a negotiated settlement. The process is designed to bring parties together and establish common ground to build trust and seek solutions. Conciliation is an important first step in conflict resolution, and the earlier it is established, the sooner misconceptions and a further escalation of conflict may be avoided.

Negotiation

Negotiation is the most common form of ADR and involves parties attempting to arrive at a mutual settlement without intervention by a third party. Negotiation is a voluntary process in which parties engage each other to reach a desirable settlement. Views are exchanged and parties aim to realize their interests, but compromise is an important aspect of negotiation as each party attempts to reach a joint agreement. Negotiation is also an important skill that we use daily in our lives. A good negotiator is a fundamental asset to ADR at all levels.

One method of negotiation is termed the 'mutual gains' approach (Susskind *et al.*, 2000). The foundation of this style of negotiation is based on first reckoning your best estimate of how you will do if *no* negotiated settlement is reached. It requires parties to understand how they will do without negotiation, and sets the foundation for mutual gains by each party. The primary steps include (after Susskind *et al.*): (i) analysing and trying to improve your position if no negotiation occurs; (ii) probing to clarify both parties' interests; (iii) inventing options that meet mutual interests; (iv) using objective criteria to argue for a package you favour; and (v) negotiating as if relationships matter.

Mediation

Mediation is distinguished from negotiation by the intervention of an independent, neutral third party. A concise definition of mediation is difficult to pin down because of the breadth of its use in the field of conflict resolution. Boule and Nesic (2001) suggest that mediation be defined according to a conceptual approach, i.e. in ideal terms, and a descriptive approach, which deals with what actually happens in practice. A conceptual definition identifies mediation as a voluntary process and usually the mediator does not have the authority to impose a settlement. The mediator is a 'go-between' who provides assistance between disputing parties by convening dialogue, clarifying issues, improving communication and providing a healthy atmosphere for parties to explore their similarities and differences. The dispute is resolved when the parties decide that they have a consensual and workable solution (Cormick, 1982).

According to Glavovic *et al.* (1997), mediation offers particular promise in environmental disputes for a variety of reasons:

> First, it fosters a mutual learning process through which parties can advance their understanding of the issues stemming from the complex interconnections between natural and human systems. Second, it seeks to engage all interested and affected

> parties and explicitly promotes better understanding and accommodation of their individual and collective interests. Third, it is not confined to jurisdictional domains and therefore transcends the cross-cutting boundary conditions of environmental disputes.
>
> (p. 273)

Mediation has grown as a recognized field of practice over the last 30 years, and a considerable literature deals with case studies and skills involved in environmental mediation. Boule and Nesic (2001) provide a comprehensive guide to the principles and practice of mediation.

Mini-trial (case presentation)

This form of ADR is a hybrid combination of dispute methods that provides a structured exchange of information. Moore and Priscoli (1989, p. 81) note that 'it involves a data presentation component similar to that in litigation, a negotiation component and the potential for third-party mediation and an advisory opinion'. The mini-trial requires a third party who usually presides over the disputants as a mediator or neutral party. The individual does not usually deliver a judgement, but on the request of the participants may offer advice as to how a court might decide the case. It is a voluntary process that provides a forum for structured negotiation and exploration of scientific or cultural issues relevant to the dispute. Often, cross-examination is included between parties, involving experts or lawyers, and a time limit is imposed on the exploration of issues. However, no binding judgement follows. Rather, the senior management or individuals responsible for decision making negotiate a settlement based on the presentations and examination of evidence.

Mini-trials have been used successfully in commercial disputes, conflict between large companies and with large government agencies. The objective of the mini-trial is to prevent business/agency disputes from becoming costly legal disputes and moving the issues to the courtroom. Mini-trials are most often used for large-value commercial cases because it is a more expensive process compared with other forms of ADR methods (Bevan, 1992).

Arbitration

Arbitration is a more formalized and established approach compared with the previous ADR methods. Arbitration may be initiated as a result of contract clauses, legislation requirements, or be voluntary once conflict has originated. Similar to mini-trials, arbitration involves third-party decision making with respect to the outcome of the dispute; however, the third party is neutral to the process and may not be selected by the disputants. The third party may consist of a single arbitrator or a panel of varying numbers.

Arbitration can be non-binding, where parties do not have to agree to the judgement of the arbitrator, or it may be binding in the decisions reached. In non-binding arbitration, disputing parties control the design of the process and must approve the outcome of the decision. Disputants can determine the flexibility of the process by agreeing on the rules for entering evidence, cross-examination procedures, choosing the arbitrator or deciding on levels of confidentiality (Neumann, 2001). Binding arbitration tends to be less flexible with respect to the parties' input, and requires the disputants to consent to the final decision of the arbitrator.

Arbitration is used most frequently where the arbitrator(s) knows the technical, cultural or legal issues associated with the dispute. Arbitration is based on the merits of the case and evidence presented. The format for the presentation of evidence can be varied, from an informal presentation of substantive issues to a structured setting where evidence and cross-examination are given as if in a courtroom.

Public inquiries/panels

The use of public inquiries and panels is common in environmental disputes which have escalated to a state or provincial level. Often, they are '*ad hoc*' and set up to deal with a specific conflict or resource management issue (Neumann, 2001). For example, in Australia, the Fitzgerald Inquiry investigated the land use issues with respect to mining and conservation on Fraser Island, the largest sand island in the world. In Canada and the USA, public inquiries are commonly used in controversial contexts. Heated forestry-related conflicts on Vancouver Island have raged for decades and, most recently, a public advisory panel was employed to resolve ecological and economic interests in the Clayoquot Sound area.

Panels or boards can also be used as a permanent adjudicative forum for resolving differences. Environmental assessment legislation, worldwide, often uses panels to evaluate the scope of impact assessment, assess the impacts of proposed developments, advise whether a project should proceed, or evaluate the mitigation required for project development. Boards are also established as permanent conflict resolution venues where harvesting practices, impacts of development or grievances about the process can be resolved. In Ontario, the Ontario Municipal Board, a semi-judicial and independent tribunal, has functioned since 1932 to adjudicate appeals under various statutes. Washington State and British Columbia use a Forest Practices Board to review timber harvesting practices.

Summary and Conclusions

Conflict is an essential component of the human condition and often an agent of change. With environmental conflict, change has occurred dramatically over the last several decades to reflect the changing values in society with respect to how resources and the environment should be developed. A fundamental shift has developed in how we view our rights to property and how certain property rights should be distributed. Different environmental values provide alternative views of human rights to the environment and how we determine our impacts on ecosystems. Increased public input and a wider audience have changed the scope of decision making in resource management. Alternative methods of dispute resolution have been developed to resolve and manage environmental conflict to supplement traditional court-based approaches. IREM is a decision-making process that recognizes conflict and manages it as a positive force. ADR approaches can be integrated into IREM as a means of dealing with different scales of conflict and improving communication and relationships between stakeholders.

The following case study provides an example of diverse interests getting together to solve environmental problems in a large, isolated river basin in northern Australia. The stakeholders have a diverse set of values and have interests that conflict with respect to how the watershed should be managed. What is the mechanism that brings the different values together? How is it managed?

Case Study
Stomping Grounds: IREM of the Mitchell River Watershed in Australia

ANNA CARR

Preamble. This case study provides a unique view of IREM being applied to a large, remote watershed with a relatively small population. The case study highlights changing values with respect to environment, diverse stakeholder groups and a wide range of environmental issues of concern. The challenge for the Mitchell River Watershed is the integration of different interests, ideologies and values to deal with complex environmental degradation. Aboriginal worldviews, private land owners, government agencies, environmentalists, miners and tourist interests all combine to provide a management direction. Conflict and decision making are facilitated through a voluntary working group that brings together interests that have no tradition of working together in this area. The working group consists of 50 members (30 from the community and 20 from the government). A facilitator is used to undertake the day-to-day operations for the working group and provide technical assistance.

Introduction

The Mitchell River Watershed (MRW) covers 72,000 km^2 of Cape York Peninsula in Queensland, Australia. The river rises in the Atherton Tablelands near Mareeba, inland from Cairns, and flows west for 500 km before draining into the Gulf of Carpentaria (Fig. 5.2). In broad terms, the region can be divided into an area of uplands, which are being eroded by the river and its tributaries, and the Gulf lowlands, which are covered by sediments derived from this erosion (Withnall, 1990, p.1). The watershed has a humid to sub-humid tropical monsoonal climate with well-defined wet and dry seasons; 95% of the rainfall occurs from November to April.

From east to west, the MRW contains tropical rainforest, wet sclerophyll forest, a variety of woodland types, savanna, tidal plains, extensive wetlands, estuaries and mangroves.

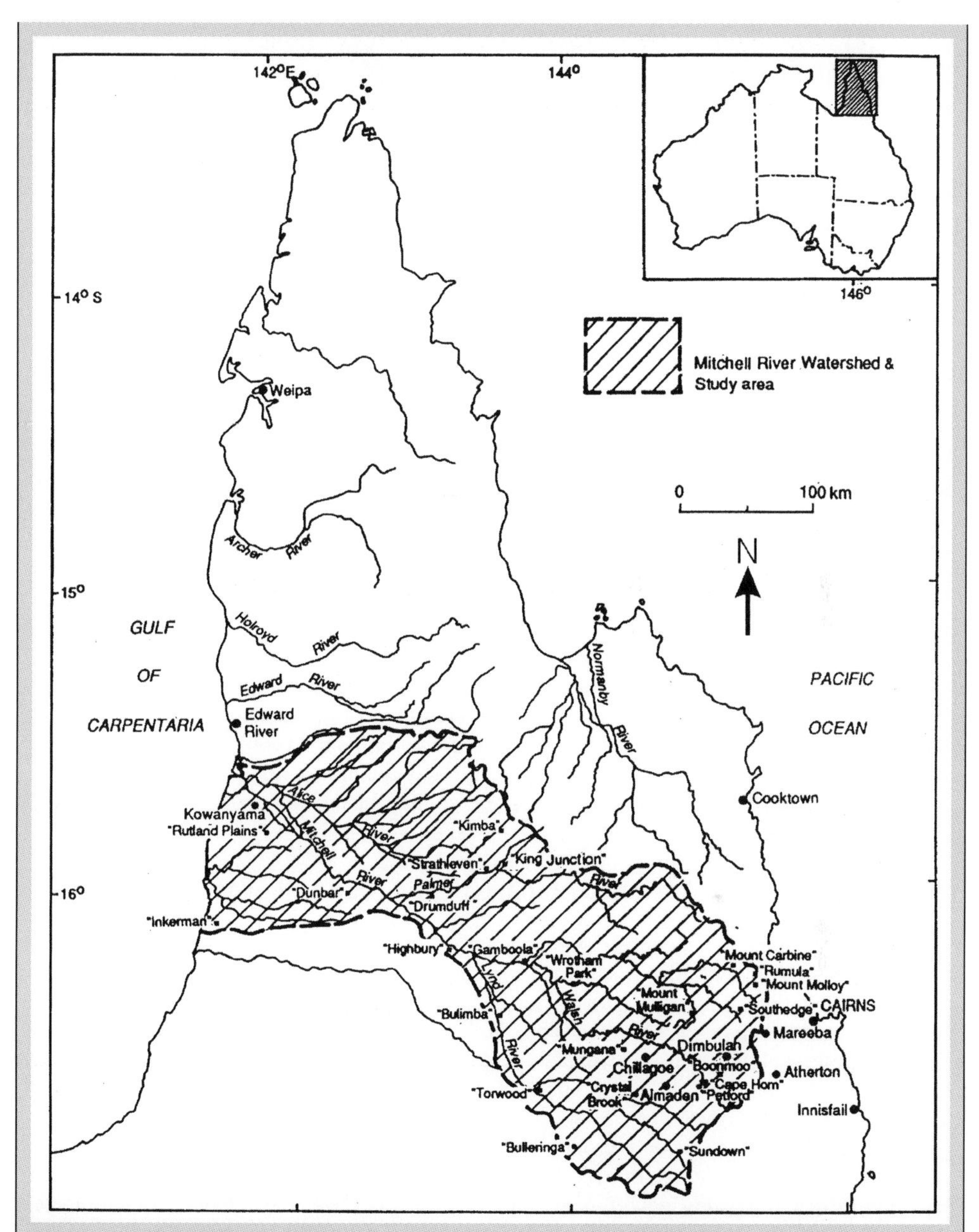

Fig. 5.2. The Mitchell River Watershed.

Woodland and open savanna are the major vegetation features of the watershed. The broad coastal delta includes an extensive wetlands system and is a significant habitat for both flora and fauna. The watershed is home to the endangered golden-shouldered parrot, the Gouldian finch and the northern bettong, with 15 other species listed as rare, vulnerable or threatened.

Grazing is the dominant land use within the MRW, with around 190,000 cattle. Arable farming is confined to small areas in the eastern parts of the watershed. With the aid of four-wheel-drive vehicles, tourists can fish, camp or investigate the culture, history and ecology of the area. The Mitchell River contributes the biggest commercial catch of barramundi in Australia and is also well known amongst recreational fishers. Gold was the largest mining industry before commodity prices fell below profitable levels. There remains considerable evidence of historical hand mining, with old shafts dotting the landscape in the upper half of the catchment.

The estimated population of approximately 7500 people makes the MRW one of the most sparsely populated areas in Australia. Young males predominate, working mainly in labouring jobs on grazing properties and in mines. The proportion of Aborigines and Torres Strait Islanders is considerably higher than the Queensland average and is extremely high in national terms. For these indigenous people, the Mitchell River holds significant spiritual and cultural heritage values.

The relative physical isolation of Cape York keeps the MRW out of the public eye. It is separated further by its position on the western side of the Great Dividing Range, outside the rainforested greenbelt. Most tourists avoid the MRW. Politicians rarely visit. IREM is not easy given the remote human influences on the landscape. It is made even more difficult by the diversity and extent of degradation. Rubber-vine, a voracious climber, has a stranglehold over huge stands of native riverine vegetation along many miles of the Mitchell. Problems of vertebrate pest invasion (notably pigs) or woody-weed eradication are among the most tenacious and far-reaching resource management issues in Australia. To many residents of the MRW, they seemed almost insurmountable.

Graziers, fishermen and Aborigines would blame the mining industry for erosion (small miners, not the large corporate structures with effective environmental control measures). The miners blamed the graziers for overgrazing and other parts of the mining industry for streambank erosion, and everyone blamed most of these environmental problems on the spread of rubber-vine and other noxious weeds. While the environment continued to degrade, some residents recognized that this cycle of blame and counter-blame was proving unproductive.

The Mitchell River Watershed Management Group (MRWMG)

The MRWMG started at a meeting organized by the Kowanyama Aboriginal Land and Natural Resources Management Office (KALNRMO) in June of 1990. There was widespread concern about the condition of the river and for watershed management generally. Arising from the meeting was initial agreement on the vision, goals and objectives for the group. Underpinning each of these was the desire that issues would be addressed *proactively*, that *cooperative* management systems would predominate and that all MRWMG activities would reflect *community ownership* (MRWMP, 2000, p. iv).

The MRWMG is a diverse group, including approximately 30 community members (Aborigines, farmers, graziers, miners, tourism representatives, environmentalists and fishermen) and 20 participating government officers from local, state and federal departments. Although there are more women than there used to be, men still comprise the majority of MRWMG membership. Membership fees are restricted to US$5.50 to encourage widespread participation. The group meets quarterly in conjunction with a field-based workshop which concentrates on one particular area, project or industry.

A facilitator, housed within the Department of Primary Industries in Mareeba, handles much of the detailed coordinating and administrative functions of the group.[1] The job includes

[1] Since my first contact with the Mitchell River Watershed Group, there have been several changes in personnel; however, the tasks associated with the position remain constant.

arranging meetings, writing submissions, administering grants, preparing research and promotional publications, community liaison, representing the MRWMG at conferences, and networking with government natural resource management agencies throughout the region. The facilitator is also a source of technical skills and advice on environmental restoration.

Issues

The big problem in the MRW is that its geographical size is not matched by its sociopolitical stature. Distance and isolation affect not only its political visibility and economic viability, but also the health, education and welfare of its residents. There are simply too many resource management problems for the sparse population to tackle. Moreover, what is a significant problem for one resident may not match others' priorities and opportunities for remediation. In addition, the sheer scale of the catchment and significance of the wet season, which makes east–west road travel almost impossible, make face to face contact difficult among stakeholder groups.

Members of the MRWMG are quick to point out that, compared with other river systems, the Mitchell is still one of the largest, most unspoiled, natural, wild rivers in Australia. Loyalty to the Mitchell River specifically and to Cape York in general is valued highly. While not wishing to have problems like those of more southern river systems, several group members admitted that there were environmental problems arising in the Mitchell (Table 5.2). Aboriginal

Table 5.2. Environmental problems in the Mitchell River Watershed.

Biophysical problems	
Weeds	Rubber-vine
	Chiney apple
Mining	Stream-bank erosion
	Environmental disturbance
	Heavy metal leaching
Ecological	Bushfires
	Loss of biodiversity
Agricultural	Feral species (telapia, pigs)
	Salinity issues
	Water quality issues
	Fragile soils
Socioeconomic and cultural problems	
Economic	Over-grazing
	Drought policies
Management	Lack of coordination
	Conflicting government policies
	Lack of data and knowledge
Demographic	Poor communication
	Lack of cultural respect
	Lack of community awareness
Tourism	Over-fishing
	Litter

members were particularly concerned that Mitchell River country be cared for now, such that future generations may benefit. One said that he was motivated to work cooperatively by the 'lack of recognition of the cumulative and long term effects of doing the bad thing'.

The significance of Aboriginal culture and connection to the environment coupled with the extent of their feelings of pride and loss is clearly illustrated by Victor,[2] ex-director of the KALNRMO. He spoke about the 'development' of the red-lotus swamp at Kowanyama:

> A 'dozer come in and took the seed bed out – all the long-neck turtles couldn't hide in the peat and the pigs came in and ate the lot. Now there is only one lotus bed out of four left.

Callum, an Aboriginal elder at Kowanyama, felt that loss acutely. Given that the Red Lily Creek dam was put in just after the MRWMG began, he spoke of his fears for the future and his anger at not being consulted over decisions which affect Kowanyama.

> That place was a story ground and now, no more red lotus lilies. That lagoon was there since [the] very beginning of creation and was looked after by those Kokoberra people.

An authority from the Council of Elders, Callum felt insulted about not being consulted by white authorities who made decisions about his country. He wondered what the purpose of the group was if it was not a consultation mechanism.

> I'd like to see the National Parks and Wildlife mob come to our community and talk to the people, and explain what they're doing . . . and maybe they could ask us what we want to do. We don't want them to tell us what to do.

Other members, too, were firmly attached to the notion that 'we know what is best for our community – not government' and 'we don't want any government people interfering in our affairs'. Self-reliance is a large part of the ideology of life on the Cape. Born out of historic need, it is a strong motivating ethic in the group. So it was particularly surprising when, in 1993, the community-based group decided to invite government officers to become members with the same voting rights and conditions as others. Yet, as Victor reflected, it took a long time before community members and government members of the group relaxed in one another's company. 'We likened it to two dogs smelling each other regarding the [government's] mistrust and jealousy of our [community-based] autonomy.'

From the outset, the relationship between local residents and those in government departments, who often live outside the area, has been important to negotiate. The traditional role for government on the Cape has been seen as enforcement. Government officers were mistrusted and disliked, seen as spoilers or police. One government officer, Peter, advised others to 'tread cautiously – be aware of the fear of government'. Another said: 'Wear your uniform, but leave your big boots at home'.

Because of the group's self-proclaimed grass-roots style, most government members were particularly cautious initially not to steer or direct the group overtly. They chose merely to provide advice upon request. However, given that government officers were now legitimate members, able to voice opinions and ideas of their own, confusion was bound to erupt. As Gareth said: 'It depends on the group and whether they want government officers to be advisory or right in with it boots and all'. At an institutional level, the government's and community's roles in management of the MRW evolved quite separately.

At a personal level, trust and effective communication were vital to sorting out government's and community's roles in the MRW. Importantly, it is the individual officers personally known to the community who are sought out for advice and information. Sometimes a

[2] All names of individuals have been changed to protect their anonymity.

resident may not recognize which department an officer represents or which department to go to for information. However, when individuals are named, the response is often one of recognition and familiarity.

Like the watershed itself, its inhabitants tend to be larger than life, diverse and extreme, making communication unusually tiring and contact with government authorities sporadic. Therefore, perhaps more than anywhere else, using everyday words is important in the MRW. To communicate across physical boundaries, diverse interests, values and attitudes, and taking into account different cultural and educational backgrounds, use of simple language is imperative. Trust has slowly developed and is apparent only now that members are confident that there will be ongoing talks and discussion.

If catchment management depends on effective cooperation between users, and cooperation depends on mutual trust and understanding, then it is in social gatherings and informal settings where trust is nourished. These social occasions before or after official MRWMG meetings have provided opportunities for members to get to know each other.

Strengthening relationships is critical for the MRWMG, given the diversity of core values within the group. Covert or latent conflict is always present over external political issues such as indigenous land rights, dairy deregulation or the collapse of the local tobacco industry. Sometimes it erupts in meetings between representatives of different factions who push their own position on issues at the expense of listening to others. A proxy representative once came to a meeting on behalf of the usual environmental group delegate. He was not familiar with the group's personalities or style of operation. The newcomer upset established relationships and impeded group progress by not being prepared to leave his position to look for alternative solutions when negotiating a difficult environmental decision.

Summary

Of all the issues facing the MRWMG, the most apparent is that of scale. Apart from the sheer physical enormity of the watershed and its ecological diversity, the personal and metaphorical space between people was substantial. The mining, fishing, agricultural, grazing and tourism sectors were all quite distinct. 'Boundaries were ridden' and not just on cattle properties where graziers built grand houses to enjoy their glorious isolation. 'Small miners' are still socially isolated in their bid to ward off potential competitors. Commercial fishers usually fly-in to the Gulf for the season, then retreat to Cairns or the Atherton Tablelands for the rest of the year. There, tobacco, avocado, sugar cane, dairy production and all kinds of horticultural pursuits keep farmers too busy to interact. Moreover, they often speak different languages and engage in their own cultural affairs to the exclusion of others. At the other end of the watershed, indigenous peoples have always derived their health and well-being from the river and its surrounding country, but opportunities to interact with those in Mareeba in the east are few and far between.

Overcoming their social isolation had to happen for MRWMG members before the biophysical problems were acknowledged and identified, so as to begin addressing the various watershed management issues. While several government-appointed facilitators have come and gone, some officers have remained, changing their approach to suit conditions. Some assumptions have been revised and some remain unresolved. Over time, however, much conflict, suspicion and mistrust has given way to greater cooperation.

What the group has achieved thus far is nothing short of remarkable. They have practised integrated watershed management for more than 10 years, being one of the earliest of Queensland's catchment management committees. They have recently completed an environmental audit of the stock-routes[3] and reserves in the Mitchell River's headwaters. There are now numerous publications on the watershed's native birds and animals. Issues of cultural sustainability and indigenous land management now rate highly on the 'to do' list. The group has a clear vision of the future mapped out within its Mitchell River Management Plan and an impressive website for visitors and local stakeholders alike (www.mitchell-river.com.au). Of course there are still conflicts and many areas in which discussion has not yet established clear outcomes. Unresolved issues include:

- What constitutes overgrazing?
- Whether to eradicate feral pigs, thus eliminating an important protein reserve for Aboriginal communities.
- What is 'appropriate fire management' and who will this benefit.
- Whether or not to promote urban development.

[3] Stock-routes are vegetation corridors historically set aside to allow the transportation of cattle between regions. Given their infrequent use, they are important sites for biodiversity assessment.

Discussion Questions

1. Examine a local case study in the newspapers or on television that deals with environmental conflict. What are the issues? Who are the dominant stakeholders? What is the major issue with respect to resource use? How are individuals justifying their positions?
2. Use a natural resource, such as fish, timber or coal, and provide an overview of the property rights affecting the management of that resource. How are they implemented? Have the property rights changed over time?
3. What is your view on the environmental justification to property rights? Should we have an instrumental or intrinsic approach to our justifications in resource management?
4. List three different approaches to resolving environmental conflict. What are the advantages and disadvantages of each approach?
5. Research a recent public inquiry with respect to natural resources in your state or province. Why was the inquiry created? What were the results of the inquiry?

Case Study Discussion Questions

The Mitchell River Watershed process raises many questions including:
6. Relating to watershed management: How can reasonable management decisions be made about the seemingly intractable issues of weeds and vertebrate pests without sufficient scientific data concerning species composition and workable resource management options within the catchment?
7. With respect to decision making and the MRWMG: How can the longer term dilemmas facing the watershed such as intergeneration equity, indigenous land management and ecological sustainability be added to the immediate action agenda?

8. Relating to the broader IREM agenda: To what extent is controversy over issues healthy for a group such as this? What strategies should a facilitator or chairperson take to prevent controversies over resource management issues escalating to intractable conflicts of interest? Should interests situated outside the watershed have influence over those living within the watershed? How can these influences be balanced appropriately?

Acknowledgements

Anna Carr acknowledges those friends and colleagues on the Mitchell River Watershed Management Group for their support and advice. She wishes to thank the Elders at Kowanyama for their insightful comments. Acknowledgement is also due to the Rural Industries R & D Corporation, who contributed funds to this research.

References

Albrecht, D., Bultena, G., Hoiberg, E. and Nowak, P. (1982) The new environmental paradigm scale. *Journal of Environmental Education* 13, 39–43.

Becker, L. (1977) *Property Rights – Philosophic Foundations*. Routledge and Kegan Paul, London.

Bevan, A. (1992) *Alternative Dispute Resolution*. Sweet and Maxwell, London.

Bingham, J. (1986) *Resolving Environmental Disputes: a Decade of Experience*. The Conservation Foundation, Washington, DC.

Boule, L. and Nesic, M. (2001) *Mediation: Principles, Process, and Practice*. Butterworths, London.

Bromley, D.W. (1989) *Economic Interests and Institutions: the Conceptual Foundations of Public Policy*. Basil Blackwell, Oxford, UK.

Brown, H. and Marriot, A. (1993) *Alternative Dispute Resolution Principles and Practice*. Sweet and Maxwell, London.

Buck, S. (1989) Multi-jurisdictional resources: testing a typology for problem-structuring. In: Berkes, F. (ed.) *Common Property Resources: Ecology and Community-based Sustainable Development*. Belhaven Press, London, pp. 127–147.

Chatwin, B. (1987) *The Songlines*. Viking Press, New York.

Copp, D. (1986) Some positions and issues in environmental ethics. In: Hanson, P. (ed.) *Environmental Ethics: Philosophical and Policy Perspectives*. Institute for the Humanities, Simon Fraser University, Burnaby, British Columbia, Canada, pp. 181–195.

Cormick, G. (1982, September) The myth, the reality and the future of environmental mediation. *Environment* 24, 15–39.

Coser, L.A. (1956) *The Functions of Social Conflict*. Free Press, New York.

Demsetz, H. (1967) Toward a theory of property rights. *American Economic Review* 57, 347–359.

Deutsch, M. (1987) A theoretical perspective on conflict and conflict resolution. In: Sandole, D. and Sandole-Staroste, I. (eds) *Conflict Management and Problem Solving: Interpersonal to International Applications*. Frances Pinter, London, pp. 38–49.

Devall, W. and Sessions, G. (1985) *Deep Ecology*. Peregrine Smith, Layton, Utah.

Dorcey, A.J. (1986) *Bargaining in the Governance of Pacific Coastal Resources: Research and Reform*. Westwater Research Centre, University of British Columbia, Vancouver, British Columbia, Canada.

Dorcey, A.J. and Riek, C. (1989) Negotiation-based approaches to the settlement of environmental disputes in Canada. In: *The Place of Negotiation in Environment Assessment*. Canadian Environment

Assessment Research Council, Hull, Canada, pp. 7–36.

Dukes, E.F. (1996) *Resolving Public Conflict*. Manchester University Press, Manchester, UK.

Dunlap, R.E. and Van Liere, K. (1978) The new environmental paradigm. *Journal of Environmental Education* 9, 10–19.

Elridge, A.F. (1979) *Images of Conflict*. St Martin's Press, New York.

Ely, R.T. (1914) *Property and Contract*. Macmillan, New York.

Emel, J. and Brooks, E. (1988) Changes in form and function of property rights institutions under threatened resource scarcity. *Annals of the Association of American Geographers* 78(2), 241–252.

Estrin, D. and Swaigen, J. (1978) *Environment on Trial*. Canadian Environmental Law Research Foundation, Toronto, Ontario, Canada.

Filley, A. (1975) *Interpersonal Conflict Resolution*. Scott, Foresman and Company, Glenview, Illinois.

Geller, J.M. and Lasley, P. (1985). The new environmental paradigm scale: a reexamination. *Journal of Environmental Education* 17, 9–12.

Glasbergen, P. (1995) *Managing Environmental Disputes*. Kluwer Academic Publishers, Dordrecht, The Netherlands.

Glavovic, B., Dukes, E. and Lynott, J. (1997) Training and educating environmental mediators: lessons from experience in the United States. *Mediation Quarterly* 14(4), 269–291.

Harrison, M.L. (1987) Property rights, philosophies, and the justification of planning control. In: Harrison, M.L. and Mordey, R. (eds) *Planning Control: Philosophies, Prospects, and Practice*. Croom Helm, London, pp. 32–58.

Himes, J.S. (1980) *Conflict and Conflict Management*. University of Georgia Press, Athens, Georgia.

Laue, J. (1987) The emergence and institutionalization of third party roles. In: Sandole, D. and Sandole-Staroste, I. (eds) *Conflict Management and Problem Solving: Interpersonal to International Applications*. Frances Pinter. London, pp. 17–29.

Leopold, A. (1949) *A Sand County Almanac*. Ballantine Books, New York.

Macpherson, C.B. (1978) *Property: Mainstream and Critical Positions*. University of Toronto Press, Toronto, Ontario, Canada.

McEnery, J.H. (1985) Toward a new concept of conflict evaluation. *Conflict* 6(1), 37–72.

Mitchell River Watershed Management Group. (2000) *Mitchell River Watershed Management Plan*. Mitchell River Watershed Management Group, Mareeba, Queensland, Australia.

Moore, C. and Priscoli, J. (1989) *The Executive Seminar on Alternative Dispute Resolution Procedures*. US Army Corps of Engineers, Virginia.

Nash, R.F. (1989) *The Rights of Nature*. University of Wisconsin Press, Madison, Wisconsin.

Neumann, R. (2001) *Resolving Environmental Disputes: Principles and Practice Study Guide*. Australian School of Environmental Studies, Griffith University, Brisbane, Queensland, Australia.

O'Riordan, T. (1989) The challenge for environmentalism. In: Peet, R. and Thrift, N. (eds) *New Models in Geography*. Unwin Hyman, London, p. 1.

Pirages, D.C. and Ehrlich, P.R. (1974) *Ark 2: Social Response to Environmental Imperatives*. W.H. Freeman, San Francisco.

Randall, A. (1987) *Resource Economics: an Economic Approach to Natural Resource and Environmental Policy*, 2nd edn. John Wiley & Sons, New York.

Ryan, A. (1984) *Property and Political Theory*. Basil Blackwell, Oxford, UK.

Ryan, A. (1987) *Property*. Open University Press, Milton Keynes, UK.

Schellenberg, J.A. (1982) *The Science of Conflict*. Oxford University Press, Oxford, UK.

Schmid, A.A. (1978) *Property, Power, and Public Choice: an Inquiry into Law and Economics*. Praeger, New York.

Scott, A. (1983) Property rights and property wrongs. *Canadian Journal of Economics* 16(4), 555–573.

Scott, A. and Johnson, J. (1983) *Property Rights: Developing the Characteristics of Interests in Natural Resources* (Resource Paper No. 88). University of British

Columbia, Department of Economics, Vancouver, British Columbia, Canada.
Simmel, G. (1955) *Conflict*. Free Press, Glencoe, Illinois.
Spitler, G. (1988) Seeking common ground for environmental ethics. *Environmental Professional* 10, 1–7.
Susskind, L., Levy, P. and Thomas-Larmer, J. (2000) *Negotiating Environmental Agreements*. MIT-Harvard Disputes Program, Island Press, Washington, DC.
Tillet, G. (1991) *Resolving Conflicts – a Practical Approach*. Sydney University Press, Sydney, Australia.
Usher, P. (1984) Property rights: the basis of wildlife management. *National and Regional Interests in the North: Third Annual Workshop on People, Resources, and the Environment North of 60°*. Canadian Arctic Resources Committee, Ottawa, Ontario, Canada.
Wehr, P. (1979) *Conflict Regulation*. Westview Press, Boulder, Colorado.
Wildavsky, A. (1987) Choosing preferences by constructing institutions: a cultural theory of preference formation. *American Political Science Review* 81(1), 3–21.
Withnall, I.W. (1990) Geology and mining in the Mitchell River watershed. In: *Proceedings of the Mitchell River Watershed Management Conference*. Kowanyama Land and Natural Resource Management Office, Kowanyama, Queensland, Australia, pp. 14–32.

6 The Role of the Social Sciences in IREM

Introduction

In 1968, Eric Forsman, a young biologist working for the US Forest Service as a summer fire guard, could scarcely believe his eyes when a northern spotted owl landed in the front of his ranger cabin. Who could have foreseen that the perfect 'ecological representative' had just landed in his yard? Just as assuredly, who could have forecast that this seemingly harmless creature would so profoundly impact the Pacific Northwest's biggest industry (timber) in less than 25 years?

It is almost as if the entire natural resource profession was blindsided by an insignificant little bird that few people even knew existed, and fewer still who really cared. Obviously, how an individual perceives issues such as the spotted owl will be influenced by one's area of responsibility and interest: the 'where you stand is where you sit' phenomenon. Some timber people will see a significant reduction in harvest levels; professional ecologists will see a threat to the fabric of the ecosystem; environmentalists will perceive a worldwide loss of biodiversity, and the list goes on.

The emerging message is clear. There is another dimension to consider from the current debate, namely that the social sciences can and must play a more important role in the development of our natural resource management systems. This is particularly significant given the importance that 'social' factors such as public perceptions, preferred management of natural resources, values, desired life-styles and citizen involvement in the decision-making process will play in the future of natural resource management (Stein *et al.*, 1999). Without understanding and accounting for the human dimension of natural resource management, any potential will probably be short term and narrowly accepted by different segments of society (Kessler *et al.*, 1992; Ewert, 1996; Margerum, 1999).

In addition, science will be instrumental in producing new information useful in management decision making, understanding public perceptions of natural resource use and determining useful managerial options. Botkin (1990) suggests that developing this new information has become more important than previously thought, primarily to offset the often mistaken belief that 'nature knows best' and if just left alone, rehabilitation and restoration will occur. We now know that nature will probably not be left alone and, as a result, resource managers are faced with the task of making decisions for both the long and short term, thereby creating a need for integrating science with management decision making. Before discussing the role of management, however, a short discussion regarding some of the current trends facing natural resource management may be instructive.

Current Trends

A number of trends have emerged concerning science and natural resource management. These trends include the growing political power of science, changing management expectations and needs, an expanding diversity of values and human impacts.

There are good reasons for this increased political power, such as the relatively objective nature of the research endeavour (Dietrich, 1992). It should be noted, however, that the social sciences are as vulnerable to scientific shortsightedness as are other disciplines. We trivialize science that we do not like or understand, and automatically assume validity and worth in the science that results in findings we agree with or which support our personal or professional goals (Buttel and Taylor, 1992). Hutchings *et al.* (1997) go further by suggesting that decision makers often overlook significant attributes commonly associated with scientific investigation, such as the presence of uncertainty, differing interpretations of the observed phenomenon, and drawing conclusions beyond the data.

Other reasons for the growing influence of science in the political process include: the increasing complexity of the problems, the different scales of effects (e.g. site, forest, landscape, ecosystem and global) and the overall lack of comprehensive databases. In the latter case, the lack of comprehensive data sets allows the manager and social scientist to develop premature closure on specific issues. For example, believing that over-harvesting is simply a result of maximizing profits and other economic forces within the timber industry can lead the public to oversimplify what is often a complex decision phenomenon. In another example, Dixon and Fallon (1989) point to the numerous, often conflicting, interpretations of sustainability that reflect complexity and ambiguity, in understanding both what the concept means and what data to collect.

In a comparable example, professional status or responsibility can also influence how the data are dealt with. Brown and Harris (1998) found that those resource managers associated with commodity production, such as timber harvesting or mineral extraction, tended to take a more traditional perspective on acceptable uses of the resources. Conversely, those professionals affiliated with the natural and social sciences tended to take a less utilitarian perspective.

Management expectations and needs

McKibben (1989) suggests that nature is partially a set of human ideas about the world and about our place in it. One of the many things that science can do is provide for a greater perceived understanding of that world, how it works, what makes sense and what things we can predict about it. In natural resource management, this 'making sense' has become more problematic. In part, this is due to the fact that the number of situations now confronting natural resource management far exceeds the capability of any one scientific group or governmental organization to deal with these issues adequately. A sampling of these issues would include the following:

- Global deforestation and environmental degradation
- Global climate change
- Loss of biological diversity
- Changing demands for forest products
- Wilderness preservation and the proper role of reserve areas
- Watershed management and allocation
- Production and harvesting practices that are sustainable
- Forest health
- Conflicting demands from society for preservation, recreation and commodity production

Scientists are increasingly asked to make their 'best guess' regarding a natural resource management issue, with complete and historic data often lacking. Stankey *et al.* (2003) suggest that one of the keys to effective natural resource management is the production of new understanding based on systematic assessment and feedback. Accordingly, we can expect that natural resource management will insist that science do the following things:

- Provide a foundation for the development of policy by defining the various alternatives. This implies that research serves an 'up-front' role in decision making instead of merely providing 'backfill', i.e. research-generated information used to develop policy rather than support for a decision already made.
- Provide monitoring information about the outcomes and quality of the decisions and policies implemented.
- Maintain an air of objectivity despite the pressure from the political process.
- Develop multiapproach and multidiscipline predictions rather than unidimensional solutions.

Diversity of values

As reviewed in Chapter 5, there is a growing diversity of values that people place on natural resources. This diversity of values is not only situational but is also a function of space and time. For example, when a person is building a house, the wood products used and price of those products are of critical importance. Most of us, however, may only build one house in our lifetime and, in the interim, other values tend to take precedence in our value system. These other values might include wanting an intact forest in order to escape the noise and congestion of the urban environment, or cherishing the recreational activities offered by the forest environment. Others would place a high degree of value on large wilderness landscapes as a way of experiencing adventure and challenge. Still others would desire to build their house near abundant water supplies (Radeloff, 2000) and, as a result, would be extremely concerned with any interruption of that water supply. In another example, Keenan *et al.* (1999) found that attitudes toward water allocation varied as a result of where one lived and the anticipated social and environmental impacts from transferring water. Wondolleck (1992) points out that understanding this diversity of values held by various stakeholders is often key to developing effective and long-term solutions to natural resource challenges.

Westman (1993) has previously distinguished between the goods and services produced by the natural environment. *Goods* include marketable products such as timber or forage, or even the use of the environment for recreation. *Services*, on the other hand, are the functions of an ecosystem and how these various functions interact. These services include: the absorption and breakdown of pollutants; the cycling of nutrients; and the fixation of solar energy. One example of this is the buffering effect that coastal wetlands serve.

Clearly, related to the issue of goods and services is the concept of values. A number of authors have identified a wide range of values associated with the natural environment (Roston, 1985). As shown in Table 6.1, these values include scientific, therapeutic and recreational entities (Ewert, 1990). Associated with all of these values are differing levels of potential conflict. For example, aesthetic values have a high potential for conflict because of the individual nature of aesthetics. One person's beautiful setting is another person's boring scene.

Human impacts

Another trend influencing the interface of science and management is the growing presence and power of human impacts upon the earth's landscape. There can be little doubt that few landscapes or sites now exist free from the influence of man. Most scientists agree that the net loss of the world's forests from human activity, since pre-agricultural times, is of the order of 8 million km^2 or an area about the size of the continental USA. Of this amount, more than three-quarters has been cleared since 1680. In addition, the annual human withdrawal of water from natural circulation is now about 3600 km^3, or an amount exceeding the volume of Lake Huron. In 1680, the annual withdrawal was less than 100 km^3. It is estimated that we exceeded the Earth's capacity to assimilate pollution in the early 1970s and that we would need four planet Earths if each individual in the world was to produce waste at the level of the average North American (Wilson, 2002). There are a number of other statistics, such as air quality, arable land

Table 6.1. Selected values associated with natural environments.[a,b]

Values	Level of potential conflict	Comments
Scientific	Low	Not well advanced; loss of wildlands is outstripping the ability to collect information
Therapeutic	Low	Many acknowledge the cathartic and rehabilitation qualities of wildland environments
Ecological/ biodiversity	Low	The importance of saving gene pools for future generations is widely recognized
Recreational	Medium	May conflict with other values such as scientific. As a highly personal quality of life issue, these values often invoke high levels of emotion
Symbolic/ cultural identity	Medium	Symbols from wildland areas such as the bald eagle or bison represent certain societal and national values (e.g. freedom, strength, 'rugged individualism')
Aesthetic	High	The intangible and subjective nature of these values often leads to disagreement as to worth and value
Inherent worth	High	For many, wildlands have an intrinsic value in just being there Others feel that wildlands should be more 'productive' for the good of society
Market	High	Usually are extractive and compete with most other values. This exclusivity creates high levels of emotion and conflict

Adapted from [a]Roston (1985) and [b]Ewert (1990).

and amount of potable water, which also point to the decline of global and environmental health (Postel, 1992).

When considering these trends, one fact becomes increasingly clear – people need to be considered in any long-term management strategy. It would be a challenge for the research community to describe any major scientific advancement that ultimately did not involve a human dimension. Reidel (1992) puts forward the idea that in natural resource policy, perhaps management has been asking the wrong questions. The research community could also be asked the same questions. What, then, would be the right questions and how can information be generated toward answering those questions?

Wolosoff and Endreny (2002) point out that managers and policy makers are often faced with fundamentally different situations from those faced by researchers and scientists, i.e. managers and policy makers often work in situations that demand relatively shorter response times than their counterparts in science or research. Furthermore, they are often faced with formulating one-sided policy that is applicable across large geopolitical landscapes, whereas the scientist is more often concerned with understanding the particular phenomenon and its complexity. Given these differences, it is useful to develop an understanding of the role that science and, in particular, the social sciences can play in the decision-making process.

The Role of the Social Sciences in Natural Resource Decision Making

Bormann (1993) suggests that concepts of the environment such as sustainability, forest health, biodiversity and ecosystem management are essentially human constructs that serve as expressions of human values. If we believe that natural resource management is one manifestation of the society in which we live, what type of scientific structure must be in place to provide the type of information necessary for effective natural resource decision making? Machlis (1992) makes the observation that biologists, ecologists and other natural science professionals are now faced with a hard reality: ultimate solutions to natural resource problems lie in social, cultural, economic and political systems – the very systems that are the focus of the social sciences. While, traditionally, the social science disciplines have included political science, geography, anthropology, sociology, psychology, economics and philosophy, more recent areas of inquiry could include the recreation and leisure sciences, education, demography, human ecology and environmental psychology (Ewert and Williams, 1994; Williams and Patterson, 1996).

What role can the social science disciplines play in the formation of natural resource policy? Global climate change presents one scenario that is both timely and of profound importance. This is particularly true because, as Kempton (1997) reports, the public's knowledge regarding global climate change is often based on 'cultural models' which represent fundamental ways of understanding regarding the environment that are shared by most members of that culture, and are often incorrect. For example, global climate change is caused by air pollution.

While in the case of global climate change, research has been defined primarily in terms of meteorological and chemical processes, the causes are almost exclusively human. Indeed, Maloney and Ward (1973) suggest that most environmental crises facing our society, and the world, are really 'crises of maladaptive [human] behavior'. For example, consider the following social sciences and the various roles they could, and do, play in the development of a comprehensive solution to any emerging challenges associated with climate change.

1. **Anthropology**. What have been the patterns of human adaptation in response to historical changes in the climate? Did communities develop large-scale adaptations, migrate or simply die out? Knowing how our ancestors reacted may provide some insight into how humans generally behave in this type of crisis.
2. **Political science**. What political and/or governmental institutional structures have been effective in producing global awareness, monitoring and enforcement procedures? As a global community, we already have some examples of international discussion and action on far-reaching environmental issues including nuclear weapons, regulating the use of the oceans, and international cooperation on issues such as illegal trade of threatened and endangered species (Feldman, 1991).
3. **Economics**. What mixes of economic incentives would be most effective in altering behaviours to produce a more environmentally friendly set of actions?

4. Education. What educational vehicles would be most influential in modifying the behaviours of individuals? What will be the most effective mechanisms for methods of education that can be translated into behaviours and knowledge which are not harmful to the global environment?
5. Psychology/sociology. How can the individual and, collectively, society take more responsibility in modifying their behaviours to lessen the overall impact upon the natural resource base? What specific cues 'tell us' that there is a threat to global health?
6. Recreation and park management. As outdoor recreation is often the primary avenue from which a large segment of the population experiences direct contact with the natural environment, can the outdoor recreation experience be managed in such a way as to increase the individual's sensitivity and willingness to act in an environmentally conscious way (Ewert, 1991)?

These are a few of the potential types of information that would add to the overall mix of solutions to global climate change. This multidisciplinary approach could and should be applied to other issues in management and policy making for our natural resources. This is particularly true if one considers the importance of the roles that public perceptions, cultural values, history and individual expectations play in the natural resource decision-making process. Unfortunately, the importance of these social-based issues is not always recognized by managing agencies or other scientific disciplines. Holden (1988) has argued that:

> The social sciences have lagged far behind in assessing the interactions between physical changes and human activities. Far more is known about the processes of global warming, deforestation, resource depletion, and pollution than about the processes of the human institutions that create these effects.
>
> (p. 663)

A growing body of literature now speaks to the need for integrating the social, physical and biological sciences (Heberlein, 1988). The fact that our research community has failed to do so points to a message of disinterest and a lack of willingness on the part of the scientific institutions to cooperate. Heberlein suggests that possible explanations as to why the social sciences have not been widely integrated with the other sciences lie in the following areas:

- The social sciences are less developed and are considered to be of lower status than the other disciplines.
- There is a lack of tangible products from social science research.
- There exists a multitude of theoretical perspectives, or paradigms, that result in a myriad of explanatory theories, often in conflict with one another. Thus, it becomes more difficult to ascertain what the real explanation is.

Field and Burch (1990) argue that social sciences have provided an essential body of knowledge to the forestry profession for a substantial period of time. Likewise, others point out that, while under-utilized, the social sciences can and should play an integral part in the development of numerous predictive tools, such as biodiversity risk assessment.

Machlis (1999) suggests that several patterns have emerged from the interaction between the social sciences, natural resource management and policy development. These patterns include: (i) a significant amount of overlap between disciplines exists with no clear distinction between them; (ii) the social sciences can and have been effective in developing an understanding of the complexity of human systems such as tradition, power, values and perceptions; and (iii) global and ecosystem level scales have been problematic for social science methods. As scientists and managers, we often overlook the 'cultural cornerstones' that guide our behaviours and the way we collect and filter information. A number of authors now suggest that these cultural cornerstones not only reflect multiple methods and ways of interpreting data, but also reflect ways of understanding the natural environment and interpreting outcomes, such as ecosystem health (Dixon and Fallon, 1989; Hetherington *et al*., 1994; Patterson and Williams, 1998). For example, there is a growing understanding that information gained through traditional scientific methods can be augmented and made more comprehensive when combined with the 'traditional environmental knowledge' (TEK) typically held by aboriginal and native groups (Sherry and Myers, 2002).

From a management perspective, Hutchings *et al*. (1997) question whether scientific inquiry can fully coexist with governmental control and political decision making. They point out that, too often, research information that is contradictory or unsupportive of government policy and/or political expedience is suppressed or disregarded.

However, merely rallying against the status quo can be counter-productive. As we seek to bring about a greater awareness of the need for the social sciences in the context of natural resource research and decision making, there are a number of points to consider.

1. Including incentives for the integration of the biological and physical sciences, i.e. competitive grants, research proposals and workshops, could frame the questions in such a way as to be of interest to the other sciences. For example, assessing the use of user fees at a particular location could also include the anticipated physical and biological impacts upon the resource based on the different use levels.
2. Social science research programmes need to be multiscale, including individuals, groups, communities, landscape and counties (human-equivalent landscape level), ecosystems, biomes and global systems.
3. To the extent possible, recreation and similar disciplines should be linked with the mainstream social science literature, as well as the biological sciences. Failure to do so tends to marginalize the information and can ultimately downgrade the information our scientists generate in the eyes of other disciplines and the courts.
4. From a funding agency perspective, we need to design our cooperative agreements to bring the output more in line with natural resource management needs, both in terms of the actual science and also with respect to the overall visibility of the research. In addition, priority should be given to cooperative agreements that incorporate a group of universities, and other research institutions, in order to bring a variety of ideas and approaches to the issue under study.

5. Increasing the awareness that many of our managers and public are not always cognizant of the human dimensions component of natural resource management. The social sciences need to pay more attention to defining what the human dimension in natural resource management is, in addition to identifying potential research questions.

While this list is incomplete, the emerging scientific challenge is to incorporate the social sciences into the policy and decision-making agenda, as they bring a scientific focus to the human dimensions aspects of natural resource use. Non-traditional disciplines, such as recreation and leisure, need to redefine their research role in this developing scenario. The issues now go far beyond visitor and experience characteristics and often include the very fabric of many of our social systems. From an ecosystem management perspective, issues related to social structure and natural resource use, the changing dynamics between social organizations and natural resources, and the challenge of dealing with uncertainty will, in part, determine the success of these disciplines in natural resource management (Ecological Society of America, 1995; Cordell and Bergstrom, 1999).

Dale (2001) reminds us that the scientific approaches needed in natural resource management must be both sophisticated and multidimensional in order to capture the information required for contemporary decision making in natural resources. In addition, research and science should be accessible to the general public as part of an overall decision-making process. Lee (1994) coined this idea as 'civic science' and believes this approach will help reduce public misunderstanding about natural resource issues and encourage more public involvement in the decision-making process.

Finally, while reflecting on the case of global climate change, the spotted owl, old-growth forests and any number of other issues previously mentioned, one is reminded of the Yukon traveller in Jack London's *To Build a Fire* (London, 1908). The man was wise in the ways of the world but not in their significance. He never saw the clump of snow hanging from the tree, directly over his fire. The clump of snow that would eventually, but surely, kill him. Our failure to carefully build in the human component of our environmental decision making may doom us to a similar fate. People make the problems but they also create the solutions. At this point, however, we still have a choice about what the future holds for some of our natural resources and for society.

Summary and Conclusions

In this chapter, we have examined the role that science and, more specifically, the social sciences can play in the natural resource decision-making process. Also discussed in this chapter were issues such as the current trends surrounding the use of science in natural resources. Another issue explored was the presence of management expectations and needs regarding what science and research can provide. Adding to this challenge is the diversity of values associated with natural resources and the increasing presence of human impacts. The chapter concluded with an overview of what role the social sciences can play, now and in the future. It was determined that the social sciences face a number of challenges associated with research in natural resources, but the areas of academic and scientific disciplines can also contribute to our understanding of the subtleties of natural resource management issues, in ways the biological or physical sciences cannot.

In the following case study, we look at a 'real-life' example of how several of the social sciences were, and are, being used to revitalize and rehabilitate a heavily impacted area near Chicago. Doubtless, the examples discussed in this case study have applicability in numerous other situations at locations across the globe.

Case Study
Restoring the Rustbelt: Social Science to Support Calumet's Ecological and Economic Revitalization

LYNNE WESTPHAL

Preamble. This case study presents an example of how various social science disciplines, such as environmental psychology, anthropology and economics, have been useful in developing information for informed decision making. Another point illustrated in the case study is the strong connection between the current health of this local ecosystem and the human history that has impacted it.

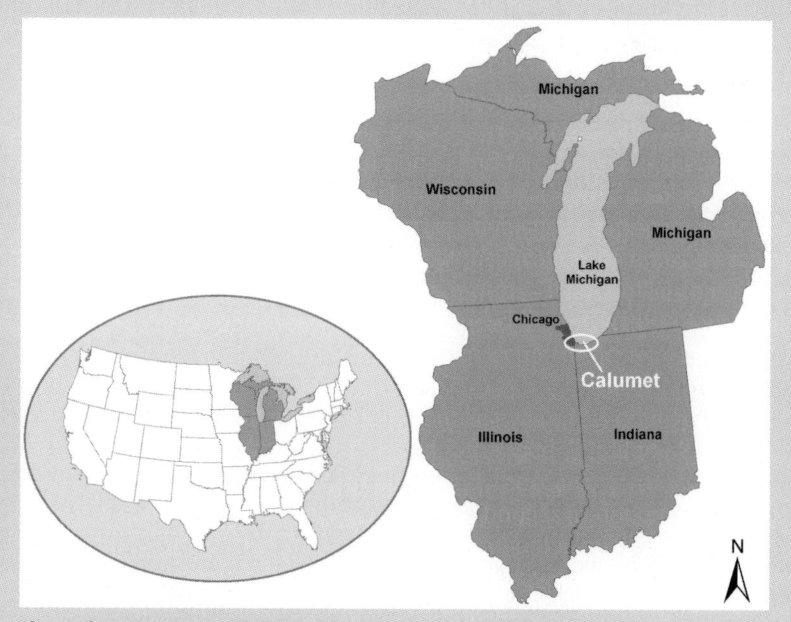

The Calumet Region of Illinois and Indiana, USA.

Introduction

Can endangered species and heavy industry thrive in the same place? Can we have jobs, vibrant communities and wildlife habitat coexisting in major metropolitan areas? Can we move beyond jobs versus the environment to a society where these issues are both, and not either/or? The majority of Americans believe this can happen, and yet its society often ends up facing logjams where the environment is pitted against economic vitality. In the rustbelt landscape of Calumet, Indiana, USA, a broad coalition of government, industry, environmentalists and local citizenry are trying to make the coexistence of a healthy economy and healthy environment a reality.

Background

The Calumet area runs along the southwest shore of Lake Michigan, including 10% of the City of Chicago, cities and towns in northwest Indiana, and the Indiana Dunes National Lakeshore. Once one of the largest wetland complexes in the Midwest USA, the Calumet region remains ecologically complex. Several major ecosystems overlap in Calumet, including northern boreal forest, prairie, savanna and dunes. The Calumet region is considered by many to be the birthplace of ecology. In 1898, Henry Cowles, the noted ecologist from the University of Chicago, first outlined the theory of succession in Calumet's Indiana dunes.

Calumet is also the early heart of US industry. The steel industry, in particular, has dominated the region and dramatically transformed the landscape, beginning in the late 1800s. By the 1920s, the Calumet area was surpassing Pittsburgh in steel production, and remains the largest steel-producing region in the USA today. The steel plants created many jobs, both in the mills and secondary jobs in the surrounding communities.

Besides shaping the local economy and the skyline with smokestacks, the steel industry dramatically changed the land. Slag, a rock-like by-product of steel making, was disposed of as landfill in Lake Michigan and in low-lying areas, 'improving' what was seen as disease-breeding swamps and creating 'useful' land. The promise of well-paid jobs in the steel and other manufacturing industries drew migrants from Europe, Mexico and the southern USA. The steel mills also shaped communities, creating virtual company towns, where taverns, churches, schools and other components of community life were supported by workers from the local mill (Jones, 1998).

With the 1970s and 1980s came a drastic global restructuring of the steel industry, and mill after mill closed. Other industries felt these economic shifts too, as the region haemorrhaged jobs, with communities as well as families facing hard times. Over 40% of the jobs in southeast Chicago were lost between 1970 and 1990 (Jones, 1998). Hundreds of acres of industrial land were left abandoned or idle, in what became known in the USA as 'brownfields' (elsewhere, these lands are called derelict or contaminated land).

Steel was not the only industry in Calumet. Meat packing, petroleum processing, and automobile and railcar manufacturing thrived in the region. Transportation was key to this industrial success, with easy access to water, rail and roads for shipping raw materials and finished product. Many of the labour struggles in the late 1800s and early 1900s could be found in Calumet too, including efforts to unionize and to create the 8 h day. From Pullman's company town (now a Chicago neighbourhood and historic district) to the steel mills, the Calumet region is rich in industrial and labour history.

Through the region's industrial growth and decline, hunting and fishing remained popular activities in the Calumet region. Many famous Chicagoans maintained lodges along Lake Calumet in the late 1800s and early 1900s. Today, Calumet hosts the only legal hunting grounds in Chicago. Calumet's natural areas provide habitat for many kinds of birds. Bird watchers flock to the area to see them, including state-threatened and endangered birds such as the black-crowned night heron and snowy egret.

By the end of the 1980s, Calumet clearly needed help. In 1990, newly elected Mayor Richard M. Daley proposed a sweeping redevelopment plan for Chicago's portion of Calumet: a third Chicago airport. This plan called for levelling landfills, filling wetlands, moving rivers and eliminating an entire neighbourhood of homes and businesses. The local feelings about the airport plan were mixed. Some saw it as the only hope for a return of jobs to the region; others saw it as a threat to a way of life and to the rich ecological gems that survived in the area. Citizens organized against the new airport, advocating for the threatened and endangered species in the wetlands and open spaces of Calumet, and for the strong communities threatened by the airport. The airport eventually was withdrawn from consideration, and an understanding and appreciation of the ecological treasures of Calumet began to spread.

Citizen action did not rest with the decision against building the airport. They felt that the wetlands and other prized natural areas were still threatened. Local activists formed the Calumet Ecological Park Association (CEPA) and continued lobbying for recognition and protection of these remaining fragments of Calumet's past ecological riches. CEPA wanted the region declared an 'ecological park'. In the mid-1990s, CEPA persuaded local congressional representatives to fund a National Park Service feasibility study for a Calumet Ecological Park. The study determined that Calumet's mixture of historical, cultural and natural features met the criteria for a National Heritage Corridor. Initial attempts to designate the area a National Heritage Corridor failed for a variety of reasons, but the dream of the designation is still alive and being pursued.

In the 1990s, mayors of cities across the USA identified brownfields as a major issue needing new policy directions. Existing federal laws created liability nightmares for interested buyers and developers of brownfields (Rafson and Rafson, 1999; Trumbull, 1999). These issues have been addressed by individual states through Voluntary Clean-up Programmes. In these programmes, the state negotiates the clean-up standards with owners, basing them on the intended future use (Trumbull, 1999). After clean-up, states assured new owners and developers that they would not be held liable for existing, known contamination on the site. These and other changes opened the door for these sites to be redeveloped.

The Current Issue

Calumet is poised for a comeback. In 2000, the City of Chicago and State of Illinois announced a major initiative to revitalize the economy and the ecology of Illinois' part of the Calumet region. Chicago's Mayor Daley paints a picture of what Calumet can become:

> Good environmental management is good for business, and good business development can also benefit the environment. Nowhere is this more true than for the Calumet region on Chicago's southeast side.
>
> Today, the era of decline is ending and it is possible to see what a new era will look like. Chicagoans will regain access to wildlands and restored landscapes that were unavailable for public use for half a century. New industries will spring up in the Lake Calumet area, bringing new jobs and tax revenue.
>
> With careful planning and management, many believe that Calumet's natural beauty and industrial strength can be brought back (City of Chicago Department of Planning and Development, 2002, p.1).

Thus, Chicago's Department of Planning and Development, in partnership with the Chicago Department of Environment, the Southeast Chicago Development Commission,

Openlands Project and the Calumet Area Industrial Council, developed the *Calumet Land Use Plan*. This plan outlines the ideas of ecological and economic growth and redevelopment, designating a significant amount of land for each use. Plans for the natural areas are outlined further in the *Calumet Open Space Reserve Plan* and the *Calumet Area Ecological Management Strategy*. Funding for the initiative comes from many sources, but the Calumet Tax Increment Financing District promises to be a significant source for both industrial and ecological redevelopment.

The vision many now have for Calumet is a tall order, and it can be difficult to imagine how to get there from here. Succeeding with an initiative on this scale takes high quality information from the physical, biological and social sciences to support planning and implementation of these plans. Dozens of past and present research projects are helping to provide this information. Because the issues cross disciplines, so too does the research. Biologists are looking at species and habitat. Chemists are looking at water quality and pollution issues. However, along with information about the biological and physical components of Calumet, decision makers need information about people – information that social scientists can bring to the table. Let us take a look at three existing social science research projects that were started to support the Calumet Initiative. These studies draw on different disciplines in the social sciences, and use different methods to conduct social science research.

Project 1: Ecologically friendly industrial landscapes

One goal in Calumet is to increase the ecological integrity of the entire landscape; to blur the lines between 'industrial area' and 'nature area' without impinging on the very different requirements of each land use. How can we increase the ecological health of redeveloped brownfields and existing industrial sites, and do so in a way that is seen as positive by industry owners, workers and visitors? This research project seeks to answer this question.

Joan Nassauer, a landscape architect at the University of Michigan, has been researching how to design and implement landscapes with greater ecological health in various human-dominated settings, and Calumet is her latest study area. Native plants increase the habitat quality and other aspects of ecological health, but often people do not like them, seeing them as 'messy' (Nassauer, 1997). Various design techniques and careful plant selection can increase the acceptability of landscapes using native, messy plants. This is important because, in a human-dominated landscape, what people do not like they do not take care of, or they change it – disliked landscapes are not sustainable. In earlier work in residential and agricultural settings, Nassauer found that 'cues to care', or signs that people care about and maintain a place, can turn a disliked landscape into a beautiful one. 'Cues to care' can be as simple as a birdhouse in a field of native grasses, or a small patch of neatly mown lawn bordering a native wildflower garden.

Nassauer tests various design options through the use of photosimulations, altering photographs of an existing landscape to indicate how it might look with different native landscaping designs, and varying amounts of maintenance. These simulations are rated by a representative group of stakeholders – in Calumet this would be industry owners, workers, local residents and site visitors – to understand how the landscape options are perceived and appreciated. This information, then, can generate landscape guidelines that will result in industrial area landscapes that provide sustainable habitat and an appealing view for the people who work there. By providing landscape guidelines that use native plants in designs that people like, habitat quality improves, business costs will probably go down and the environment will be more sustainable in the long run.

Project 2: Asset-based understanding of Calumet communities

Revitalizing the economy and ecology of Calumet is not just in the hands of agency employees. In fact, if it were, the chances of success would be much lower. Local residents are central to the success of this initiative, for several reasons. First, the impact from changes will be greatest for local residents, just as the overwhelming impact of the changes over the past decades has been greatest on them. Therefore, it is important that local residents have a say in what happens, when, where and how. Secondly, there is more work to be done than any one agency, or even a partnership of government and non-profit agencies, can do. Local resident involvement in restoration and rehabilitation of the environment, from backyards to wetland restoration, is critical to long-term success of this initiative. Thirdly, residents in the Calumet region have successfully blocked projects in the past. From a pragmatic perspective, it is better to include local residents in decision making in order to minimize the potential for protracted court and other battles. Finally, local residents have been advocating on behalf of their local environment for decades. Their commitment is clear, and it is important to respect this active history in the current initiative. Therefore, it is important to understand the variety of local communities, their interests and their perceptions of the environment. One way to gain such an understanding is through applied anthropological research.

Alaka Wali and a team of ethnographers from Chicago's Field Museum of Natural History are conducting applied anthropological research in several of Calumet's communities. This research is aimed at understanding the assets (e.g. skills, social networks, existing volunteer efforts and other strengths) that these communities have brought, and could bring, to the initiative (Kretzmann and McKnight, 1993; Wali *et al.*, 2003). Researchers visited neighbourhoods, attended public meetings and interviewed residents, workers, clergy, business owners and others in the local community. Photographs, local newspapers and other artefacts filled-out the observational and interview data to help the researchers paint pictures of each neighbourhood, its unique characteristics, views and strengths.

The Field Museum researchers found many assets in the communities of Calumet. For example, one key strength in the Hispanic communities is the kinship networks. With regular meetings in homes or parks, it is the kinship network that fosters social networks and civic activism in the Hispanic community. This means that agency representatives who want to reach out to the Hispanic community need to find ways to work within these kinship networks, just as they might work with church groups or Rotary clubs to reach other segments of the population.

Another asset the Field Museum researchers identified in one neighbourhood is the popularity of gardening – just about everyone has a garden at home and there are community gardens that celebrate the local culture. The popularity of gardening could be a point of common interest between local residents and outside agencies, and gardening skills are an asset that could be applied to local ecological restoration projects. The Field Museum project is designed to provide an understanding of community assets and to provide an entrée for agencies to partner with new community groups to meet both locally identified and regionally identified environmental goals.

Project 3: Special places of Calumet

People are often very attached to certain places – places where they can get away, where they can experience beauty, solitude and nature. Imagine such a place, a special place for you. What is it like? How would you feel if your special place was in an area where plans were being made for major changes.

Herb Schroeder, an environmental psychologist with the USDA Forest Service, has developed a research method to uncover what is important about people's special places and translate this into useful information for natural resource planners and decision makers. Most of his work has been conducted in the North Woods of Wisconsin and Michigan. These are the kinds of places where we expect people to have attachments to special places – places where free-flowing water, natural forests and wildlife abound. The same research process was followed in the Calumet area, and people's responses indicate that these same attachments are alive and well in the midst of one of the largest industrial areas in the world.

In a written survey, people were asked to describe, in their own words, what makes their special place special. The answers were analysed for common themes and interesting differences. The Calumet area has many diverse landscapes, with natural areas next door to industrial areas. These natural places provide many of the same important experiences as do wilder, more remote settings. Looking across all of the survey responses, from the pristine North Woods to the Calumet region, the findings suggest that solitude, beauty and refuge are important in all of these settings. For example, compare two descriptions of special places:

> A river with a unique eddy creating a hole for brook trout. A mile walk through wet cedar swamp and tag alder. Occasional sightings of raccoon, bear, deer, heron and hawks. . . . No easy spot to find, but is visited four times per trout season on the average. A spot discovered alone but since have found others know of it and have fished it. Have encountered only one other party there in 8 years. Complete privacy, solitude is relaxing.

> Although I passed by frequently, I never noticed a big swamp through the trees and down the hill until a friend showed me an obscure path down to it. Now, throughout the year, I sit immobile on a fallen tree and watch deer, muskrats and beavers. . . . I never encounter another soul there, yet friends tell me they have visited. The people who go there treat the site with awe and respect. . . . It's a tiny, private undiscovered place where I can go all by myself to chill out and get reconnected to that which is important in my life.

The first place is in Michigan's Upper Peninsula, the second is in Calumet. What does this mean for managers and decision makers interested in improving Calumet's natural areas? Many of these sites will be moving into public ownership, and may be developed for more access. Maps, tours and new trails will introduce these hidden places to more and more people. It is important to take the experiences these places currently provide into account when making changes, to ensure that the opportunity for solitude, refuge, beauty and other experiences is not lost in the rush for improvement.

Discussion Questions

1. Discuss the 'power' of science and how this power has developed and grown, particularly in natural resource management. Provide an explanation as to why this has occurred.
2. Why have so many natural resource issues become so difficult to solve? Are these issues primarily a manifestation of a lack of scientific understanding, management confusion or the lack of a clear political direction?
3. Differentiate between nature's 'goods' and 'services.' Using a local setting, identify some of the goods and services present at that location.
4. Select a natural resource issue not discussed in the chapter, and describe how the various social sciences might contribute to a better understanding of this issue.
5. Peruse your local newspaper and identify a natural resource issue. Discuss the ways in which various social science applications and disciplines, as presented in this chapter, can help further our understanding regarding that issue.

Case Study Discussion Questions

6. Compare and contrast the kinds of information from the Calumet studies. What kinds of project planning and implementation questions can this information help to answer? How do you think managers can apply the information from these research projects?

7. Imagine yourself as an industrialist in Calumet. What social science research information do you need to be an effective part of this initiative? (Check: are you assuming all industrialists are anti-environment?)

8. Imagine yourself as an environmentalist. What social science research do you need to be an effective part of this initiative? (Check: are you assuming all environmentalists are anti-industry?)

9. Social science research is important in its own right, but must also be integrated with biological and physical science in initiatives such as at Calumet. How are these social science research projects integrated with biological and physical science information?

10. Not everyone shares the common belief in the USA that economic and environmental issues can be balanced successfully. What do you think? What do you think those who *disagree* base their opinion on? What can social science contribute to this debate?

References

Bormann, B. (1993) Is there a social basis for biological measures of ecosystem sustainability? *Natural Resource News* 3, 1–2.

Botkin, D.B. (1990) *Discordant Harmonies: a New Ecology for the Twenty-first Century*. Oxford University Press, New York.

Brown, G. and Harris, C. (1998) Professional foresters and the land ethic, revisited. *Journal of Forestry* 96(1), 4–12.

Buttel, F. and Taylor, P. (1992) Environmental sociology and global environmental change: a critical assessment. *Society and Natural Resources* 5, 211–230.

City of Chicago Department of Planning and Development. (2002) *Calumet Land Use Plan*. City of Chicago, Chicago.

Cordell, H.K. and Bergstrom, J.C. (1999) *Integrating Social Sciences with Ecosystem Management*. Sagamore, Champaign, Illinois.

Dale, A. (2001) *At the Edge: Sustainable Development in the 21st Century*. UBC Press, Vancouver, British Columbia, Canada.

Dietrich, W. (1992) *The Final Forest*. Simon and Schuster, New York.

Dixon, J.A. and Fallon, L.A. (1989) The concept of sustainability: origins, extensions, and usefulness for policy. *Society and Natural Resources* 2, 73–84.

Ecological Society of America (1995) *The Scientific Basis for Ecosystem Management*. The Ecological Society of America, Washington, DC.

Ewert, A. (1990) Wildland resource values: a struggle for balance. *Society and Natural Resources* 3, 385–393.

Ewert, A. (1991) Outdoor recreation and global climate change: resource management implications for behaviors, planning, and management. *Society and Natural Resources* 4, 365–377.

Ewert, A. (ed.) (1996) *Natural Resource Management: the Human Dimension*. Westview Press, Boulder, Colorado.

Ewert, A. and Williams, G.W. (1994, September) *Getting Alice Through the Door: Social Science and Natural Resources Management*. Presented at the Social Dimensions session at the meeting of the Society of American Foresters National Convention, Anchorage, Alaska.

Feldman, D. (1991) International decision-making for global climate change. *Society and Natural Resources* 4, 379–396.

Field, D.R. and Burch, W.R. (1990) Social science and forestry. *Society and Natural Resources* 3, 187–191.

Heberlein, T.A. (1988) Improving interdisciplinary research: integrating the social and natural sciences. *Society and Natural Resources* 1, 5–16.

Hetherington, J., Daniel, T.C. and Brown, T.C. (1994) Anything goes means everything stays: the perils of uncritical pluralism in the study of ecosystem values. *Society and Natural Resources* 7(6), 535–546.

Holden, C. (1988) The ecosystem and human behavior. *Science* 242, 663.

Hutchings, J.A., Walters, C. and Haedrich, R.L. (1997) Is scientific inquiry incompatible with government information control? *Canadian Journal of Fisheries and Aquatic Sciences* 54, 1198–1210.

Jones, E.L. (1998) From steel town to 'ghosttown': a qualitative study of community change in southeast Chicago. Unpublished master's thesis, Loyola University, Chicago.

Keenan, S.P., Krannich, R.S. and Walker, M.S. (1999) Public perceptions of water transfers and markets: describing differences in water use communities. *Society and Natural Resources* 12, 279–292.

Kempton, W. (1997) How the public views climate change. *Environment* 39(9), 12–21.

Kessler, W.B., Salwasser, H., Cartwright, C.W. and Caplan, J.A. (1992) New perspectives for sustainable natural resources management. *Ecological Applications* 2(3), 221–225.

Kretzmann, J.P. and McKnight, J.L. (1993) *Building Communities from the Inside Out: a Path Toward Finding and Mobilizing a Community's Assets*. Asset-based Community Development Institute, Northwestern University, Evanston, Illinois.

Lee, K. (1994) *Compass and Gyroscope: Integrating Science and Politics for the Environment*. Island Press, Washington, DC.

London, J. (1908) To build a fire. *Century Magazine* 76, 525–534.

Machlis, G. (1992) The contribution of sociology to biodiversity research and management. *Biological Conservation* 62, 161–170.

Machlis, G. (1999) New forestry, neopolitics, and voodoo economics: research needs for biodiversity management. In: Aley, J., Burch, W., Conover, B. and Field, D. (eds) *Ecosystem Management: Adaptive Strategies for Natural Resources Organizations in the 21st Century*. Taylor and Francis, Philadelphia, Pennsylvania, pp. 5–16.

Maloney, M. and Ward, M. (1973) Ecology: let's hear from the people. *American Psychologist* 28, 583–586.

Margerum, R.D. (1999) Integrated environmental management: the foundations for successful practice. *Environmental Management* 24(2), 151–166.

McKibben, B. (1989) *The End of Nature*. Random House, New York.

Nassauer, J.I. (1997) Cultural sustainability: aligning aesthetics and ecology. In: Nassauer, J.I. (ed.) *Placing Nature: Culture and Landscape Ecology*. Island Press, Washington, DC, pp. 65–84

Patterson, M.E. and Williams, D.R. (1998) Paradigms and problems: the practice of social science in natural resource management. *Society and Natural Resources* 11, 279–295.

Postel, S. (1992) Denial in the decisive decade. In: Brown, L. (ed.) *State of the World*. W.W. Norton and Company, New York, pp. 3–8.

Radeloff, V.C. (2000) Exploring the spatial relationship between census and land-cover data. *Society and Natural Resources* 13(6), 599–612.

Rafson, H.J. and Rafson, R.N. (eds) (1999) *Brownfields: Redeveloping Environmentally Distressed Properties*. McGraw-Hill, New York.

Reidel, C. (1992) Asking the right questions. *Journal of Forestry* 90(10), 14–19.

Roston, H. (1985) Valuing wildlands. *Environmental Ethics* 7, 23–48.

Sherry, E. and Myers, H. (2002) Traditional environmental knowledge in practice. *Society and Natural Resources* 15, 345–358.

Stankey, G.H., Bormann, B.T., Ryan, C., Shindler, B., Sturtevant, V., Clark, R.N. and Philpot, C. (2003) Adaptive management and the Northwest forest plan. *Journal of Forestry* 101(1), 40–46.

Stein, T.V., Anderson, D.H. and Kelly, T. (1999) Using stakeholders' values to apply ecosystem management in an upper Midwest landscape. *Environmental Management* 24(3), 399–413.

Trumbull, W. (1999) State and county programs. In: Rafson, H.J. and Rafson, R.N. (eds) *Brownfields: Redeveloping Environmentally Distressed Properties*. McGraw-Hill, New York, pp. 102–108.

Wali, A., Darlow, G., Fialkowski, C., Tudor, M., del Campo, H. and Stotz, D. (2003) New methodologies for interdisciplinary research and action in an urban ecosystem in Chicago. *Conservation Ecology* 7(3). Retrieved 12 December 2003 from http://www.consecol.org/vol7/iss3/art2

Westman, W. (1993) How much are nature's services worth? *Science* 197, 960–964.

Williams, D.R. and Patterson, M.E. (1996) Environmental meaning and ecosystem management: perspectives from environmental psychology and human geography. *Society and Natural Resources* 9, 507–521.

Wilson, E.O. (2002, February) The bottleneck. *Scientific American*, 82–91.

Wolosoff, S.E. and Endreny, T.A. (2002) Scientist and policy-maker response types and times in suburban watersheds. *Environmental Management*, 29(6), 729–735.

Wondolleck, J.M. (1992) Resource management in the 1990s. *Forest Perspectives* 2(2), 19–21.

7 Power and Decision Making in Natural Resource Management

Introduction

This chapter addresses three specific questions essential for the effective management of natural resources and the implementation of IREM processes. (i) How are decisions made in natural resource management agencies and organizations, and how are these processes likely to affect the way that IREM is adopted and implemented? (ii) Broadly speaking, how do power and influence affect the natural resource management policy process? (iii) What does the policy process look like and in what ways do various policy actors and policy circumstances influence that process? The first question is addressed by discussing the attributes of well-acknowledged decision-making models; the second question by examining the most popular theories of power and also by examining the more modern policy literature to see how power and influence have been examined; and the third is scrutinized by tracing the dynamics of the different phases of the policy process. Each of these discussions draws on theory from various social science fields such as public administration, political science and social policy, and places these in the context of natural resource management.

Decision-making Models for Natural Resource Agencies

Four models of decision making are presented here that help explain how decisions are made within public bureaucracies and similarly in private and commercial organizations (see Table 7.1). The first model is the basic rational model, the second is the bounded rationality model, the third is the organizational processes model, and the fourth is known as the political bargaining model. The first model is regarded as a normative model, a model that reflects how experts believe decisions in organizations should be made, while the other three are more descriptive models – models that try to explain how organizational decisions are

Table 7.1. Decision-making models: general concepts.

Rational models	Procedural rational models	Organizational processes models	Political bargaining models
Singular coherent unit	Rational within bounds	Emphasizes differences between individuals and organizations	Implicit rules of bargaining
Purposeful behaviour	Objective setting fundamentally subjective	Assumes no 'super-individual'; no single definable goal	Search for politically viable solutions
Maximizes benefits – minimizes costs	Goals are dynamic	Decisions are fragmented	Intricately linked issues
Includes reason why, what and how	Decision makers 'satisfice'	Profoundly different from individual decision making	Informal power
Disregards constraints	Adverse to risk	Emphasizes importance of groups	Outcome generally supported by elites
Groups simply redefined as if an individual		Subunit complexity	Neither rational nor routine
Conscious choice for a calculated objective		Recognizes individual goals	
Prioritization of objectives		Subunit control of resources	
		Disjointedness and bias affect feasibility	
		Appears to be non-rational	

actually made. Initially, the three latter models were promoted in the scientific literature as competing models to explain the decision-making process. More contemporary theorists, however, believe that they are most usefully employed as complementary models. Each model helps in explaining particular aspects of the decision-making process or provides a different but supporting perspective.

The basic rational decision model

This model suggests a linear, step-by-step process for decision making, where one step is completed and logically leads to the next. It begins with a clear definition of the problem and the stating of an equally unambiguous set of objectives. This process largely assumes an ideal world where all the necessary information is available at the right time and that this information is in an appropriate and usable form. It further assumes that the decision maker has a broad range of skills and strategies available to solve the problem and that she/he is clearly able to rank all aims objectively and rationally. This model takes for granted that all those involved in the decision think and act in the same way as a single individual does; all buy into the chosen decision and work in a coordinated manner to implement it. In summary, decisions tend to be explained as a unified conscious choice to attain a single calculated and agreed upon objective (McGrew and Wilson, 1982, p. 8).

Table 7.2. Comparison of decision-making characteristics.

	Model type			
	Rational models	Procedural rational models	Organizational processes models	Political bargaining models
Key variables	Characteristics			
Unit of analysis	Individual	Individual	Organizational structures	Power
Key behaviour determinants	Objectively set goals	Bounded goals	Individual and organizational survival	Self-interests
Nature of decision-making process	Goal directed	Goal directed	Fragmented	Concentrated within elites
Role of Individuals	Professionalism emphasized	Blend of profes-sionalism and bureaucratism	Bureaucratism predominates	Elites predominate
Level of routines	High	Moderately high	Moderate	Low
Information management	Available and open	Generally open and available	Selective disclosure and gathering	Used as basis for bargaining
Inducement for change	Radical	Radical–conservative	Conservative	Conservative
Prevailing ideology	Professional	Realism	Bureaucratic	Market driven
Non-competing models	Many features are incommensurate			

Evidence from decision-making research suggests that such an idealized decision process is impossible to achieve in practice. In reality, even when natural resource managers consciously try to follow a rational decision-making process, they resort to circular or iterative processes where the linear, step-by-step process is broken up by frequent returns to earlier stages in the decision-making process (called feedback loops). Here, earlier 'decisions' are refined in the light of changing needs, evolving objectives, new information and the recognition of various barriers to earlier plans. In most IREM decision-making situations, it has to be recognized that the decision-making process is normally a highly complex affair involving a broad range of management players, agencies and decision influences. In such circumstances, any evidence of a pattern of a rational decision-making process is often limited. As will be seen, the decision models described below generally have greater efficacy in the real world of natural resource management, where many different values and priorities abut (see Table 7.2).

The bounded rationality models

For decades, two public policy analysts, Simon (1947) and Lindblom (1980), debated the true nature of bureaucratic decision making. Despite their differences, each, nevertheless, dismissed the usefulness of the rational model and in its place advocated what are now termed procedural or bounded rationality models. While Simon emphasized that goals and expectations were lowered or shifted with policy experience (lowering or shifting the goal posts), Lindblom argued that policy makers simply muddled through without much thought for goals or objectives. In general, each agreed that decision makers' skills, technical knowledge and habitual ways of addressing problems and solutions 'bounded' or limited rational or truly reasoned approaches to decision making. They also agreed that goals were forever evolving as a result of continually changing circumstances and they were set subjectively rather than objectively.

While the idealized basic rational model implies that a broad range of possible solutions are considered before a final decision is made, the procedural rationality models stress that only a narrow range of options are ever considered in practice. Not only is a rather narrow range of options considered, but also the decision maker's training, areas of interest and the standard decision-making ways that the decision maker has used in the past limit those options. Bounded rational models also reflect situations where risk is avoided if at all possible, and the potential consequences of a policy decision are regularly ignored or underestimated. This is particularly relevant to the field of natural resource management as environmental consequences are often ignored or underestimated; such practices reinforce the need for IREM. In a world of bounded rational decision making, typically politicians or policy makers advocate only small incremental changes to present practice, while professional experts rely on rules of thumb and established ways of addressing problems. Whether policy maker or professional expert, each normally seeks compromise to limit the type of solutions implemented. They tend to favour those processes and procedures with which they are most familiar and comfortable.

In the resource management field, such limiting behaviour tends to favour physical solutions to most social problems, for example building larger landfills, rather than considering the possibility of behavioural changes such as reducing solid waste in the first place. Scientific observation of policy decision-making practice suggests that only limited effort is made to find the best possible solution; the search for solutions is more often confined to selecting a solution that 'will do under the prevailing circumstances'. This has been termed as 'satisficing' (this word is a contraction of satisfactory and sacrificing). Satisficing implies sacrificing a search for an optimal solution for one that is merely satisfactory given all the time pressures and limited financial and information resources available (Rees, 1990).

Organizational models

Organizational models of decision making first and foremost recognize that a group of individuals in an organization involved in a collective decision do not make decisions in the same way that a single individual does. There is no one individual working within organizations who is capable of handling and sifting through huge amounts of information. These models recognize the frailties of humans working in organizations who have communication failures, harbour self-interests such as maintaining or expanding power, have differing objectives and priorities from their colleagues and the organization, and have differing capacities in information processing. As a result, organizational decisions are often observed to be an aggregate of disjointed rather than coordinated actions throughout an agency, which lead to very different kinds of decisions from those conceptualized by more rational models. For example, in this way, a decision to build an access road into a sensitive ecological area may be conceived at head office involving a whole host of environmentally sound practices, but may be executed in the field quite differently because of the lack of training and expertise, or simply a lack of conviction.

It is important to understand that organizations make most of the important decisions in our society (this is very true in natural resource management), and as a result it is critical to distinguish between the ways in which individuals and groups, different individuals within groups, and different groups or organizations within larger entities such as professions or industrial sectors, make decisions. Generally, as group complexity increases, through either greater numbers or organizational intricacy, so it becomes more unlikely that subunits will agree on policy objectives, priorities or methods. Organizational staff such as senior managers, field supervisors and ground level workers tend to value more personal goals, such as job security, professional status, peer recognition and professional networking as well as personal risk avoidance, above the goals of the organization. These personal motivations often interfere with and distort organizational decisions. This is particularly important in the implementation of IREM strategies that often require considerable cooperation between agencies, units within agencies and individuals within subunits. Unfortunately, bureaucracies and large corporations rarely outwardly acknowledge such decision-making influences although they routinely appear to accommodate such interference on a day-to-day basis. In attempting to formulate and implement natural resource management policy, therefore, it is important to recognize the capacity of individuals and subunits within an organization to control, manage, suppress and otherwise distort decision-making processes.

Political bargaining models

Although, on the surface, organizations may seem to be rational and working towards a common and well-articulated goal, this is regularly some way from the truth. Often organizational decisions can best be described as a process of bargaining between individuals and units based on the power and influence that each holds, and their commitment to reaching the best possible outcome for themselves or their units. Although individuals or units initially may define their position by determining a reasonably rational solution to a resource management problem, the decision that results rarely takes into consideration the best available information, considers the optimum evaluation criteria or uses the most advantageous methods to reach the agency's explicit policy objectives. The final outcome of a collective decision is dependent on the interplay of power among those with influence and what in the end amounts to a politically viable solution. An organizational decision in this case represents the outcome of a bargaining process where, at least temporarily, a mutually satisfactory outcome among those negotiating the decision has been reached or for the time being some actors involved in the decision-making process do not expect to get a better arrangement.

The underlying principle of the political bargaining decision-making process, whether concerned with individuals, groups, organizations or nations, is that each has self-defined interests to protect. In this decision-making process, bargaining continues at all levels for as long as possible to ensure that individual or unit interests are least compromised by the final decision. The major forces acting on the decision process and its final outcome are then the underlying informal structure of power, the resources that players are able to devote to the issue and the negotiating skills that each player possesses. The effectiveness of each decision is thus defined by what individual actors and units consider 'key issues' to protect their own well-being as much as it is about the well-being of the agency.

Political bargaining in the resource management context is rarely seen – once the analysis is broadened beyond the scope of the individual or isolated decision – as a 'rational' decision process, nor does it appear to follow an established rational or reasoned routine (McGrew and Wilson, 1982). Political bargaining clearly emphasizes a new dimension to the decision process, placing considerable emphasis on individuals within an organization or unit and the informal power that an individual or unit holds. Because of the continued bargaining process that takes place until a decision is considered irreversible, all manner of distortions can be expected to affect a decision throughout its life. Such bargaining, based on evolving interests, frequently leads to concerns about natural resource management policy or decision-making inconsistency.

Multiagency decision making

The idea of organizational processes and political bargaining already apparent and influential within the boundaries of an agency draws attention to the increased challenges of coordinating and implementing decisions in the multi-agency and multi-interest resource management situation. As will be seen later in this book, especially in Chapter 10, the challenges of implementing IREM at the landscape scale, where numerous agencies with broad-ranging and differing resource management objectives may be invovled, are especially difficult. Also working to influence resource and environmental management in this broader organizational context is the interplay of government (the state), the market including various resource management sectors such as the mining industry, and civil society such as NGOs. While the relationship of market, state and civil society (which is of considerable importance in the implementation of resource management policy) is dealt with at some length in the next chapter, the following section examines various theories of power that are typically applied at the regional or national levels but also have importance for understanding the dynamics of the policy process on a smaller scale.

Power and Influence Shaping Natural Resource Management Policy

In the natural resource management field, some individuals and individual organizations, such as a large multinational organization like Standard Oil, have considerable power over public policy, and other individuals or individual organizations have little or no influence. Why is this? Why in a democratic society where each adult typically has one vote can someone or some group be very powerful and another be quite powerless when it comes to shaping and influencing public policy? This section attempts to answer these questions by reviewing what now might be considered the classic theories of power and influence in public policy development.

Three groups of theories are examined. The first is the pluralist theory of power, the second is elitism and the third is structuralism. Pluralism and elitism are considered to be subjective and descriptive theories, as they deal with how individual preferences aggregate into power and influence, while structuralism is considered to be more objective and normative. The focus of its analysis is on the results of public policy rather than expressed preferences, and it is normative because it suggests or implies prescriptions or solutions to what is typically observed as a gross inequity in the apportionment of public policy benefits among the under-classes such as the poor. As with the decision-making models, theorists first advocated these individual theories as competing explanations of power, but they have been advanced more recently as different dimensions and complementary theories that help explain the way in which power and influence impact the policy development process (Blowers, 1984).

Pluralism

The pluralist theory of power was postulated by Dahl (1984), Polsby (1980), McFarland (1969) and others as the most appropriate theory to explain how power and influence aggregate in a democratic society to influence and shape public policy. Power is seen as the result of the aggregation of the expressed views and preferences of numerous people in society. These preferences are communicated through the typical channels of democratic processes such as elections, opinion polls, public meetings, and phone calls to political constituency offices. This view of the policy-shaping process assumes that individuals, groups or coalitions, and various social movements have open access to policy makers and that policy makers are responsive to the public's views and inclinations. Such an expression of power and influence might be seen in the situating of a county landfill. Individuals express their preferences for one landfill site over another by forming pressure groups, informing their elected representatives, and to some extent through their voting patterns at election time. To come to a final decision, policy makers weigh the support and the viability of the various opposing alternatives and make a decision that is said to be in the public interest.

Elitism

In returning to the discussion in Chapter 4 on the 'driving forces' underlying IREM and concerning environmental justice, we learned that unpopular or unhealthy resource development initiatives were almost invariably located in the backyards of the poor or racial minorities. Part of the explanation for this lies in the theory of elitism. Elitism, although still focusing on the subjective preferences of policy actors, argues that effective power and policy influence is concentrated among powerful groups in society (Mills, 1959). These groups typically include the wealthy, the well educated and those who belong to the upper classes. These groups are able to get preferred access to policy makers to influence their decisions. As a consequence and as a general rule, coalitions of these influential groups are able to keep unwelcome natural resource developments out of their neighbourhoods.

This is an example of the NIMBY (not in my backyard) phenomenon. One interesting point about elitism is that elite groups in society often claim to represent the broader interests of society, including the under-classes, although this is rarely substantiated by the situation's facts. The evidence suggests instead that elites, once they gain power, are rarely able or willing to act in the interests of the under-classes.

Structuralism

In contrast to pluralist and elitist theories, structuralism examines power and influence from the perspective of the results of the policy process rather than from an accumulation of subjective preferences. Structuralism, as mentioned earlier, is not only a descriptive theory explaining how power impacts the policy process; it is also meant to be prescriptive. By drawing attention to the inequitable way that policy mechanisms usually favour society's elites and habitually deprive the poor of tangible benefits, this theory focuses its analyses on policy outcomes. In addition, structuralists argue that policy mechanisms must be reorganized to ensure that the under-classes receive their fair share of policy benefits (Sandbach, 1980). Structuralism, both as a descriptive theory and as a policy prescription, is concerned with how various ruling-class interests are self-serving by manipulating policy machinery to meet their own needs. Structuralism as a diagnostic tool is especially interested in the way that power permeates the capitalists' system from international, national, regional and local political economies to ensure that society's most powerful continually receive more than their 'reasonable' share of public policy benefits.

In the case of situating a landfill discussed above, structuralists will argue that while all segments may benefit from the development of a modern sanitary landfill that might replace an old one, the odds are that the landfill will finish up in the backyards of the under-classes. Such a landfill, no matter what level of technological sophistication is used, comes with residual negative effects such as smell, increased truck traffic, the perceived risk of pollution and property devaluation.

Idioms of Analysis

The three theories previously examined have undergone various metamorphoses in the policy literature to give special attention to particular aspects of policy concern. Weale (1992) made an extensive review of analytical trends in the natural resource management literature to discover four basic and broad-brushed analytical approaches that he called 'idioms of analysis':

- Rational choice idiom
- Social systems idiom
- Idiom of institutions
- Policy discourse idiom

The *rational choice* idiom focuses on the individual unit, whether that is an individual such as a woodlot owner, a solitary firm such as a commercial fishing company, a single administrative unit such as regional government, or a nation. This idiom recognizes that each 'individual unit' has specific preferences and acts more or less rationally to optimize its individual welfare while others attempt to rationalize theirs. A rational choice in the context of this book is seen as that action that most efficiently attains the individual's desired natural resource policy outcomes. Interestingly, if an individual unit acts rationally in a resource management situation requiring communal agreement and support, such action can give rise to the free rider effect. The free rider effect recognizes that many members of a community will act to curb water pollution, for example, and will voluntarily incur the pollution abatement costs – if not always with great enthusiasm. The problem is, however, that some and possibly a critical mass of the community will avoid such deeds in the hope or in the belief that enough others will comply to attain a satisfactory clean-up. Under such conditions, left to individual rational choice alone, everyone will suffer the consequences of insufficient pollution abatement, unless political action is taken to encourage or force everyone to comply.

The *social systems* idiom examines the complex interactions of various social subsystems such as the economic, political/administrative and civic subsystems, and also examines their interactions as they impact the public policy decision process. In this regard, a more in-depth explanation of the interrelationships of the state, market and civil society is given in the next chapter. Whereas the rational choice analysis focuses on individuals and their interactions, and the systems approach takes a broad view of the whole social system as well as examines its inter-related components, the institutional analysis provides an intermediate examination or view. According to Weale (1992), institutions are considered in the rational choice and social systems analyses but appear 'obliquely'. For rational choice theorists, as with pluralist and elitist interpretations of power, institutions are seen as the outcomes of society's 'aggregated preferences'. In this way, expressed preferences serve to both guide and constrain the options available to politicians, bureaucrats and industry in natural resource and environmental management development.

From the systems perspective, *idiom of institutions* are seen as the component parts of a broader set of organizational relationships. Their particular significance is explained best by their role in that wider political system. In the case of forest management, for example, land ownership may be seen as an over-arching institution, while small woodlot owners, multinationals, commercial forest owners and public land managers may be seen as a lower-order subset with its own set of roles and influences.

The fourth and final idiom is that of *policy discourse*. This analytical perspective views the world from a cognitive developments (key understandings and social learnings) perspective that leads to new or different approaches to problem solving in natural resource management. Within discourse theory, there will be policy actors with different points of view, but their motivations will centre on their beliefs, substantiated by empirical and theoretical evidence, rather than identifiable material self-interests.

In this sense, we see policy solutions explained best by the dominant belief systems of the era. This is frequently referred to as paradigm shifts. Neo-classical theory, for example, strongly influenced economic policy between the two World Wars, and in environmental management, end-of-pipe pollution abatement strategies guided policy of the late 1960s. In contrast, more integrated approaches to resource management were adopted in the late 1980s and in the late 1990s; and ecosystem and IREM approaches began increasingly to influence resource management policy and land management. This was largely because, as this idiom suggests, of a better understanding of the importance of biodiversity, landscape management approaches to sustainable development and globally scaled pollution problems.

The Policy Process

Having discussed the ways in which decisions are made within organizations and the way that power and influence shape policy, it is now appropriate to examine the various phases of the policy development process and consider how these various decision-making models and theories of power discussed above actually play out in the policy process. This section begins with an overview of the various dimensions of the policy development process and the major influences on its evolution. A walk through the various phases of the policy process follows this discussion in order to examine the various transformations of the key variables impacting each policy phase. As we move through each phase, it will be possible to see how each impacts the potential dynamics of each subsequent phase, and it will also be possible to see how all the phases combine to form ever-recurring policy cycles that create new policy needs, generate new policy ideas, offer new policy implementation possibilities and suggest new ways to assess and evaluate various policy activities.

According to Mitchell (1979), natural resource policy has three basic dimensions, a time dimension, a perspective dimension (see social systems idiom) and a spatial dimension (see Fig. 7.1). In terms of a time dimension, resource policy must be considered from a historical perspective, from the perspective of the present and for their ramifications on the future. A temporal dimension is particularly important, for example, when considering fisheries policy. In the

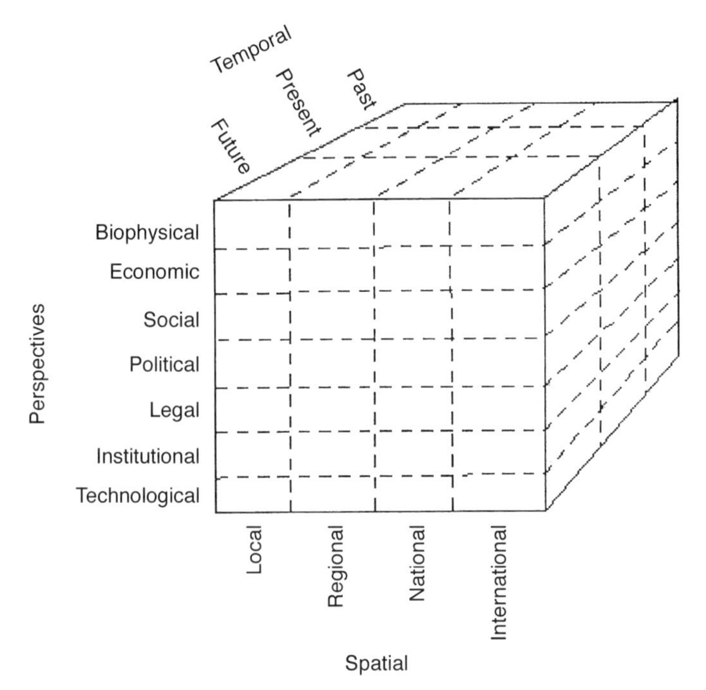

Fig. 7.1. Dimensions of policy analysis. Source: Krueger and Mitchell (1977), cited in Mitchell (1979).

case of the North Atlantic fishery, for instance, we know from reports of New World explorers that the ground fishery was abundant then. We now know, however, that fishing stocks are substantially diminished and we deduce from this that, unless drastic measures are implemented now and maintained in the foreseeable future, the health of the fishery will be jeopardized for many generations.

Mitchell (1979) argues, which is consistent with the social systems idiom discussed earlier, that policy can be considered from broad multi- and interdisciplinary perspectives such as considering the economic, political and technological systems at play in any policy arena. If we are to implement a truly integrated resource and environmental management regime successfully, we must find ways to account for the various impacts and interactions of one system on another. If we fundamentally alter our economy, for example, we need to ask how this will impact, among other things, our social system, our legal system and our political system. The impacts of such interdependencies were most evident in the collapse of the North Atlantic ground fish stocks that devastated the fishing industry as well as the economies and the social fabric of many commercial fishing-dependent communities. This industrial adjustment, in turn, led to changes in commercial and recreational fishing regulations and to social policy.

The impact and severity of impact of an economic change will depend on its spatial coverage as well as its intensity within a particular area. Inside a small fishing village, for example, a walk down one street might reveal both affluence and poverty side by side as a result of the particular fishing licences held by its residents. The owner of a crab-fishing licence, for instance, may convey affluence, whereas owning a ground fishery licence will probably reflect financial hardship. At the broader village scale, the overall picture may be one of financial hardship and social decay. In contrast, when a wider net is cast, more varied industrial and commercial activity throughout the broader region may mean a more diversified economy and a more resilient social fabric that tempers the direct impact of the fishery collapse. As we move from regional to national and to international scales, so the impacts will vary according to the significance of the natural resource management sector that is under scrutiny. In terms of intensity, it is useful to compare the potential impact of a similar number of job losses experienced in the fishing industry with that in the car manufacturing industry. Instead of having thousands of jobs scattered in small towns and villages along an extensive shoreline, as in the case of the fishery, where the political impact is likely to be muted, it is interesting to consider whether similar job losses concentrated in a single town would have more political fallout, such as if the jobs in an automobile manufacturing industry were transferred to another region or country.

As useful as these dimensions are in examining the inter-relationships of natural resource policy, O'Riordan (1981) suggests a rather different set of lenses to view this process, which provides an alternative way of examining the policy process. This set of lenses considers actor personalities, the institutional environment and the resource policy issue's particular characteristics or nature (see Fig. 7.2). The first of O'Riordan's analytical dimensions is in many ways analogous to the three descriptive models of decision making (see above); he looks to the problems that policy actors have in integrating all aspects of the policy development process, how they deal with internal inconsistency, and how they react to continually changing external pressures. O'Riordan is also concerned with resource management as an institution. What are the various agencies' mandates, how are resource management goals set, what are the standard ways of dealing with problems and issues and, leading from the first dimension, how do agencies deal with policy inconsistencies? Finally, O'Riordan focuses squarely on the issues specific to the particular natural resource development context. In this way, he draws attention to the very different problems that arise in renewable compared with non-renewable resource development practices and the different challenges that might arise in the petroleum industry as compared with coal mining.

The Policy Cycle

Whether using Mitchell's (1979) or O'Riordan's (1981) analytical framework, neither provides sufficient insights to examine the various influences over the life of a policy process. These frameworks tend to take snapshots of the policy process rather that make full-length, feature movies. As made clear earlier, rarely if ever does a policy solve a resource management problem in perpetuity. More probably, a policy initiative may resolve matters for a while and then changing

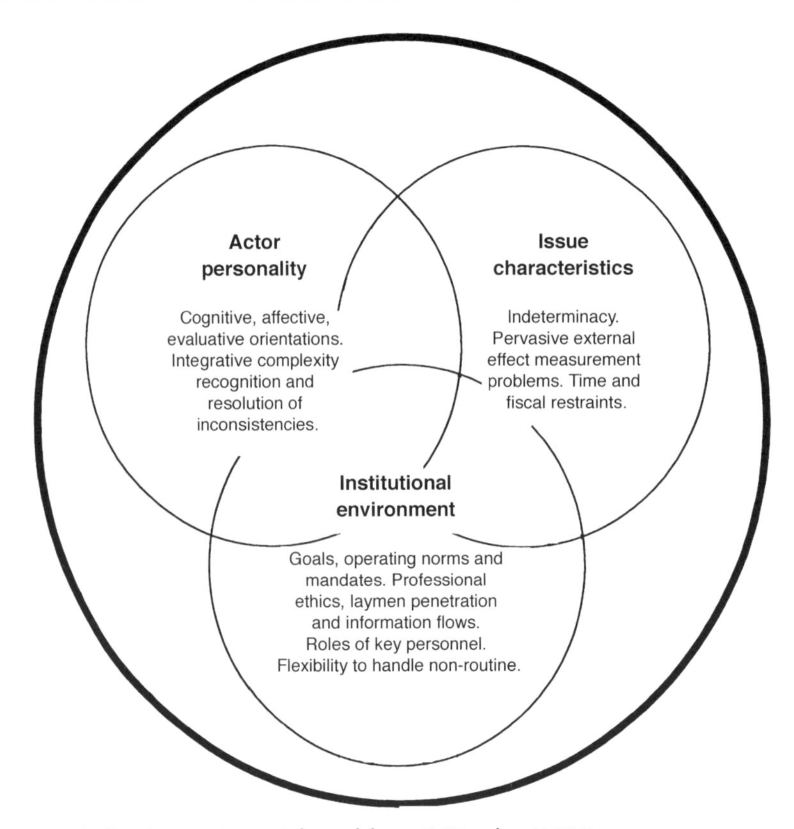

Fig. 7.2. Policy interactions. Adapted from O'Riordan (1981).

conditions will demand a new policy approach. The policy development process is thus, when viewed over an extended period, a recurring process that sees one policy phase evolving into the next or is replaced by a completely different policy initiative. While in reality it is difficult to differentiate clearly the various stages or phases of a policy process, as they often merge, it is possible to recognize four inter-related stages, each with its own characteristics and dynamics.

The first phase is the *gestation stage* where the need for a new policy or a revised policy is acknowledged and articulated. The second is where *policy is formulated* and ratified, while the third phase is where *policy is implemented*. The fourth phase is where *policy impacts are actually felt*, where they are assessed either formally or informally, and where they are evaluated. Depending upon how it is assessed and the influence that disaffected policy actors have, the implementation phase is either extended or efforts are increased to revise it, leading back to a renewed gestation phase (see Fig. 7.3).

Policy gestation

It is important to understand from the outset that there is no set prescription to take a policy idea and see it through to implementation. For example, over a

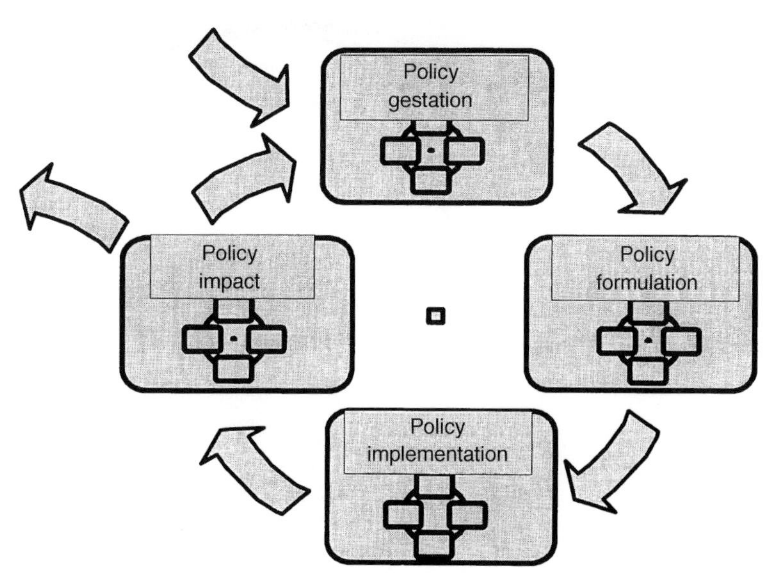

Fig. 7.3. The policy cycle.

period of 4 years, one of this text's authors worked intensively as part of an interdepartmental government team to have a single paragraph, based on considerable background analysis, adopted as a guiding resource management principle. In another instance, the same individual was consulted for 20 minutes by a lobby group. The recommendation that emerged from this consultation was adopted by government within a couple of days as legislation – word for word. Despite this wide disparity in preparation, influence and time over public policy formulation and adoption, there remain some key policy considerations that help explain the adoption process, which can be helpful in influencing the policy adoption process.

Hall *et al.* (1972) identified a set of general and characteristic variables from their analysis of the policy adoption process (see Fig. 7.4). They suggested that the first group of variables were almost invariably key in the policy adoption process, while the characteristic variables were often important but less predictably so. The important general variables were legitimacy, feasibility and support.

- Do policy makers see this as a legitimate need?
- Do policy makers consider this policy idea as feasible, even if technically it is not?
- Does this idea have a critical support – perhaps from an influential group such as a trade association?

The characteristic variables include:

- Association and scope – who will this policy adoption process affect and how? Is this idea linked to other important policy initiatives?
- The impact of a policy crisis – a crisis will sometimes put an idea on the fast track policy agenda but at other times displace it forever.

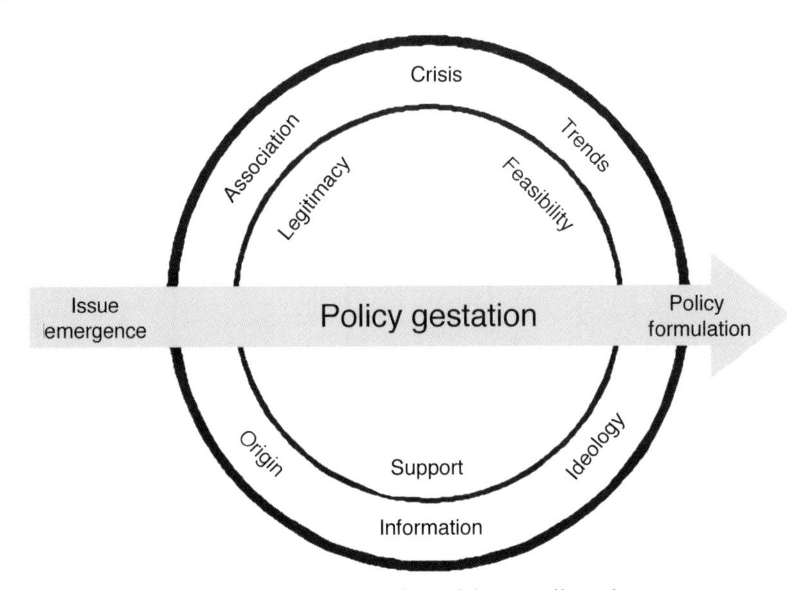

Fig. 7.4. The policy gestation process. Adapted from Hall *et al.* (1972).

- Trend expectation – refers to whether something is likely to gain or lessen in influence or popularity.
- Issue origin – who is seen as the source of the idea and is that group or person respected?
- Policy information – this does not necessarily have to be true but it generally has to be believed.
- Management ideology – does this idea support the political beliefs and managerial style of the policy agency?

Each of these variables interplay in the policy process that either leads to a policy idea being adopted or, as is more often the case (given that our policy system is designed to be conservative rather than radical), sees the policy idea flounder and get lost in a sea of competing priorities.

Policy formulation

One reason that our policy adoption system seems to be conservative and resistant to change is that since the Second World War, public policy has become much more pervasive and intrusive on the private sector and, as a result, it is much more convoluted (see Mayntz, 1983). Given this increased complexity, it is difficult to predict with much precision the outcome of any single policy initiative because of numerous potential knock-on effects. A change in one policy area, for instance, may well have unintended influences on a wide and sometimes unpredictable spectrum of policy issues. Adding to this is the question of which policy form is most appropriate. There is a broad range of possible policy forms, which makes it difficult to select the most effective. Several examples exist.

- Regulatory norms – you must do this or be sanctioned
- Financial incentives – do this and receive a monetary reward or subsidy
- Procedural regulations – follow these prescribed steps and get appropriate authorizations before proceeding with an intended project
- Education – persuade using information
- Public sector provision – when the private sector is unable or unwilling

As inferred by the aforementioned examples, the time frame and the care which a public agency takes in formulating and ratifying a public policy varies widely depending on, among others, the nature of the resource development initiative, the sense of urgency, the level of perceived support, whether the initiative is driven by a crisis or not, and whether it appears politically expedient. In working through the policy development process, it is important to understand that just because a policy has been officially ratified, this is no guarantee that it will be implemented or that it will accomplish its intended objectives.

Policy implementation

To understand the policy implementation process, Rees (1990) suggests focusing on three inter-related factors about a natural resource agency that is charged with implementing natural resource policy. The first relates to past performance or, as Rees refers to it, pre-conditioning elements. The question is: How have natural resource agencies handled similar problems in the past? To understand an agency's past performance, it is useful to look at its character, its conventional operating methods and its span of influence and accountability. The second concern, 'internal factors', refers to an agency's administrative processes such as its degree of centralized and decentralized decision making and its decision-making style – is it able to make decisions on its own or does it have to continually refer to outside authority? Furthermore, this notion is concerned with the way an organization intervenes in the policy arena – does it act without regard to its negative effect on others, and does it act now and apologize later? The third influence, external factors that Rees (1990) calls '*ex-post* changes', includes the biophysical variables that impact the resource sector and the social, political, economic and technological factors that impinge more directly on the policy process (see Fig. 7.5).

Policy impact

Whether or not a policy is implemented, there is always a policy impact, or consequence. This may be the result of no policy intervention when one is necessary or when not implemented (non-implementation), or may be the direct result of policy outputs. Such outputs may begin to resolve an identified and targeted problem, or they may create negative (or positive) unintended outcomes. Surprisingly few public policy programmes, especially when one considers the costs of many natural resource management initiatives, are formally evaluated or

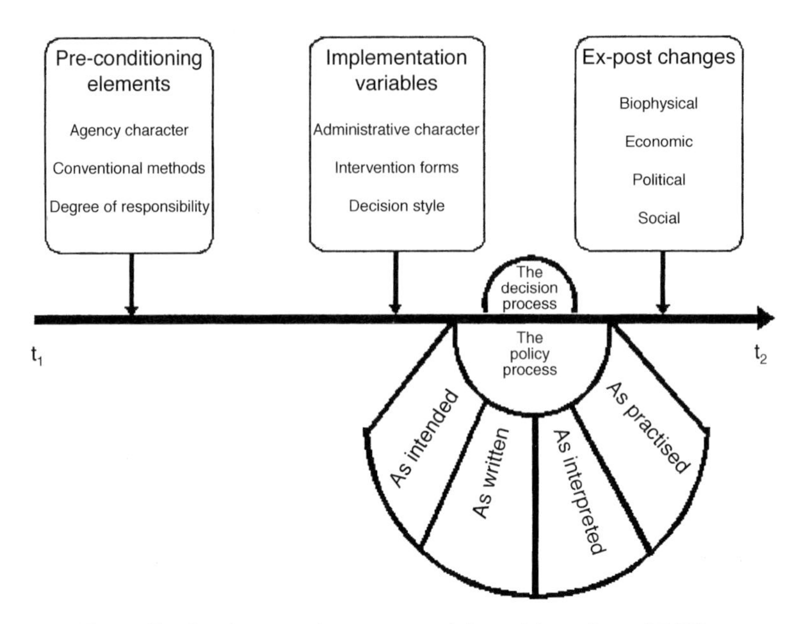

Fig. 7.5. The policy implementation process. Adapted from Rees (1985).

they are rarely assessed in the context of initially formulated objectives. As Lindblom (1980) and Simon (1947) suggest (see earlier in this chapter), the goalposts are often lowered or moved, or policy implementers simply muddle through until a new round of policy initiatives is instigated.

Recognizing the complexity and uncertainty of implementing most natural resource policies, some policy architects have advocated an adaptive and iterative approach to policy implementation (see Walters, 1986). This adaptive approach assumes that each policy iteration is considered to be an applied experiment, to be assessed and evaluated and, where appropriate, adjusted so as to be consistent with any new reality. As will be seen in the following chapter regarding the more specific workings of natural resource management, the relationship between the state and the market with regard to resource and environmental management is a convoluted one that often results in both state failure and market failure.

Summary and Conclusions

This chapter has introduced the reader to the complexity of decision making in the natural resource management policy context and has shown how power and influence permeate that process sometimes to distort and, in other situations, confound a policy maker's true intentions. This chapter has brought to focus the idea that public policy and industrial decision making is rarely a mechanistic process dominated by strict rules and protocols but is made by people with differing values, capacities and motivations. These human dimensions of the natural resource management process strongly influence decision making in a way which strains even the best efforts of the most dedicated and seemingly rational practitioners.

As was seen, the decision-making and policy account described in this chapter provides a rather perplexing and challenging but nevertheless realistic platform upon which to build a new natural resource management paradigm. It suggests a rather problematic policy inertia, one dominated by conformism on the one hand and muddling through on the other. To be successful, then, policy makers must somehow cut through this complexity to build a public policy and industrial management regime that is both responsive to environmental imperatives and mindful of the broad range of resource values that the public now clearly asks for. In this regard, the next chapter builds on the conceptual foundation introduced here to examine the intricacies and nuances of natural resource management. In combination, these chapters provide the theoretical and conceptual foundations for building a realistic conceptual and procedural framework for IREM which is expanded upon in following chapters.

Case Study
The Macal River Upper Storage Facility: Chalillo Dam, Belize

CHRIS RODRI

Preamble. In the summer of 2002, a Supreme Court Justice decided on the fate of the remote Macal River Valley, Belize, which is under threat from flooding by a hydroelectric project. This pristine rainforest-covered valley is among the most ecologically diverse areas on the planet and home to one of the largest jaguar populations in Central America. It is also the only known nesting site in Belize for the rare scarlet macaw, and shelters tapirs, howler monkeys and a host of other threatened and endangered creatures. This case study examines the Chalillo Dam Project in Belize from the perspectives of political power and natural resource decision making.

Introduction

In order to meet the rapidly growing energy demand in Belize, the Belize Electric Company Limited (BECOL) proposed to construct a dam that would flood the Macal River Valley. Proponents of the Macal River Upper Storage Facility (MRUSF), for which an approach road was already laid, hoped soon to begin construction in this remote area of Central America. The river drains the heart of the Maya Mountains, a large tropical rainforest with high relief and high seasonal rainfall. The proposed dam would hold back the river in a 9.53 km^2 storage reservoir that would be released in order to feed an existing dam further downstream, thus maintaining an adequate river flow during the dry season when river levels are too low to operate the turbines.

Its supporters tout it as a cornerstone of the government's strategy to make electricity generation self-sufficient in the country and not reliant on imported power from Mexico or costly diesel-driven generators. The aim ultimately is to initiate socioeconomic development by reducing the cost of electricity and improving efficiency of the supply. On the surface, this seems a laudable and timely project, Belizeans already pay considerably more for electricity than many of their neighbours, and the current supply is often sporadic and unreliable. For some, however, the big question is not whether the project can fulfil its ambitious economic promise and help alleviate poverty, but rather if it can do so without irreversible and unreasonable impacts to the pristine environment.

When one begins to scratch below the surface of this initiative and its underlying policy-making influences, it becomes clear that a highly convoluted set of issues conspire to question whether this project is more about political and corporate power than about electrical power.

Background

Belize

Belize is a small country about the size of the State of New Hampshire, situated on the Caribbean coast of the Yucatán Peninsula, Central America. Bounded to the north by Mexico and to the west and south by Guatemala, it has a multiethnic population of approximately 250,000.

The Yucatán region was once part of a proud ancient Mayan civilization that extended from the southern states of Mexico in the north to El Salvador in the west, and which reached its peak population around AD 900 before rapidly declining. More recently, the area of Belize has experienced a turbulent colonial history, first being claimed by the Spanish as early as 1638 and then by British pirates in the 17th century and official colonization by the British in 1862, when the land became known as British Honduras. Since the mid-1980s, the country's natural environment and ancient Mayan ruins have formed the basis of a rapidly expanding ecotourism industry. In 1981, Belize became an independent state within the Commonwealth, despite continued claims on its territory from Guatemala, a dispute that dates back to the 17th century and which has resurfaced periodically over the years.

The Chiquibul Forest Reserve and National Park

The Chiquibul Forest Reserve and National Park (CFR and NP) is a multifunctional extractive forest as well as a protected area in the Maya Mountains of Belize. It covers approximately 0.5 M ha and stands at an altitude of between 300 and 900 m above sea-level. The area, like much of Belize, contains numerous archaeological remains from the ancient Mayan civilization and includes the extensive ruins of Caracol and Nohoch Chen. Wildlife is abundant in the area, including many endangered and threatened species, such as scarlet macaws, jaguars, Baird's tapir, Morelet's crocodiles and ocelots, scarce in many other parts of Central America. The 1900 mm of rainfall per year is unevenly distributed through the year, meaning that the dry months have virtually no rainfall at all. The complete profile of the Raspaculo and Macal watershed on the northern border of the CFR and NP therefore becomes a critical habitat for much of the wildlife during the dry season.

The proposed Chalillo Dam project

The proposed MRUSF will be located on the border between the Mountain Pine Ridge Forest (MPR) and the CFR and NP, some 12 km downstream of the confluence of the Macal and Raspaculo rivers. With a maximum depth of impounded water of approximately 35 m and a maximum surface reservoir of 9.5 km^2, the resulting reservoir will extend 20 and 10 km upstream of both rivers, respectively (see Fig. 7.6).

Chalillo will supply water to the existing run-of-river dam at Mollejon, which produces 25.2 megawatts (MW) of electricity. Water releases from Chalillo will be timed to reach Mollejon at peak hours during the dry season to maximize electrical generation when it is most needed and at the same time generate approximately 7.3 MW itself through its own base turbine. The Mollejon Dam was built in 1994, and is owned and operated by Fortis Inc., a billion-dollar Canadian corporation that owns real estate, hotels and utilities as well as being the majority owner of Belize Electricity Limited (BEL) and BECOL. This dam has been beset with problems since its commissioning, is barely able to operate during the dry season and, consequently, has not been cost-efficient. According to Fortis, the Mollejon Dam generates only about 30% of Belize's present electricity demand.

The Chalillo Region

The geology of the Chalillo Region within the CFR and NP is characterized by meta-sedimentary rock, granite and limestone karst topography. The karst that occurs throughout the area of reservoir impoundment is interspersed with a vast array of uncharted subterranean

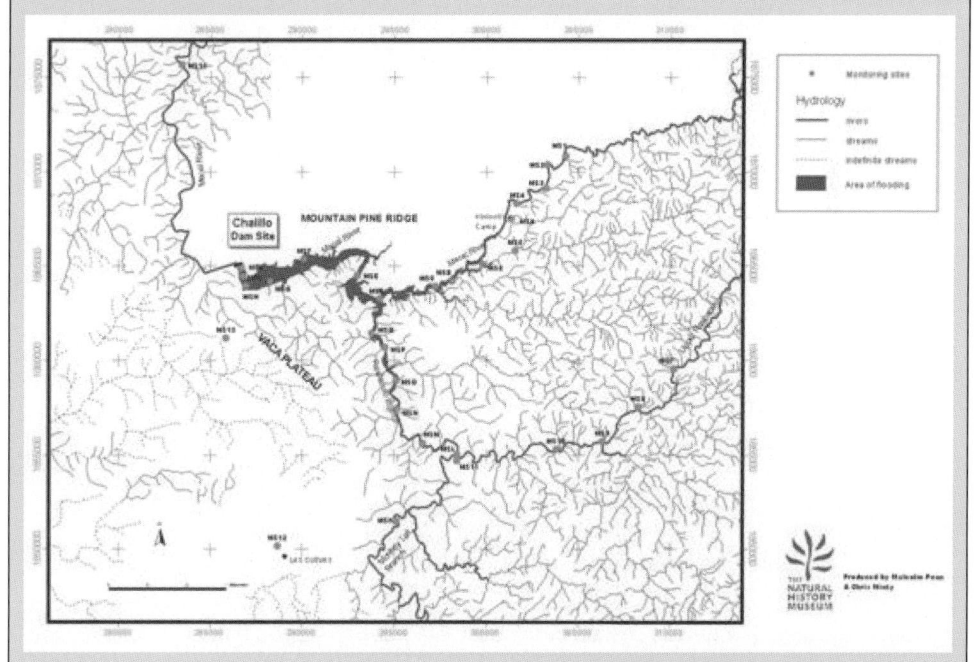

Fig. 7.6. Map to show the proposed Chalillo Dam impoundment. Source: Minty *et al.* (2001).

Fig. 7.7. The Chalillo Dam site. Photo by Glyn Bissix.

caves and underground chambers. This rock type generally supports lime-loving broad-leaved, subtropical moist forests, whilst granite locations north of the river support the distinctive Caribbean pine flora on acidic soils of the MPR (see Fig. 7.7).

Environmental Impact Assessment (EIA)

Under the Belize Environmental Protection Act (BEPA), it is a requirement of the proponents of such a project to submit an EIA. AMEC, a British-based construction company with offices in Canada, was contracted to produce the five-volume EIA, which they completed for their client, BECOL, in August 2001. The EIA was then assessed by the National Environmental Appraisal Committee (NEAC), a Government-appointed committee. One of the many supporting documents contained within the EIA was a wildlife impact assessment (Minty *et al.*, 2001) carried out by the Natural History Museum (NHM) of London. The report concludes:

> That this area (to be flooded) is a rare and discrete floral floodplain habitat, classified as 'riparian shrubland in hills' (Meerman, 1999 in Minty *et al.*, 2001), which acts as both a conduit and critical habitat for resident and non-resident fauna and avifauna. The maintenance of this exceptional habitat relies on the flow of oxygenated water and severe but temporary seasonal flooding of the river system following intense rain. This habitat, which does not occur elsewhere in Belize, will suffer approximately 80% permanent inundation if the project proceeds as planned, a figure now supported by interpretation of satellite images.

Based on intensive field work and data collected over a 10-year period (Rogers and Sutton, 1991, 1993, 1995, 1997, 2000 in Minty *et al.*, 2001), the report further concludes that:

> the remoteness and strongly seasonal hydrodynamics of the Macal and Raspaculo Watershed make it one of the most biological rich and diverse regions remaining in Central America.

> A seasonal shortage of food in the region for both herbivores and carnivores means that the relatively rich floodplain habitat becomes a critical seasonal food source for many species during the dry season. It is also a key staging post for many bird species migrating to and from North America; the watershed provides an important habitat for many endangered vertebrates species of international value.

The report continues that the construction of the dam would be likely to cause:

> Significant and irreversible reduction of biological diversity in Belize, initially at the population level but potentially at the species level, some of the species affected being of international importance.

Fragmentation of the proposed Meso-American Biological Corridor, and

> Rapid reduction in the already endangered population of Scarlet Macaw subspecies (*Ara macao cyanoptera*), leading to population inviability and probable eventual extirpation from Belize.

While AMEC did not completely hide the above findings of the NHM, they did relegate them to an Appendix and decided to publish only a 'Draft' version of the report, despite the Final Report being submitted by the NHM in good faith before the AMEC-imposed deadline.

AMEC further tried to discredit the report by stating that there were some 'some errors in fact' contained in the 'Draft Report', especially respecting the conservation status of certain wildlife species discussed (p. 15)'. The point of contention was with the scarlet macaw subspecies *Ara macao cyanoptera*, which is not listed on the Convention on International Trade in Endangered Species (CITES) website, although *Ara macao* is, as only full species are listed. In any event, the subspecies is considered by the IUCN Parrot Action Group as 'Critically Endangered'. AMEC, as lead consultants, evidently failed to recognize or acknowledge this fact when passing judgement on the wildlife report.

In an attempt to find out why the NHM's report was relegated to an Appendix and dismissed rather than used as part of the decision-making calculation, the Canadian Broadcasting Corporation (CBC), in an interview for the *Disclosure* programme, questioned the CEO of AMEC. He conceded in this interview for the first time that there will be environmental damage in building the dam, but claimed that the NHM 'did not do what they were asked to' in documenting that damage. Despite the concerns expressed by the NHM over biodiversity impact, the EIA received clearance in Belize by the NEAC in October 2001, a decision that was challenged in Belize's highest tribunal.

The dam proponents

The Government of Belize (GOB)

In their election manifesto of 1998, the People's United Party (PUP) identified agriculture and tourism as the twin pillars of the economy in Belize, and specifically committed the party to promote Belize as 'an exciting ecotourism destination'. The PUP were elected in 1998 and re-elected in 2002, and have consistently supported construction of the dam in order to support the growth of tourism and agriculture, by bringing down the cost of electricity, thus facilitating socioeconomic development. Indeed, the GOB Prime Minister has stated in an interview with CBC that his government has an 'understanding' that very soon after construction of the dam, prices will begin to drop. The GOB is a minority shareholder in BEL.

Belize Electricity Limited and Fortis, Inc.

Previously, the GOB operated the Belize Electricity Board (BEB), but electricity distribution was privatized in 1992 and in 1999 Fortis Inc. became the majority shareholder in BECOL and BEL, making them the monopoly provider and generator of electricity in Belize. In the Fortis

Project Summary, it is stated that one benefit among others of the Chalillo Dam 'project will [be to] help Belize attain greater autonomy in its electricity sector by increasing its capacity to generate renewable, stably priced energy (p. 15)'. However, this was not corroborated in an interview with the CEO of Fortis by the CBC *Disclosure* programme, when he made it very clear that electricity prices would continue to rise after construction as a result of Fortis having to finance the new infrastructure.

The Belize Audubon Society (BAS)

The BAS represents one of the largest conservation groups in Belize and, as a result of funding from the European Community (EU), operates seven protected areas in Belize as co-managers with the GOB. Despite this ongoing relationship with government, this organization prides itself on its ideological autonomy. BAS openly reacted to criticism of being rather quiet on the Chalillo Dam issue in its August–September 2000 newsletter, when somewhat surprisingly – given its strong nature-conservation mandate – it endorsed the construction of the dam, referring to its mission of balancing environmental and socioeconomic development concerns.

The dam opponents

Official GOB opposition

In the Belize Assembly on 21 January 2002, the leader of the United Democratic Party (UDP), the official opposition, the Right Honourable Dean Barrow, stated his party's opposition to the dam.

The Belize Zoo

The Belize Zoo has been a leading opponent of the dam's construction. Supported by funding from the Wildlife Trust, teams from the Zoo have made numerous expeditions to both the Macal and Raspaculo rivers and the countless unnamed tributaries that are to be flooded. Together they discovered previously unknown scarlet macaw nesting sites and have documented the importance of the river system to this and other endangered species. The Director of the Belize Zoo is an American national.

Local journalist

A well-respected local journalist, writing for the Belizean newspaper, *The Reporter*, commented on the PUP´s propaganda media:

> It is revealing but profoundly depressing that any party in government could hold the intelligence of its electorate in such low esteem that it would put forward these proposals as serious constitutional reform measures. There is no more classic illustration of this than the Chalillo dam debate with the full force of the government media focused on trying to discredit all or anyone who dares to question the folly of the original Mollejon dam or the continuing folly of Chalillo.

He added:

> The technique of the government, through its media, is to color ANY critic of Chalillo as an interfering foreigner being paid from abroad. In other words, if you criticize the Chalillo Project you are somehow 'unBelizean', acting against the state.

The report continues

> The Chalillo dam is NOT, as EXPERT after EXPERT has testified, either economically feasible, a potential source of cheap electricity or without environmental impact. It is this bunch of lies that

government keeps pushing that reveals exactly the extent of the fraud being perpetrated against the true interests of all Belizeans – even the as yet unborn, for they too will inherit the enormous burden of debt caused by the stupidity of placing hydro dams on a river which runs almost dry in the dry season and which, in some years, has hardly any rainfall at all. And of course the cost of the inevitable increase of pollution of both the Macal River and the Belize River is incalculable.

The Mollejon dam on the Macal River was built against all experts' advice at the time. It was promoted and pushed through by our present minister in charge of finance. Chalillo is no more than a further affront to logic to try and prove that he was right and also because there is a lot of money in and around dams. The Vaca dam which follows Chalillo will be the same story, and by then Belizeans will have some $250,000,000 to $300,000,000 dollars to pay back over the next 30 years at an astronomical 12% to 15% interest rate – and they STILL won't have cheap electricity, just a billion dollar debt!

Local NGOs

The Belize Ecotourism Association (BETA), through its Board of Directors, formally stated its position on the project in a press release dated 22 October 2002 stating:

> The World Commission on Dams has issued a report that has brought international attention to the numerous downfalls of dams, pointing out that the mitigation factors have been largely unsuccessful. Nor do dams provide flood control, in reality increasing the devastation, as did the Patuca Dam in neighboring Honduras. We are deeply concerned that Fortis and BEL are misleading the public and our government with their claims of the benefits of Chalillo. With serious consideration, and using all of the information and resources at our disposal, we have come to the following conclusions:
>
> There is insufficient data available to determine that the Chalillo Dam would be able to perform as projected. Given the inaccurate projections of the Mollejon Hydroelectric Project by the same consultants, we do not have confidence that the Chalillo Project will succeed in providing the additional power. A proposed third dam at Vaca Falls and the accompanying roads and infrastructure would further jeopardize the increasingly valuable ecotourism potential of the entire area.

International NGOs

One very well-resourced and influential USA-based environmental advocacy group, the Natural Resource Defence Council (NRDC), joined the fight in 1999, adding the Macal River to its list of 'Biogems' – environmentally critical regions threatened by development. The GOB-backed newspaper called the NRDC and opponents of the project 'lawbreakers' and 'terrorists'. Responding, the famous Robert F. Kennedy, Jr, an environmental lawyer with NRDC, said: 'This is one of the worst boondoggles I've ever seen in nearly two decades as an environmental lawyer, it will make a few Canadian businessmen wealthier and impoverish the people of Belize for a generation. This is globalization at its worst'. In addition, Probe International, a self-appointed watchdog of the Canadian government's international development performance, exposed a seemingly cosy relationship between the Canadian International Development Agency (CIDA) and AMEC, the company responsible for the EIA. Over the years, CIDA has awarded AMEC several contracts worth in excess of Can$46million to support various engineering and environmental studies associated with hydroelectricity projects. In this case, contrary to the usual tenure of an EIA to provide a neutral assessment of the costs and benefits of the dam's construction and operation, a rider was included within the EIA contract that encouraged AMEC to recommend construction.

More than a dozen advocacy groups in the USA, Canada and Belize organized a letter-writing campaign that delivered tens of thousands of opposition faxes and e-mails to Fortis, with celebrities such as Harrison Ford lending their names to the cause. Harrison Ford in a letter to the Canadian *Globe and Mail* newspaper highlighted the connections:

> Here's how Fortis set it up: to prepare an environmental assessment to support Fortis' dam, the Canadian government's foreign aid arm, CIDA, hired AMEC, a multinational engineering company that also happens to be a major dam developer. CIDA cloaked its assessment of the dam in secrecy; no documents were made public and no public input was solicited. If that isn't bad enough, Canadian taxpayers are footing the $250,000 bill. Fortis' perfect environmental scheme hit a snag, however. AMEC subcontracted a piece of the job to the Natural History Museum of London. Unfortunately for Fortis, the Natural History Museum wasn't prepared to mince words about the dam's devastating effects on Belize's wildlife. The Museum's scathing report says that the dam would cause 'significant and irreversible reduction of biological diversity' and fragment the proposed MesoAmerican Biological Corridor – an international effort to maintain the connections between the few remaining forested areas of Central America.

Probe International and the NRDC say that people downstream from the dam would be threatened. According to the director of NRDC's Biogems Programme, 'Fortis' geological studies state that the site is granite when it's really sandstone and shale. The worst case scenario – the dam breaks, floods communities downstream and kills people. Fortis' contract guarantees that they can sell the dam to the government for $1 without any liability.'

The office of IUCN-Meso-America reviewed the pertinent reports and bibliography and evaluated the general scope of the EIA. The interest of IUCN in this evaluation is based on the resolution adopted on 10 October 2000 by the World Conservation Congress, urging the developers of the project: 'to conduct a fully transparent and participatory environmental impact assessment of the proposed hydroelectric facility . . .' and to the Belizean NEAC 'to follow and apply the laws of Belize on environment and environmental impact assessment on the proposed project, with special attention to participatory processes and consistent with the best international practices'.

Conclusion

On 10 June 2002, the Supreme Justice Court in Belize began hearings on two lawsuits filed by BACONGO, a coalition of Belizean environmental and business groups challenging the GOB's approval of the US$30 million project. The suits charged that the project will destroy crucial habitat and raise electricity bills by passing construction costs to customers and that insufficient public consultations were given during the review of the EIA prior to NEAC's decision to grant environmental clearance to the project. BACONGO attorney Young-Barrow appeared before the Chief Justice to present an overview of the laws governing the approval of EIA by the Department of the Environment and the NEAC. Young-Barrow emphasized the importance of public consultation, a point expected to be of major significance in the course of the trial. Despite the fact that the approach road to the dam was already well under construction, the decision to build the dam could have been reversed by the Chief Justice and, even if the dam was built, the water storage facility need not necessarily be filled.

In December 2003, this case was before the Privy Council in London, UK, the highest court of appeal in the British Commonwealth. Regardless of how this case unfolds in this court of appeal, this case is fundamentally about policy power and really more about influence over fundamental natural resource policy decision making. It is a question of who has the power, who should have the power and who ultimately has the power to make the key decisions in this situation that clearly pits natural resource development against the interests of maintaining environmental quality.

Discussion Questions

1. Blowers (1984) argues that we should treat the various models of organizational decision making such as bounded rational, organizational and political bargaining models as non-competing and complementary explanations of the decision-making process. How might such an analytical approach differ from, say, one that relies only on a single model to explain the organizational decision-making process?
2. Pluralist and elitist theories of power in policy making are said to be concerned with subjective preferences (mainly wants and demands) and are generally descriptive, while structuralist theories focus more on objective matters (such as measurable deprivation) and are descriptive and normative. Based on your knowledge and experience of natural resource management, provide an example where you believe pluralist influences prevailed in natural resource policy making, where elitist influences succeeded, and where a class struggle led to a fair or unfair policy outcome.
3. The rational choice idiom is based on the premise or assumption that each 'individual' has specific preferences and acts rationally to optimize her/his individual welfare. Think of an example in natural resource management where you believe that the manager or the agency has made a decision about the management of the natural resources under its control that appears to be primarily motivated by some other factor such as compassion for others.
4. What is the free rider effect and how will it affect the efficient and effective application of IREM in an ecosystem made up of many owners.

Case Study Discussion Questions

5. Based on the evidence from the wildlife impact assessment, what socioeconomic benefits to the people of Belize might justify the construction of the dam and balance the environmental costs?
6. In your opinion, do you think it reasonable or unjustified that international conservation interests should remain quiet on the issue of dam construction in developing countries? Explain your answer.
7. It could be argued that the opponents of dam construction were masquerading as legitimate opponents so as to satisfy their own 'financially driven' agendas. What rules should a policy decision maker invoke to decide who has a legitimate interest in this case and those who do not?
8. Fortis included two 'unproven or may be ineffective' habitat mitigation methods in its Project Summary. Is it reasonable for Fortis to include predictions in what is supposed to be a factual information document concerning the proposed dam? What do you consider their motivation was for doing so?
9. What is your opinion and, looking at the available evidence, how should the Privy Council in London rule? What is the basis for your argument?

References

Blowers, A. (1984) *Something in the Air: Corporate Power and the Environment*. Harper and Row, London.

Dahl, R.A. (1984) *Modern Political Analysis*, 4th edn. Prentice-Hall, Englewood Cliffs, New Jersey.

Hall, P., Land, H., Parker, R. and Webb, A. (1972) *Change, Choice and Conflict in Social Policy*. Heinemann, London.

Lindblom, C.E. (1980) *The Policy Making Process*, 2nd edn. Prentice-Hall, New York.

Mayntz, R. (1983) The conditions of effective public policy: a new challenge for policy analysis. *Policy and Politics* 11(2), 123–143.

McFarland, A.S. (1969) *Power and Leadership in Pluralist Systems*. Stanford University Press, Stanford, California.

McGrew, A.G. and Wilson, M.J. (eds) (1982) *Decision-making: Approaches and Analysis*. Manchester University Press, Manchester, UK.

Mills, C.W. (1959) *The Power Elite*. Oxford University Press, New York.

Minty, C.D., Sutton, D.A. and Rogers, A.D.F. (2001) *A Wildlife Impact Assessment for the Proposed Macal River Upper Storage Facility*. Natural History Museum, London.

Mitchell, B. (1979) *Geography and Resource Analysis*. Longman, London.

O'Riordan, T. (1981) *Environmentalism*, 2nd edn. Pion, London.

Polsby, N. (1980) *Community Power and Political Theory*. Yale University Press, New Haven, Connecticut.

Rees, J.A. (1985) *Natural Resources: Allocation, Economics and Policy*. Methuen, London.

Rees, J.A. (1990) *Natural Resources: Allocation, Economics and Policy*, 2nd edn. Methuen, London.

Sandbach, F. (1980) *Environment, Ideology and Policy*. Blackwell, Oxford, UK.

Simon, H.A. (1947) Administrative Behavior. MacMillan, New York.

Walters, C. (1986) *Adaptive Management and Renewable Resources*. MacMillan, New York.

Weale, A. (1992) *The Politics of Pollution*. Manchester University Press, Manchester, UK.

8 The Theoretical Foundations of Natural Resource Management

Introduction

This chapter examines the meaning and implications of various dimensions of 'natural resource management' and also considers a number of key ideas and theories that explain the background for natural resource policy decision making. Initially, this chapter explains the fundamental concepts and processes of natural resource management and then links these to environmentalism and multiple-objective resource management. A discussion of the connections of sustainability and ecological modernization to natural resource management follows, along with the weighing of the impacts of globalization on good governance and sustainable development. Specifically, this chapter provides explanations and overviews of the following concepts related to natural resource management.

- Conceptualizing natural resource management
- Environmentalism
- Multiple-use/multiple-objective resource management
- Sustainability and ecological modernization
- Decoupling development from environmental degradation
- Market and state functions
- Globalization
- Good governance and sustainable development
- Environmental subsidies and green taxes

In conclusion, this chapter ties together the contributions of the social sciences, and the policy and decision-making literature discussed in earlier chapters to illustrate further the need for an IREM process.

Conceptualizing Natural Resource Management

Natural things or parts of nature are not considered to be 'natural resources' until humans place value, usually utility or monetary value, on them. For example, colonialists initially considered the North American wilderness as a barrier to human development, and this only changed for parts of it when direct value was placed on what was usually considered 'neutral stuff'. For example, and until relatively recently, the northern aspen forests of Alberta, Canada were rarely thought of as a natural resource until Japanese pulp and paper manufacturing companies began to see their economic value in the pulping process. In both cases, nothing physically or objectively changed with the wilderness or forests; it was rather their subjective valuation or the way people viewed them that changed (see Zimmerman, 1933, cited in Mitchell, 1979).

Given this important distinction as to what constitutes or does not constitute a natural resource, it is useful to examine what the process of 'natural resource management' entails. Natural resource management is considered to be all those decisions concerning the exploration of, the allocation of, and the processing, manufacturing and marketing of raw natural materials into usable commodities and products. This process includes the broad economic, social, environmental and technical considerations that influence natural resource management decision making (Mitchell, 1979). Especially important in the modern context of renewable natural resources is the notion of sustainability. This aspect refers to those decisions that ensure, or jeopardize, the supply of natural resources for the future.

Natural resource development, as one facet of the broader process of natural resource management, refers to the transformation of neutral stuff into a commodity or service that attends to human needs (Mitchell, 1979). In the context of both services to humans and sustainability, this includes the conservation of biodiversity, aesthetics and nature as amenity.

Environmentalism

Environmentalism as an area of academic interest focuses on the interface of two largely abutting ideologies that influence decision making in natural resource management. They are the *technocentric* and the *ecocentric* ideologies. Natural resource managers who operate largely within the *technocentric* realm are profoundly optimistic about humans' ability and ingenuity to continually solve problems of natural resource scarcity and the ongoing problem of environmental degradation. Technocentrists generally believe that environmental problems created in the past and the present will be solved in the future by the legacy of knowledge and expertise conveyed to future generations and future generations' resourcefulness. In contrast, *ecocentrists* are much more pessimistic about the potential of human inventiveness. Their thinking and behaviour are tempered largely by the belief that all things are constrained by natural limits, and moving beyond these limits has dire consequences in the long term for humankind by upsetting the natural balance of ecological processes (O'Riordan, 1981).

Although these two ideologies are best understood to be at the opposite ends of a continuum (see Fig. 8.1), natural resource managers and analysts may be seen to fall along that continuum (O'Riordan, 1981). At one end of the continuum are the 'deep ecologists' who argue that we should abandon much of our industrial world and return to a far more simple way of life that operates safely within natural limits. Less radical in their thinking are the 'soft technologists' who see humankind returning to a somewhat more simple life but who also see the advantages of new technologies that can be utilized to reduce the human footprint. The accommodators represent thinking and behaviour that largely rely on science and technology to solve environmental problems and maintain our present standard of living, but are nevertheless mindful of the role of nature in setting limits to our use and abuse of natural processes. The cornucopians, on the other hand, who are at the opposite end of the continuum from the deep ecologists, regard problems of natural resource scarcity and environmental degradation simply as challenges to be overcome and opportunities for the application of science and technology.

Environmentalism is not simply of academic interest, as the tension between these two viewpoints, of how the world is seen and how it should play out, impacts natural resource management and managers in their everyday lives. These two opposing ideologies and their various manifestations such as soft technologists and accommodators appear to have driven the debate, for example, over the Kyoto Accord. There are, for instance, those who trust that science and technology will prevail in solving the problem of global warming (Essex and McKitrick, 2002), and there are those who believe we must quickly curb human extravagance to live within the limits of our environmental means (Brown, 2001).

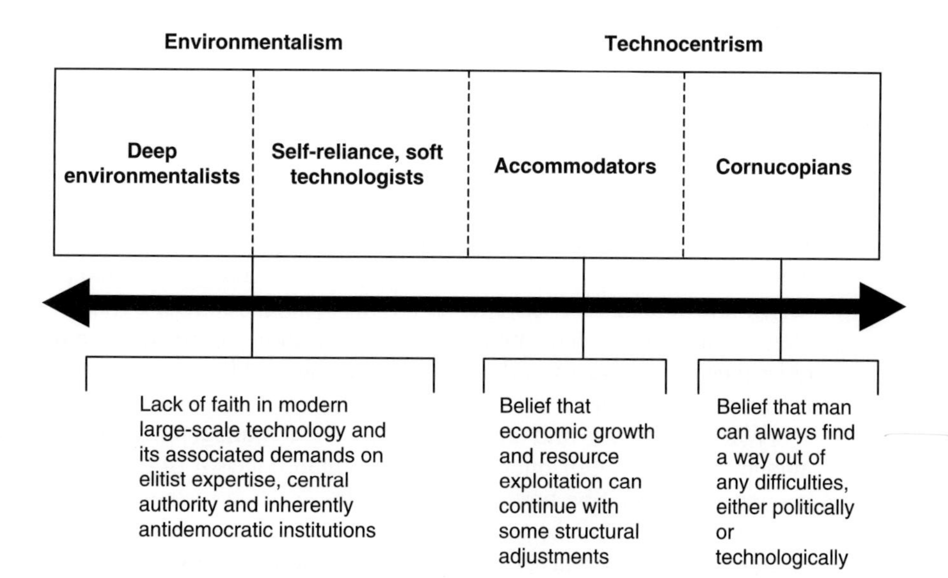

Fig. 8.1. Concepts of environmentalism. Adapted from O'Riordan (1981).

Multiple-use and Multiple-objective Resource Management

Multiple use in resource management may be the result of formal and prescribed resource policy decision making or, alternatively, can be seen as the outcome of passive, permissive or residual use of a natural resource. For example, in Northern European countries, such as in Sweden and Norway, the *allemansrätt* (Sandell, 1998) entrenches a tradition of public use of private lands for collecting wild foods such as mushrooms and berries as well as recreational access. Such use is generally considered permissive rather prescriptive management and a residual rather than primary use of a natural resource such as a forest. Of note is the fact that in Scotland such recreational access has recently been formalized in legal statute (Bissix, 2002/2003). Similar informal use of the Pacific Northwest public and private forests has evolved over the years and, interestingly, recent immigrants, bringing with them rather different traditions of public use of forests, have created novel challenges for forest managers in this region.

An important forerunner of the IREM process is multiple-objective resource management (as distinguished from multiple use). In its idealized or theoretical form (Van Maaran, 1984), it entails a more formal approach to natural resource management than multiple use implies. Applied to watershed management, it considers, among others, a broad range of resource values such as wood fibre production, clean air protection, potable water supply, aggregate and gravel production, mining, hydropower development, agriculture, and wildlife production and protection, as well as housing, industry, recreation and tourism. Once the potential of each of these values in the watershed has been calculated theoretically, Van Maaran prescribes a process of ascertaining the public will concerning the balance of these different values. This then becomes the basis for public policy. An important aspect of this process is measuring the public's valuation of outputs not normally traded in the marketplace, such as aesthetics, clean air and informal recreation, so that their value can be adequately and fairly weighted alongside marketable commodities such as timber or hydroelectricity. The final stage in this process is to persuade natural resource managers to adopt a multiple-objective regime that works in the public interest.

As a deliberate natural resource management strategy, an attempt is made to apportion the watershed's natural resources based on well-articulated and publicly endorsed objectives. Such allocation sometimes necessitates the spatial zoning of land and water (see the case study at the end of this chapter) and at times requires temporal zoning such as allowing generally unabated recreational use of the forests until scheduled for harvesting. Multiple-objective resource management sometimes involves the integration of resource management objectives in both space and time, such as when cattle grazing and recreational use of rangelands occur simultaneously.

While in theory and on the surface such management approaches are attractive, Rees (1990) offers a rather scathing criticism of this approach as it is often practised. She argues that multiple-objective resource management is conceived fundamentally as a process for augmenting resource flow decisions that enhance conservation. Rees contends, however, that the 'economic net benefit maximization' concept implied in multiple-objective theory:

- relies too heavily on abstract models of perfect competition that do not work well in practice;
- suggests that recreation and aesthetic values prevalent today will persist into the future, which cannot be supported by empirical evidence;
- argues that discounting the practice of translating future economic values into present-day values unfairly skews decision calculations to favour consumption in the present rather than conserving resources for the future;
- translates all decision calculations about resource values into monetary terms, which favours those who have the ability to pay; and
- reserves recreational and wilderness space for future generations unfairly and penalizes those under-classes who might benefit today if, for example, present-day supply of resource commodities was increased, with a resulting drop in prices.

In practice, Rees argues that multiple-objective resource management regularly is reduced to a process of natural resource exploitation where the dominant resource extraction activity takes precedent over those resource values such as clean air, potable water and aesthetics that cannot easily be quantified or traded in the marketplace.

Sustainability and Ecological Modernization

This section considers the relatively recent rise in importance of sustainability in shaping public policy, and relates this discussion to the evolution of different approaches to sustainable resource and environmental management, and the formalized consideration of intergenerational equity.

Concepts of sustainability

The Brundtland Report defined sustainable development as 'development that meets the needs of the present without compromising the ability of future generations to meet their own needs' (World Commission on Environment and Development, 1987, p. 43). According to O'Riordan, sustainability can be considered 'most conveniently' as the 'replenishable use of renewable resources' where the 'rate of 'take' equals the rate of renewal, restoration or replenishment' (cited in Turner, 1993, p. 43). Although O'Riordan suggests that 'it is tempting to dismiss the term sustainable development as an impossible ideal . . . [however,] the phrase has stuck. . . . Like it or not, sustainable development is with us for all time' (cited in Turner, 1993, p. 37). Turner, who along with several colleagues examined the variation in use of this term among various jurisdictions throughout the world, suggests that sustainable development has been used to categorize a wide range of resource management approaches that reflect very different worldviews of sustainability. Turner arranged these approaches into a four-category typology.

It must be emphasized that none of these types is well represented on the ground at the present time. The theoretical basis for these various types is the 'sustainability inheritance asset portfolio'. This concept incorporates four broad values or asset categories that represent various types of sustainable resource management practices. The assets categories are: (i) manufactured capital (K_m); (ii) natural capital (K_n); (iii) human capital (K_h); and (iv) moral or ethical capital (K_e). In theory, each of these assets can be traded with each other to form an array of assets that creates human wealth and well-being.

More recently, Ross and Bissix (2000) expanded Turner's (1993) typology to capture the full range of natural resource exploitation practices and sustainable development possibilities found throughout the world (see Fig. 8.2). It is this expanded typology that is outlined here. The most stringent sustainability paradigm is the 'very strong sustainability' paradigm (VSS). The second most rigorous type is the 'strong sustainability' type (SS) and the third is the 'weak sustainability' (WS) paradigm. The least demanding type identified by Turner is the 'very weak sustainability' (VWS) paradigm, which has been renamed in this expanded typology as 'weak exploitation' (WE). The two types added to this continuum by Ross and Bissix are the 'strong exploitation' (SE) paradigm and the 'very strong exploitation' (VSE) type. The latter two, as inferred, were added to reflect the non-sustainability of the vast majority of present-day natural resource management practices. It should be noted that a theoretical key or dividing line in the separation of the sustainable paradigms from the exploitation types lies in the second law of thermodynamics. According to Turner, his fourth type, the VWS type, violates this second law, which recognizes that matter cannot be created or destroyed. From a strictly biophysical perspective then, this paradigm of so-called sustainable development is not sustainable at all.

The very strong sustainability paradigm

This first sustainability paradigm assumes that the global economy has already exceeded ecological limits and concerted effort is needed on a global scale to restore the world's ecological balance. Such a scheme requires strict limits on energy and resource consumption to avoid or minimize further environmental

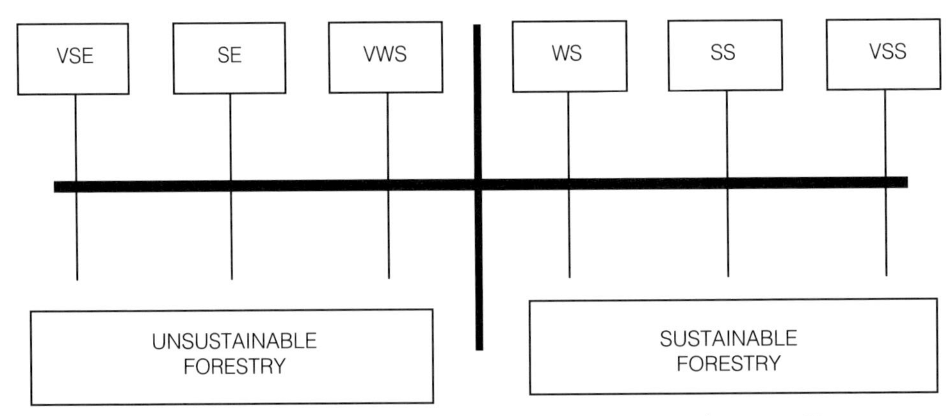

Fig. 8.2. The extended sustainability typology. VSE, very strong exploitation; SE, strong exploitation; VWS, very weak sustainable (or WE, weak exploitation); WS, weak sustainable; SS, strong sustainable; VSS, very strong sustainable.

damage. The aggregation of environmental damage throughout the globe, which gives rise to the need for a very strong sustainability paradigm, is known as the 'scale effect'. This scale or accumulative effect is most evident in the build-up of greenhouse gases, the depletion of the ozone layer and the widespread fallout of acid precipitation. Although the VSS paradigm acknowledges a serious environmental management problem, it is important to note that this paradigm does not presume the end of development but does stress its root meaning that emphasizes qualitative development over growth. Within this sustainable development paradigm, considerably more emphasis is placed on the accumulation of moral, human and ecological capital development as opposed to a concentration on the production of economic assets.

In this context, Trainer (1998) critiques present economic development practices that he suggests are 'totally indiscriminate' (p. 173). The economy as Trainer puts it is unconcerned about what is produced, and argues that much of it is pointless and extravagant. In a moral and equitable world, such a shift to a more ecologically sound developmental process can only be achieved by a substantial reduction of 'environmental' consumption by the rich countries so that the poor have sufficient room to improve their standard of living.

The strong sustainability paradigm

The SS paradigm requires that natural capital in general must be protected and is non-substitutable or tradable with other capital values. While in some circumstances fluctuations in natural capital on a local or regional scale may be acceptable, overall losses in natural capital in one area must be compensated by gains elsewhere. Similar to the VSS paradigm, this type does not in itself argue against development, especially when defined in terms of improved quality rather than quantitatively; it does suggest, however, that development must be separated from environmental degradation or is linked to meaningful gains in restoration ecology. In this circumstance, forestry, for example, is concerned with ensuring that the broad forest landscape improves in scope and vitality even while harvesting continues.

The weak sustainability paradigm

This paradigm's focus is not about the 'preservation of specific attributes of the ecological community but rather the management of the overall system to meet human needs, [to] support species and genetic diversity [but not specific species and particular genetic diversity, that] enable[s] the system to adapt to changing conditions' (Turner, 1993, p. 11). The WS paradigm provides both upper and lower limits to the exploitation of natural assets. The upper limitation specifies the non-substitutability of certain natural capital assets such as keystone species and fundamental ecological processes as well as protecting the assimilative capacity of vital life support systems such as rivers and the air we breathe. Having said this, a substantial amount of natural resource exploitation is tolerated to maintain living standards. This defines the lower limits of sustainable activity. While generally condoning resource exploitation, this approach to sustainable development nevertheless accepts some level of constraint on natural resource economic activity, which also implies limits to population growth as well as to the exploitation of natural resource stocks.

These conservation initiatives are generally consistent with the principle of 'ecosystem stability and resilience'. Turner (1993) suggests that a 'set of physical indicators will be required to monitor and measure biodiversity and ecosystem resilience', which also implies a system of safe minimum standards. Turner also points out that there is presently no scientific consensus over how biodiversity should be measured and hence what safe minimum standards are necessary. This makes the implementation of even this modest level of sustainable development problematic in the real world.

The weak exploitation paradigm

The WE paradigm (formerly VWS) necessitates only that the total amount of natural and human capital remains constant over time. This standard requires society to be as well endowed at the end of any period as it was at the start. It assumes, however, perfect substitutability between the various capital assets such as natural, manufactured, human and moral assets. Within this framework, the free-flow trading of capital assets must be sufficient to offset depreciation, but no particular focus is placed on the conservation of biophysical matter. Transforming a forest, for example, by clear-cutting and processing trees into chopsticks is justifiable as long as the social value of the chopsticks is as great as the value of the standing forest (Bissix, 1999). Although programmes that fit within the WE paradigm are held out to be models of sustainability, they do in fact violate the second law of thermodynamics, which in a practical sense renders them unsustainable.

The strong exploitation paradigm

The SE paradigm sees the short-term exploitation of natural resources as having greater benefit than any foreseeable future value. For example, the clearance of a natural ecosystem and replacing it with a short-lived cash crop system corresponds to the SE paradigm. Historically in Europe, North America, Australia, New Zealand and in Chile, as well as in many developing countries, the preponderance of monoculture plantations, where soil degradation can be safely assumed over the long term, attests to the acceptance of short-term gains as an acceptable trade-off for the degradation in natural capital over the long term (see, for example, Dudley *et al.*, 1995). In Britain and Denmark especially, the farming community has long heralded the output capacity of their agricultural systems but has largely ignored the vast inputs necessary to the system in terms of fertilizer, energy, etc., and the long-term degradation of its farm production and ecosystem potential.

The very strong exploitation paradigm

The most obvious large-scale example of the VSE paradigm is the worldwide destruction of tropical forests (see Dudley *et al.*, 1995). The clear management ideology operating here is that tropical forests are more useful as wood products rather than as fully functioning ecosystems. While poverty, inequitable land tenure, poorly educated peasants and unabated population growth contribute in part to tropical forest exploitation and the destruction of biodiversity, such societal pressures seem inconsequential compared with the damage resulting from the international forest products trade, much of which is conducted unethically if not always illegally (see Dudley *et al.*, 1995).

In general, when considering the representation of natural resource management practices within these various typologies, it is readily seen that the vast majority fall within the exploitation typologies rather than the sustainability types. This results from an overwhelming lack of incentive to reduce, conserve and recycle natural resources and the failure of the marketplace to incorporate the full environmental costs of production. Given the failure of the marketplace to make appropriate adjustments, it is seductive to think that a more substantive role for government and state intervention is necessary. The following discussion regarding the interactive roles of the marketplace and the state (government) in abating environmental damage is provided to make a more informed assessment of this suggestion.

Ecological modernization

Drawing on conclusions from previous discussions, that little if any natural resource management is conducted on a global scale in a truly sustainable way, it seems reasonable to have little hope for future sustainable natural resource and environmental management. Despite this, there is in fact both room for optimism and reason for despair in the way that Western culture has approached environmental management since the 1960s.

Many see the 1960s as the birth of the environmental movement. This was the era when Earth Day was conceived, Rachel Carson's seminal book *Silent Spring* became required reading on many university campuses throughout North America, and local, regional and federal governments began to take note of environmental concerns and initiate action. Governments took direct action on environmental problems; environmental action groups sprung up to address issues of local and national concern, and the age of so-called 'end-of-pipe' environmental clean-up strategies was born. End-of-pipe terminology refers to the set of government regulations and industry strategy used to mitigate environmental problems after their creation. Typically agencies or departments of the environment focused on remediating specific pollutants generally using command and control measures, or other regulatory strategies. For example, methods were utilized to regulate and reduce particulate matter emissions in industrial smokestacks, industrial waters were purified before release into river systems and, later, catalytic converter systems were required and added to automobile exhausts to reduce toxic air emissions. Most end-of-pipe strategies shared a common concern. The dominant theme was to deal with the problem after its creation rather than deal with the pollution production problem head-on. Many end-of-pipe strategies suffered the common concern of media transfer. Solutions directed at one form of pollution abatement such as that from smokestacks merely transferred the problem to create a landfill conundrum and, in turn, a water pollution concern. The net gains in environmental quality consequently were disappointing.

In the late 1980s, and throughout the 1990s, a different approach to environmental management was superimposed on the pollution problem. It was during this period that the global scale of the environmental problem was beginning to be better appreciated and, as a consequence, new approaches to environmental management were deemed necessary. A more integrated resource and environmental management approach was conceived whereby the pollution problem was considered at the start of the production design process and the goal of elimination or reduction was built into the complete production and product use process. Such an approach had a number of important benefits. Such comprehensive and integrative thinking and action began to lower the level of pollution per unit produced; it also reduced the overall amount of energy consumed, lowered the amount of raw materials necessary and increased, in some instances, profit margins. Instead of pollution abatement always being considered as a cost centre, it was now a possibility to consider environmental management as a profitable corporate strategy. Sawmills, for example, looked to ways of utilizing sawdust and woodchips as an energy source rather than mounding up unsightly piles that eventually created leachate problems. Woodchips were sold to pulpmills and utilized in the pulping process, and a new generation of saws, using harder thinner blades, created less waste and increased production per harvested tree.

Decoupling development from environmental degradation

Given the advances in production efficiency and environmental management at the level of the firm, it is possible to conceive of ways to combine these improvements at the local, regional, national and international levels to maintain growth and at the same time reduce pollution. This is the goal of decoupling development from environmental degradation. Pearce (Turner, 1993), for example, argues that it is possible to 'decouple' the seemingly inseparable connection between economic growth and increased environmental decline and cites the reduction in energy requirements of a number of OECD (Organization for Economic Cooperation and Development) countries that accompanied real growth in gross domestic production (GDP) between 1970 and 1987. Pearce describes several policy mechanisms useful for stimulating sustainable development and this process of decoupling development from environmental degradation. He specifically focuses on various incentive systems, pricing mechanisms and fiscal policies, as well as information and education programmes.

Market and State Functions

Unfortunately, attaining significant and consistent gains in environmental management is not as easy as we might hope. To gain a better appreciation of this, it is useful to examine the interlocking roles of the market, the state and civil society in bringing about a more environmentally just society. This overview focuses on the phenomena of market failure and state failure.

Market failure

The suggestion has already been made in this text that the market does not properly account for environmental pollution by failing to include the full social and environmental costs of pollution in the pricing structure. As a result, according to Janicke, the market has 'manifest imperfections' as society's primary socioeconomic steering mechanism (Janicke, 1990, p. 15). Such failures have a number of knock-on effects. The first is that outdated industries are able to produce goods at a profit by externalizing the full costs of production to society at large. This means that new industries or old industries with improved industrial processes are unable to compete reasonably with and eventually replace these outmoded production methods. Secondly, a recurring feature of established industries is their propensity to capture public sector subsidies at the expense of innovation and sustainable development. In Britain, prior to the Thatcher era, for example, the coal industry was substantially over-subsidized. This led to a largely obsolete energy industry spewing unnecessary air pollution. On two counts, therefore, this industry did not properly reflect the true costs of production through environmental externalities and subsidies, and as a consequence endured way beyond its usefulness to society.

Two other factors work against the market as a leader in environmental management. First, most business cycles operate on a 3–5 year cycle and are highly sensitive and responsive to shareholders' short-term investment goals. Unfortunately, most pollution abatement strategies require much longer investment periods before measurable and significant benefits to shareholders are accrued. Secondly, businesses anticipate resource supply problems and deal with long-term natural resource supply management problems ineffectively. In the North Atlantic fishery, for example, the industry chose to ignore the early signals of fish stock crises and instead chose to invest heavily in more efficient fishing gear and respond to increasing demand. Normally investment in more efficient equipment provides an environmental premium, but in this case the ground fishery was exploited to almost the very last fish. In the absence of adequate state intervention, and in this case international oversight, a tragedy of the commons (see Chapter 2) emerged where all participants were concerned with taking what they could while stocks lasted.

State failure

Given that the market is largely incapable of or unwilling to manage its own excesses, this calls for a politic of natural resource exploitation constraint and state stewardship over the environment. According to Janicke (1990), there are two key functions of the state. There are the services it provides, firstly, in advance of, and, secondly, as a consequence of industrial development. Many of these functions result directly from market failure rather than being the consequence of proactive state initiatives. Primarily the state has a regulatory function, a legitimization role, an infrastructure production function and a nuisance abatement role.

Often the government (the state) takes the blame for the failure of industry. This is largely a consequence of the lack of accountability of the market and the fact that there is no central authority to account for or accept responsibility. In response to market failure, the state will expand its regulatory function, but in doing so often oversteps its financial capacity. To address such shortages, it will often encourage industrial expansion to increase the tax base, thus encouraging more environmental degradation, and inviting additional criticism of its regulatory and environmental stewardship role. This cycle of environmental problems, state expansion and market expansion theoretically continues until, for example, a taxpayer revolt forces a rollback in government services. This can have catastrophic consequences for an expanded and unregulated market such as the failure to properly inspect drinking water supply, adequately scrutinize environmental performance generally, and, for instance, to properly certify seaworthiness of tankers, which may lead to massive oil spills.

Globalization

Despite various declarations at the United Nation's Conference on Environment and Development (UNCED) held in Rio de Janeiro in June 1992, the follow-up New York City conference in 1997, and the recent World Summit on Sustainable Development in Johannesburg in 2002, there is substantial evidence to suggest that the environment has been losing ground on a global scale to natural resource exploitation and creating net environmental degradation. Environmental advances in 'per unit' production of commodities and services have been dulled by an overall and worldwide increase in production, increases in per person consumption, and increases in total environmental damage. Given the great concern expressed by world leaders at these and other key global conferences, it is at first difficult to see how this might be. To a large extent this can be explained by the success of those driving for the liberalization of international trade.

Globalization and free trade

In itself, the lowering of trade barriers is not necessarily a driving force for increased environmental degradation. However, in the campaign to reduce so-called arbitrary protectionist strategies used by some countries, other countries such as the USA have pressed, through conventions like the Multilateral Agreement on Investment (MAI), for the removal of measures that might be construed as an impediment to free trade. In many instances, this has meant the dismantling of rules and regulations designed to protect the environment, and in others it has resulted in the removal of workers' rights and protections. To add to these extensive courses of action, there has been a rather zealous push to increase personal consumption in the quest for creating jobs theoretically to raise the standard of living. This pressure to increase world trade has also increased the overall demand for natural resources, increased environmental degradation as a result of increased extraction, processing, manufacturing and use (e.g. automobiles), and reduced the net progress towards sustainability.

The global anti-globalization movement

The world's largest corporations and banks, which seemingly stand to gain the most, have largely championed the agenda promoting free trade. There has been in recent years, however, a counter global movement. This is perhaps one of the classic examples of pluralist, grass-roots power (see Chapter 7) to shape public policy anywhere. It is a movement largely spawned on the World Wide Web. It involves a growing coalition of human rights activists, environmentalists, the labour movement, ordinary citizens and even peasants from developing countries who would not normally have a voice in public policy development. On occasion, this movement has taken to the streets to oppose the corporate agenda, and return, as they argue, decision making on important public policy issues from the corporate boardrooms, which takes place behind closed doors, to more transparent forums of national and regional legislative assemblies. The key to this movement's effectiveness has been the rapid dissemination of information that often counters unsubstantiated or disputable claims made by the most powerful elites. It side-steps equally powerful and concentrated media outlets, and makes transparent what would otherwise be hidden from the public eye. The thrust of their campaign suggests that what is good for international corporations is not necessarily good for either the people or the environment.

Good Governance and Sustainable Development

In the United Nations 'Agenda for Development' (1994), good government implies 'the wisdom and the historical responsibility to know when to let the market forces act, when to let civil society take the lead and when government should intervene directly' (cited in Ginther *et al.*, 1995, p. 4). The Graz (Austria) Seminar on Sustainable Development, Human Rights and Good Governance in 1994, also stated that:

> Good governance, meaning sustainable management of resources in legal, institutional, political, economic, social and ultimately cultural terms, is difficult to translate into a concrete operational program for the world as a whole.
>
> (Ginther *et al.*, p. 9)

The gross domestic product

One notable impediment to translating such a declaration into good governance is the trading world's focus on each nation's GDP as the primary indicator of national success. The GDP, which measures the total economic activity of a nation, was originally a stop-gap, wartime measure to monitor and stimulate production in the Allied war effort. This measure has endured, however, and has evolved to a measure or proxy for a nation's overall quality of life. There are, however, substantial and misleading notions regarding such use. For example, should a nation or region endure an environmental calamity such as a massive ice storm, hurricane or drought, then the remedial measures undertaken to compensate for and repair this damage, usually at great economic cost and human sacrifice, is included as a positive in the calculation of the GDP.

While most rational individuals would agree that the nation or region is likely to be worse off compared with the situation had it never experienced such a calamity, the GDP calculation nevertheless suggests otherwise. As a result of the remedial work carried out to clean up after the environmental incident, the GDP shows an increase as a result of the necessary economic activity. Such activity is above and beyond what might otherwise have been the case and what can be afforded. The chances are that most people directly affected would consider themselves to be worse off than before, and even those in the broader community who helped to compensate for these losses through their taxes would also consider themselves worse off. Despite what might be reasonably considered a lowering of the quality of life for all, the GDP shows a rise. In a similar way, increased exploitation of natural resources to meet increased market demand may result in substantially elevated and perhaps irreversible environmental degradation. Such irresponsible natural resource exploitation leads to a very real decrease in the quality of life; however, the GDP calculation again suggests otherwise.

The genuine progress index

Given that government action and its appropriation of scarce resources is often driven by a simple or summary indicator (the GDP), there is a clear need to dismantle the old scoreboard and replace it with a new one that more clearly and more accurately measures increases or decreases in the quality of life of a nation or region rather than its gross economic activity alone. One such initiative is the *genuine progress index* (GPI). The GPI concept was developed about a decade or so ago, and variations of the GPI have been developed at the theoretical level in a number of locations throughout the world.

The GPI is a complex measure of human wealth and well-being that includes, among others, a jurisdiction's ecological footprint, its employment status, the value of the region's voluntary capacity, its crime and sense of security, its adherence to sustainable development strategies, its population's mental and physical health status, its adherence to sustainable transportation, etc. (Cobb *et al.*, 1995). These are all measures that add up to an index of the quality of life rather than a gross indicator of economic activity. In this index, decreases in environmental quality, i.e. additions to the environmental deficit, perhaps the result of an open pit mining operation, would be registered as a negative factor rather than a positive one. The economic gains of the open pit would of course register as a positive, but only in the context of its environmental and social costs. Money necessarily spent on drug and gambling addiction that often results from failures in the fabric of society would also be calculated as a negative, while the resulting gains in community health as a direct result of such treatment would compute as a positive. The damage resulting from a major windstorm would compute as a negative, while the value added from restoration work would be seen as a positive.

While there are no national or regional jurisdictions in the world yet committed to implementing the GPI concept as a replacement or even as a parallel activity to measure human well-being, GPI-Atlantic, based in Halifax, Nova Scotia, Canada, has applied the theoretical concepts of the GPI to construct a practical application (see http://www.gpiatlantic.org). A pilot programme has been developed, focusing on a rural jurisdiction and a small industrial town to survey their human and environmental condition. Using these data, a coalition of community members, community health and environmental professionals, and academics has combined to provide timely information as a broad GPI and as various and more specific indicators to inform and drive public policy formulation. The goal is of course to demonstrate the usefulness of such an index in driving public policy at the local, regional, national and international levels, hopefully to be adopted by various levels of government to complement or replace the GDP.

Environmental Subsidies and Green Taxes

Changing the scoreboard for measuring human development is likely to be a drawn-out process, given the present widespread and uncritical use of the GDP and the political power of those who directly benefit from its continued use. There are, nevertheless, other faster acting measures that can be utilized more immediately by government to drive a more environmentally concerned agenda. Subsidies have generally been favoured by industry as they do not negatively or directly impact the corporate bottom-line of the target agency. On the contrary, over-generous subsidies can in some instances contribute to windfall profits.

Increasingly acceptable by industry and politically astute for government is the use of green taxes (see Anderson, 1994). However, the implementation of green taxes remains problematic, as it requires a rather substantial philosophical shift among politicians and bureaucrats. Both these groups in the past have favoured command and control policies rather than economic tools to influence public or industrial behaviour. Such radical measures (i.e. green taxes) must be implemented in a political system heavily influenced by those most likely to be adversely affected by such taxes. These are often the large and powerful multinational natural resource corporations who persistently wield a great deal of economic and political power.

Summary and Conclusions

As discussed in this chapter, the practice of natural resource management is as much about its human dimensions as it is about biophysical science and technology. To begin, we saw that the concept of natural resources is very much a human phenomenon; natural things do not become natural resources until humans value them as such. We also saw that attitudes to how we approach natural resource and environmental management are shaped by management ideology and competing worldviews as to how the globe functions and humans' capacity and limitations to manipulate it.

A fundamental tenet of IREM is the desire – a growing political as well as civic support for a sustainable society – to embrace meaningful sustainable development strategies that will deliver on-the-ground results rather than provide shallow rhetoric. We learned from the work of Turner, however, that there are very different conceptions of sustainable development adopted worldwide, and the weakest of these is unable to deliver a sustainable society because it violates the second law of thermodynamics. More disturbing was the assertion that the vast majority of natural resource management practices throughout the world are unsustainable and that there is little, if any, momentum on a global scale to shift to sustainable ways.

From the discussion of decoupling development from environmental degradation, we know that more sustainable natural resource management is possible, but we also saw that the relationship between the market and state, as they relate to resource and environmental management, is an uneasy one. We also saw that their interaction almost invariably finishes up in the failure of both. Finally, we saw that good governance, an effective marriage of the state, the market and civil society, is one driven by appropriate incentives. We have already seen a growing civil society rejection of unabated free trade that appears to disproportionately favour the corporate sector at the expense of average citizens and the environment. We were also introduced to the judicious use of state subsidies and green taxes as ways forward, and the adoption of a more appropriate scoreboard to guide regional, national and international use of natural resources to enhance well-being. Although perhaps a few years away, the widespread adoption of a GPI is perhaps the most exciting development that we can muster as an appropriate policy decision-making framework that will stimulate the widespread adoption of IREM strategies. Once the very real and negative consequences of most natural resource exploitation is accounted for on a continuous and widely transparent basis, the value of IREM practices will be more widely appreciated and increasingly adopted.

Case Study
Kananaskis: Oil and Gas Exploration in Paradise

MEGAN SQUIRES

Preamble. In 1997, the *Calgary Herald* newspaper issued a special report entitled, *Kananaskis: a 'country' at a crossroads*. In it, author Bruce Masterman wrote, 'paradise is under pressure' and he suggested that the inevitable *tug of war* between development, resource extraction, conservation and recreational interests had raised concerns about the future of Kananaskis Country. In this case study, Megan Squires provides a glimpse into the complex world of resource management in Canada in one of Alberta's most spectacular regions by offering examples to illustrate the challenges that decision makers face in trying to manage resources and provide opportunities for an increasingly diverse range of interests.

Introduction

Kananaskis Country is located in the Eastern Slopes region of the Canadian Rocky Mountains, covering an area of approximately 4160 km^2. Situated on the western border of Alberta, the area known as Kananaskis Country is home to alpine and subalpine environments, native grasslands, deciduous and coniferous mixed wood forests, and glacial streams and rivers. Resource planning and management in Kananaskis Country is implemented through a system of IREM – a system that has been in existence since the area was first established in 1977. Within this integrated framework, planners and managers work to provide opportunities for a broad spectrum of interests including environmental protection and conservation, outdoor recreation, mineral exploration and development, rangeland activities, energy development and forestry.

A fisherman on Upper Kananaskis Lake. Note the dam controlling flow to the Kananaskis River and subsequently to the Bow River. The Bow River flows through the city of Calgary. Photo by Glyn Bissix.

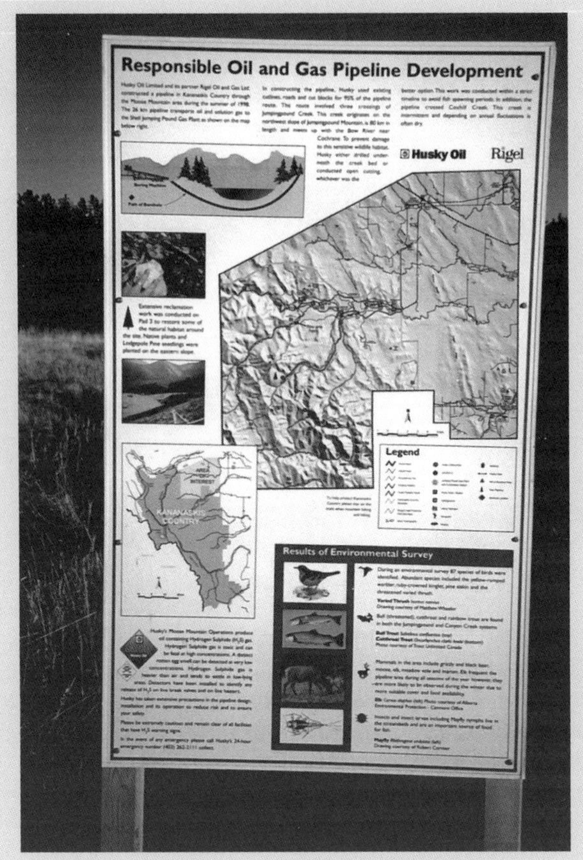

An oil company interpretative sign in Kananaskis Country emphasizing the importance of multiple-use resource management. Photo by Glyn Bissix.

Kananaskis Country is divided into five management areas: Kananaskis/Spray; Upper Elbow/Upper Sheep; Elbow/Jumping Pound; Sheep/Threepoint; and Highwood. A zoning scheme is applied within each area, and the designations, in order from the highest level of environmental protection to the lowest, include: prime protection, critical wildlife, special use, general recreation, multiple use, industry and facility.

Background

The Elbow/Jumping Pound resource management area is located along the eastern boarder of Kananaskis Country. It is part of the foothills ecoregion and contains the greater Elbow River and Jumping Pound Creek drainage. The main priorities for management within the Elbow/Jumping Pound area include the following:

- Maintenance of water quality and quantity
- Provision for recreational opportunities
- Provision for mineral resource development

- Maintenance of fish and wildlife resources for conservation and recreation
- Ecological and archaeological protection (Alberta Forestry Resource Evaluation and Planning Division, 1986, pp. 55–56)

The Elbow/Jumping Pound area contains four management zones: (i) prime protection; (ii) critical wildlife; (iii) general recreation; and (iv) multiple use. In conjunction with the identified zones, management objectives and guidelines are also required to ensure that opportunities for recreation, oil and gas development, ranching and environmental protection are maintained and enhanced.

In 1994, Grey Wolf Oil initiated a licensing application to the Provincial Resource Development Board to gain permission to drill three wells from three different surface well locations within a 6400 ha area located near Magpie Ridge in the Elbow/Jumping Pound region of Kananaskis Country. Concurrent with this application, Grey Wolf also submitted an application to Kananaskis Country for approval to develop access roads and well sites. Grey Wolf's proposed application generated a considerable amount of public interest among recreational users, ranchers, environmentalists, First Nations groups and surrounding communities. As a result of the overwhelming concern and in light of the significant opposing evidence, the Resource Development Board ordered a public hearing to review the merits of Grey Wolf's application.

The Current Issue

In 1924, the Federal Department of the Interior initiated the transfer of responsibility for mineral resources to the Province of Alberta. A 'wet gas' discovery at the Turner Valley Royalite No. 4 sparked immediate interest in mineral development, and it forever changed the face of oil and gas exploration in Alberta. However, it was not until 1930 that the province gained formal control over all of its mineral resources (Warne, 1981, p. 33). Once in charge, the provincial government and industry representatives were under pressure to devise and maintain a legislative and regulatory framework to ensure that mineral exploration and development would be efficient as well as consistent with the present and future needs of Albertans. Responding to this pressure, the Provincial Resource Development Board (PRDB) was established in 1971 to assist with legislative and regulatory responsibilities. Through its evolution, the PRDB has successively broadened its mandate to include regulation for all energy resources in Alberta including oil, natural gas, oil sands, coal and electrical energy, as well as the pipelines and transmission lines that transport energy to market. As part of its regulatory function, the PRDB handles, processes and issues decisions on new or amended applications for energy activities throughout Alberta. A fundamental aim of the process is to ensure that energy development activities are conducted in a manner that is consistent with the public interest.

In 1992/1993, Grey Wolf Oil acquired the rights to drill a discovery well in the Magpie Ridge area in Kananaskis Country. As a result of that well, Grey Wolf confirmed the presence of hydrocarbons. However, the company's inability to perform adequate testing to determine the extent of the well's productivity was the basis for this current application. Grey Wolf representatives argued before the PRDB that there was a significant need for five additional wells in order to determine adequately whether further development of the Magpie field was necessary. In theory, by the time that an application is presented in front of the PRDB, applicants should already have assessed the economic, social and environmental impacts of their proposed development. The pre-hearing preparation is viewed as an important step to ensure that companies such as Grey Wolf Oil establish their credibility and their commitment to the long-term interests of sustainability in the area where they are proposing to drill.

At its application hearing in 1995, Grey Wolf acknowledged its uncertainty regarding the commercial productivity of the proposed wells. The company argued that without clear documentation of the drilling potential, it was neither economically feasible nor procedurally efficient to initiate a detailed environmental assessment of the impacts that might ensue if extensive development were to occur. While Grey Wolf recognized the gaps in its understanding of the environmental factors, the company stated upfront that it would commit to a full-scale environmental assessment if and when the proposed wells were proven to be productive. In the assessment, Grey Wolf would address issues related to wildlife and aquatic habitats, recreational use, public safety and rangeland productivity. Prior to the hearing, Grey Wolf Oil undertook a public consultation process in order to present the details of their application and to hear the issues and concerns of key stakeholders. The elements of their consultation programme included research, distribution of an information circular, exposure through various forms of media (television, radio and newspapers), and meetings with local communities and interest groups.

The Interest Groups

Recreation

Recreation in the Elbow/Jumping Pound managing area represents a diverse range of activity groups including hiking, camping, fishing, mountain biking, horse riding, dirt biking, off-highway vehicle use and hunting (Alberta Forestry Resource Evaluation and Planning Division, 1986). The characteristics of the different recreational users including their distinct patterns – type, timing and frequency of use as well as group size and equipment requirements – are fundamental in developing an understanding about the motivations and expectations that different users have for their recreation experiences. Likewise, recognition of the characteristics of various recreational groups can provide insight into different users' levels of understanding about IREM and multiple use and the presence of oil and gas development in Kananaskis Country. Recreational representatives attending the Grey Wolf Oil hearing were tasked with the complex challenge of presenting a unified account of the issues despite concerns that many diverse and often conflicting opinions existed within the recreation community. Table 8.1 provides an overview of the recreational interest groups and their issues regarding Grey Wolf's proposal.

Environment (aquatic and terrestrial habitats)

Context plays an important role in understanding how and why certain decisions are made. Oil and gas exploration and development have long been the driving force behind Alberta's robust economy. In fact, it is widely accepted that Alberta's social, environmental and economic well-being is inherently linked to the price of oil. Therefore, it stands to reason that oil and gas companies in Alberta maintain a significant amount of power in the decision-making process. However, despite what may appear to be an uneven playing field, environmental groups have continued to fight for tougher environmental legislation, regulation and enforcement in Alberta. Presenting information before the PRDB was a welcome opportunity for these groups to give voice to the hitherto unspoken amenity values of natural environments. Several representatives from the environmental community presented information at the Grey Wolf Oil hearing. Among their concerns, the representatives expressed regret for the level of degradation that had already compromised the ecological integrity of Kananaskis Country, and specifically the Elbow/Jumping Pound management area. The environmental groups also communicated their concerns about the 'big picture' issues and the impacts of development on regional resources. A complete listing of the environmental community's concerns is documented in Table 8.1.

Table 8.1. Summary of interest group characteristics and issues.

Interest groups	Characteristics	Issues with proposed oil and gas development
Oil and Gas (Grey Wolf Oil)	Direct: Grey Wolf Oil (primary licence applicant) Indirect: Oceanic Oil (also maintain an oil and gas lease in Kananaskis Country)	Uncertainty about the commercial value of the well development, concern about the loss of revenue that the companies will suffer if the well application is denied.
Recreation	Active: hiking, camping, fishing, mountain biking, horse riding, dirt biking and off-highway vehicles Hunting: large carnivores (bears, wolves, cougars), ungulates (moose, elk, sheep, deer) and game birds (grouse and ptarmigan)	Access, future opportunities, compromised amenity values (the spiritual and emotional attributes of the landscape), crowding, conflict, and diminished quality of experience for both local and tourist populations. Potential impacts to game and habitat, and diminished wilderness experience.
Environment (terrestrial and aquatic conservation)	Non-profit environmental advocacy groups, interested public	Inadequate environment assessment outlining: impacts to sensitive wildlife populations and habitat (ecological diversity), monitoring strategies to access cumulative effects, the impacts of proposed surface disturbances (access roads and well pads), and possible impacts to habitat resulting from a sour gas leak.
Rangeland	Local stock association	Range quality and livestock health and productivity.
First Nations	Native reserve bordering Kananaskis Country and the Elbow/Jumping Pound Region	Spiritual values, traditional hunting practices, involvement in the pre-hearing public consultation process.
External communities	Residential and commercial communities bordering Kananaskis Country and the Elbow/Jumping Pound Region	Increased traffic through Bragg Creek, trans-boundary effects (air and water pollution), diminished tourist market resulting from poor visitor experiences in Kananaskis Country.
Government	Alberta Environment (water issues); forestry, lands and wildlife; parks and protected areas; Alberta Energy	Development that is consistent with: regional policy, the IRP (integrated resource plan) for Kananaskis Country, subregional management objectives; lack of interagency coordination.

Rangeland

Domestic grazing has been an active component of the Elbow/Jumping Pound management area for over 40 years (even before Kananaskis Country was formally established). Within the area, there are four complete grazing allotments and portions of two additional allotments (Alberta Forestry Resource Evaluation and Planning Division, 1986). Ranchers tend to view themselves as 'stewards of the land' whose job it is to maintain and to enhance land resources through sound management. In the event that terrestrial or aquatic resources were to be damaged by oil and gas developments, range owners would be forced to consider not only the environmental impacts and the long-term health of their cattle, but also the economic investment that would inevitably be lost. In the past, ranching interests expressed concerns about being ignored in decision making. Collectively, they represent a small number amongst the broad range of stakeholders in Kananaskis Country. As a result, many ranchers believed that despite the legitimacy of their concerns, their interests had been ignored in past decision-making processes. The PRDB hearing was an opportunity to express opinions and to discuss the potential impacts of mineral development on rangeland health and cattle productivity. The issues raised by rangeland representatives are highlighted in Table 8.1.

First Nations

First Nation representatives issued a formal request for an adjournment prior to the Grey Wolf hearing. Their rationale for the request was to give affected aboriginal groups an opportunity to study the impacts of the well application on their treaty and aboriginal rights. First Nations indicated that an investigation into the impacts of development on native spirituality and hunting rights in the Magpie Ridge area was needed and that Grey Wolf Oil should have addressed these issues during their public consultation process. First Nations groups intended to use the hearing process as a forum for discussing aboriginal rights to natural resources. The issues that First Nations groups brought to the table are also listed in Table 8.1.

External communities

Individuals from several surrounding communities attended the Grey Wolf Oil hearing. The community representatives indicated that trans-boundary issues were the primary reason for their participation. The representatives stated that the proximity of their communities to the Elbow/Jumping Pound area raised real concerns about public safety, water quality, public access and economic development through tourism as a result of the proposed oil and gas development. Again Table 8.1 documents the external communities' concerns.

Provincial government

The provincial government's role in oil and gas development within Kananaskis Country is to ensure that drilling is consistent with legislative and policy frameworks, official planning documents, and regional and subregional management objectives. Government interests are represented through multiple agencies. Also, the provincial government is responsible for considering the diverse interests of recreation, industry, ranching, First Nations and environmental protection as part of its commitment to multiple use.

Conclusions

IREM and multiple-objective resource management are tools, i.e. approaches designed to aid resource planning and management. Their role is to guide and direct planners and managers in their efforts to make decisions amidst the increasing complexity of demands from various interest groups in natural resource management. This complexity is illustrated in the Grey Wolf Oil application. Based on evidence presented at the PRDB hearing, it is clear that Grey Wolf's proposal for oil and gas development in the Magpie Ridge Region of Kananaskis Country has significant consequences for a broad range of resource interests: recreation, conservation, rangeland, First Nations, external (or surrounding) communities, and provincial government agencies (see Table 8.2). Without question, the PRDB faces an enormous challenge in reaching a decision about the proposed development.

Table 8.2. A summary of opportunities and challenges for Maple Ridge.

	Opportunities	Challenges
Recreation	Improved access for recreation to areas in the Elbow/Jumping Pound Region due to road construction and subsequent reclamation; greater likelihood that Grey Wolf will partner with land management agencies and interest groups to help maintain and enhance resource opportunities; and an increased potential for the public to learn about industrial operations and to become involved in the monitoring and regulation of industrial resource use.	Loss of access for recreation to areas within the development zones; health and safety risks for human and livestock populations resulting from water and soil contamination and air pollution; and compromised spiritual and amenity value of the area for visitors.
Environmental	Industrial development in the Elbow/Jumping Pound Region might ensure that: (i) companies comply with environmental regulations; (ii) regulating bodies are more consistent in enforcement of regulations; and (iii) public support for environmental and social advocacy increases as more people become aware of the impacts of industrial development on natural resources, wildlife and people.	Industrial development in the Magpie Ridge area might result in: (i) a loss of critical habitat for wildlife and aquatic populations; (ii) permanent damage to terrestrial and aquatic ecosystems; (iii) water and soil contamination; and (iv) air pollution.
Economic	Oil and gas development can contribute positively to Alberta's economy, which can result in increased provincial spending on health, education, environment and arts and culture. It can also result in job creation.	There are significant costs associated with the development application process. If the development proves unsuccessful, companies may find it difficult to recover their costs. Local communities could suffer losses resulting from diminished real estate values and the tourism industry could lose business because visitors may choose to travel to other locations, where development is not permitted. Similarly, environmental damage could prove costly for companies in the event of a leak or spill.

Discussion Questions

1. What does the term resource management mean? By using an example, illustrate how the term resource management differs in meaning from that of resource development?
2. Can leaving an area to natural forces be considered a form of resource management or resource development? If so, why?
3. Give an illustration as to how natural matter can have its value altered by advances in technology.
4. Using examples from your home region, describe where you believe *technocentric* thinking has dominated resource management and where *ecocentric* ideology has prevailed?
5. Explain why no truly rational solution can be developed for the so-called 'tragedy of the commons'.
6. According to Van Maaren, multiple-objective resource management requires the systematic assessment of a natural resource to deliver an optimal combination of outputs from that resource. Why does Rees argue that this is difficult or impossible to do in practice?
7. Turner's *very weak sustainability* (VWS) typology is said to violate the second law of thermodynamics, which states that matter cannot be created or destroyed. In relation to the various forms of capital, such as manufactured, natural, human and moral capital, what trading between these various forms of capital would infer a creation or destruction of matter?
8. The United Nations 'Agenda for Development' argued that we should 'know when to let market forces act, when to let civil society take the lead, and when government should intervene directly'. Give an example where market forces have worked well, where government intervention has been successful, and where civil society has provided strong leadership in resource and environmental management.
9. Give a specific example as to how environmental degradation might be usefully decoupled from development.
10. It is said that as market activities expand and production externalities increase, the need for government or state intervention increases. Provide an example of where civil society might be more successfully involved to reduce the pollution excesses of the marketplace.
11. In Western countries, it is common to find so-called gas (petrol) wars where the price of a litre or gallon of fuel changes quickly in response to local market conditions. List three difficulties policy makers might have in setting an appropriate size for a green tax designed to reduce environmental degradation under such quickly fluctuating market conditions.

Case Study Discussion Questions

12. Imagine you are a member of the PRDB and you are overseeing the public hearing in the Grey Wolf application for well licences in the Magpie Ridge area. Which criteria would you select to evaluate the information presented in this case?
13. In the event that the Grey Wolf application is approved, what are the potential sustainable development impacts including economic, social and environmental impacts of this decision?
14. IREM is an approach that recognizes that the use of one natural resource can affect the management and use of other resources. In many cases, the goal of IREM is to optimize resource use in order to achieve the maximum benefit (Alberta Forestry, Lands and Wildlife Resource Planning Branch, 1991). In accordance with this definition, what are the benefits of an integrated approach in Kananaskis Country? What are the most significant challenges?

References

Alberta Forestry Resource Evaluation and Planning Division (1986) *Kananaskis Country Sub-regional Integrated Resource Plan*. Queen's Printer, Edmonton, Alberta, Canada.

Alberta Forestry, Lands and Wildlife Resource Planning Branch (1991) *Integrated Resource Planning in Alberta*. Queen's Printer, Edmonton, Alberta, Canada.

Anderson, M.S. (1994) *Governance by Green Taxes: Making Pollution Prevention Pay*. Saint-Martin's Press, New York.

Bissix, G. (1999) Dimensions of power in forest resource decision-making: a case study of Nova Scotia's forest conservation legislation. Unpublished doctoral dissertation, Department of Geography and Environment, London School of Economics and Political Science, London.

Bissix, G. (2002/2003) Residual recreation and sustainable forestry: historic and contemporary perspectives in Nova Scotia. *Leisure/Loisir: the Journal of the Canadian Association of Leisure Research* 27(1–2), 31–50.

Brown, P.G. (2001) *The Commonwealth of Life: a Treatise on Stewardship Economics*. Black Rose Books, Montréal, Québec, Canada.

Cobb, C., Halstead, T. and Rowe, J. (1995) If the GDP is up, why is America down? *The Atlantic online*. Available at: http://www.theatlantic.com/politics/ecbig/gdp.htm Retrieved 18 December, 2002.

Dudley, N., Jeanrenaud, J.P. and Sullivan, F. (1995) *Bad Harvest? The Timber Trade and the Degradation of the World's Forests*. Earthscan, London.

Essex, C. and McKitrick, R. (2002) *Taken by Storm: the Troubled Science, Policy and Politics of Global Warming*. Key Porter Books, Toronto, Ontario, Canada.

Ginther, K., Denters, E. and De Waart, P.J.I.M. (eds) (1995) *Sustainable Development and Good Governance*. Martinus Nijhoff, Dordrecht, The Netherlands.

Janicke, M. (1990) *State Failure*. Pennsylvania State Press, University Park, Pennsylvania.

Mitchell, B. (1979) *Geography and Resource Analysis*. Longman, London.

O'Riordan, T. (1981) *Environmentalism*, 2nd edn. Pion, London.

Rees, J.A. (1990) *Natural Resources: Allocation, Economics and Policy*, 2nd edn. Methuen, London.

Ross, M. and Bissix, G. (2000) Extending Turner's spectrum of sustainable development typologies: application to global and Canadian forest management practices. *Proceedings of the 8th International Symposium on Society and Resource Management*. University of Western Washington, Bellingham, Washington.

Sandell, K. (1998) The public access dilemma: the specialization of landscape and the challenge of sustainability in outdoor recreation. In: Sandberg, L.A. and Sörlin, S. (eds) *Sustainability: the Challenge: People, Power and the Environment*. Black Rose Books, Buffalo, New York, pp. 121–129.

Trainer, T. (1998) *Towards a Sustainable Economy: the Need for Fundamental Change*. Jon Carpenter, Oxford, UK.

Turner, R.K. (ed.) (1993) *Sustainable Environmental Economics and Management: Principles and Practice*. Belhaven Press, London.

Van Maaren, A. (1984) Forests and forestry in national life. In: Hemmel, F.C. (ed.) *Forest Policy: a Contribution to Resource Development*. Martinus Nijhoff/Dr W. Junk Publishers, The Hague, The Netherlands, pp. 1–19.

Warne, G. (1981, January/March) A history of energy resource regulation in Alberta. *Journal of Canadian Petroleum Technology* 20(1), 33–34.

World Commission on Environment and Development (WCED) (1987) *Our Common Future*. Oxford University Press, Oxford, UK.

9 Theoretical Framework for IREM

Introduction

The theoretical research on IREM can be generally divided into two areas: (i) *normative* research that defines the essential elements of IREM; and (ii) *descriptive* research that establishes criteria and the context for IREM to work as a successful integrated planning process. The normative literature defines what IREM 'should be', as an ideal process, and outlines elements or components that are necessary to characterize this type of planning. The descriptive research attempts to identify what it takes for the effective implementation of an operational model of IREM.

This chapter will examine five elements from normative research that define the important processes that need to be included in planning and decision making for IREM. The second half of the chapter deals with how to improve the IREM process and identifies strategic areas for increasing the performance of this decision-making process. A case study of the Cache River watershed in southern Illinois provides an example of the complexity in integrating interests to plan at a watershed level.

Normative Characteristics: What it Takes to Make IREM

The normative conceptualization of IREM has evolved over the past two or three decades and continues to change, but has generally focused on the substantive and procedural aspects of the process. The 'substantive' refers to the *what*, or the descriptive issues associated with resource management such as the scope and scale of application, and the application of scientific methods, or what is included in the planning process. In other words, how we divide up the resource pie: in essence who gets what and how much. The 'procedural' aspects refer to the process of *how* IREM should be undertaken. The answer to how IREM is conducted centres around what is included in the decision-making process and how they are included. It must be noted that this conceptual division between the substantive and procedural components of IREM is an artificial distinction because frequently process and substance are inter-related and cannot easily be divided in practice or at ground level.

In particular, authors have sought to develop a definitional template for the integration process (Mitchell, 1986; Born and Sonzongi, 1995; Margerum, 1997) that identifies the general characteristics that are required to make IREM 'work'. One of the more useful conceptual frameworks for defining the normative components of IREM is provided by Born and Sonzongi (1995). They divide the core elements into four major divisions:

- Comprehensive/inclusive
- Interconnective
- Strategic/reductive planning
- Interactive/coordinative

The first three elements define the substantive aspects of IREM, while the fourth component defines the procedural requirements. We have added a fifth building block, *'Holistic'*, to this framework that we have borrowed from Margerum's (1997) evaluation of the IREM process. Figure 9.1 provides a conceptual overview of these five inter-related characteristics of IREM.

Comprehensive/inclusive

This composite element refers, first, to the extensive breadth of issues and considerations that are to be included (as much as possible) in the scope of the planning system. Secondly, *comprehensive* refers in more detail to the process of ensuring that the biophysical and social components associated with IREM are integrated appropriately into the initial assessment of the evaluation framework. Identifying the suitable boundaries and scale, for example, is integral to this process. In this regard, Beanlands and Duinker (1983) identify the following five crucial boundaries for comprehensive scoping in environmental assessment: *ecological*, *administrative*, *technical*, *cultural* and *temporal*. The complexity of these boundaries and the degree to which they are considered in effect delimits the realm of inquiry, but also effectively defines spatial and political concerns.

The act of being *inclusive* complements the action of comprehensiveness in that it incorporates a wide range of views, adopting multiple perspectives, and a breadth of scope to include a depth of both social and biophysical functions in the planning process. This approach takes in a broad view of human issues and ecosystem functions as an initial step in decision making.

Interconnective

One of the important tenets of IREM is that relationships among different biophysical and socioeconomic processes are recognized and integrated into the

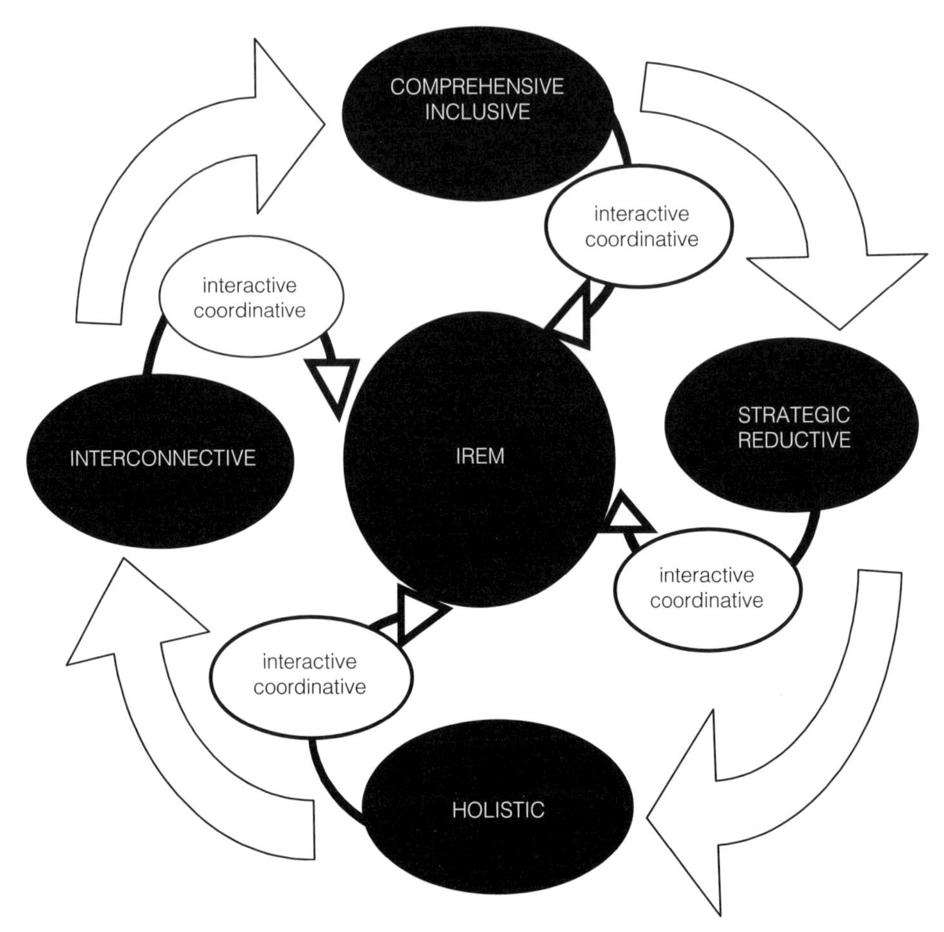

Fig. 9.1. Theoretical framework for IREM.

resource management decision-making process. The complexities of the interdependencies of a natural system, such as a mangrove ecosystem for instance, need to be recognized when we consider a management intervention (such as dredging) in that system. While trying to solve one problem, such as navigation along the river (a human dimension), a whole host of ecological problems may be introduced that have more wide-ranging human implications later on. The *interconnections* between natural and social systems need to be recognized. In a similar manner, various social systems impacting the management arena and internal interdependencies within the management system must also be considered. Social linkages that connect complex management systems are as important as the substantive issues being considered. To be effective, IREM must recognize this complexity.

Strategic/reductive planning

Strategic/reductive planning introduces the need for IREM to focus solely on key or pertinent issues that reduce or limit the focus of management action to a specific agenda and desired outcome. This orientation is frequently necessary because needed efficiencies in budget and time often constrain resource managers in their operations. Strategic and reductive elements of IREM, therefore (although seemingly contradictory to the previous two elements), are a required step in effective resource and environmental management. Once the necessary inter-relationships and the scope of analysis are recognized, then subsequent steps require more action-oriented approaches. This necessitates accurate scoping of the problem, such as the requirement to ensure that information is passed throughout the management complex in the required format and in a timely manner. This process is fundamentally an iterative process that changes with organizational learning, evolving external, environmental and organizational conditions, shifting information insights and the passage of time.

Interactive/coordinative

The fourth element of our model provides the procedural guidance for IREM and deals primarily with *how* the process is carried out. Similar to the first two substantive elements, the *interactive/coordinative* aspect of IREM recognizes the complexity of the social dimension but focuses on the diversity of different actors and interests that must be considered in the natural resource management environment. Friedmann's (1987) model of transactive planning provides a theoretical basis for normative planning that incorporates interaction and dialogue as a foundation for the transactive approach. Lang (1986) defines 'interactive' as one of the essential elements of resource planning, which moves it beyond the routines of conventional comprehensive planning. He suggests that interaction should consist of a spectrum of management practices that includes information feedback, consultation, collaboration and negotiation. At one end of the spectrum, information is disseminated and responses retrieved from participants. As managers progress along the spectrum, consultation goes one step further to inform and solicit views. More intense collaboration provides for joint problem solving and direct stakeholder involvement; and at the far end of the spectrum, negotiation resolves conflicting interests and allows for mutually beneficial bargaining amongst parties with initially competing views.

Margerum and Born (1995) define interaction as being divided into two general forms, with the general public and affected stakeholders having variable input into the decision-making process. In this context, information is disseminated on a continual basis to the general public, and there are a series of strategies to involve the public so as to include their input in the decision-making process. Interaction with the general public may take place using mail-out surveys, conducting public surveys or gaining input through the media or open houses. Identifying 'the public' and their concerns and conducting an effective public consultation programme are considered to be essential in ensuring adequate interaction at this level. As Hooper *et al.* (1999, p. 753) suggest, 'a hallmark of integrated and co-ordinated approaches is often the creation of partnerships by state agencies with local community-based groups'. Thus, integrating local interests through partnerships with state or municipal planning provides a forum for information exchange and shared decision making.

Stakeholders have a vested interest in the issues being addressed, and some are directly involved in operational decisions. Interaction with stakeholders frequently involves exchanging information and views on a resource issue and attempting to incorporate their input in the decision calculation. Stakeholder interaction may involve working groups, technical steering committees, joint decision boards and a variety of methods that can, to varying degrees, share decision-making power. In some cases, conflict resolution methods need to be employed to mediate interests.

The importance of stakeholder involvement has been well documented in the literature. Wondolleck (1988), for example, provides a comprehensive account of the US Forest Service and the conflict ensuing from the exclusion of stakeholders and the public by the Forest Service in decision making on public lands. Both changing expectations by the public and the access to information have forced public agencies to include stakeholder and public views in their management plans.

Coordination is an essential element and defines a role for how we interact with participants. Margerum and Born (1995) separate the coordinative functions into the categories of communication and conflict resolution. They note that the functions are often intertwined and 'constitute key elements or tools used to put integrated management into practice' (Margerum and Born, p. 385). The continuum from communication to conflict resolution can range from simply information sharing between parties to binding arbitration.

Holistic

The addition of the *holistic* component to the model is important because it recognizes the need for IREM to embrace the wide spectrum of management variables, rather than focus on isolated elements of the decision-making process. Holistic identifies the broadest range of physical and social factors across an area or region (Margerum, 1997) and allows for the widest scope of elements and issues feasible to be included in a substantive evaluation. This necessitates the consideration of a wide range of variables in the decision calculation such as cultural, ecological and economic issues on a broad scale, such as watersheds or river basins. A holistic approach generally provides analysis at an ecosystem level that recognizes the integrative complexity of both natural and social systems.

Descriptive Criteria: How to Make IREM Work

How should IREM work? How do we integrate interests in substantive and procedural areas to improve decision making and operational management? There has been considerable research defining the essential characteristics of IREM and the desired performance of its application; most of this empirical work has been done at the catchment or watershed level. The body of research that deals with increasing the effectiveness of this planning process can be characterized in the following categories:

- Context and culture
- Institutional arrangements
- Overcoming boundaries
- Interagency coordination
- Commitment and funding

Context and culture

Context and culture refer to the broad perspective in which IREM is applied. The historical conditions that form the social and ecological environment affect the application of resource management strategies. Walther (1987) cautions that the successful application of IREM is often a function of the historical and political context. He suggests that it cannot succeed without political commitment, or be legitimized by the culture in which it is developed. Hooper *et al.* (1999) divide context into place and time. Time frame is considered important because of the longer lengths of time it often takes to achieve integrated strategies in the resource management field – noticeable success is frequently identified in decades. Mitchell and Pigram (1989) set 'context' as the over-arching leverage point in any attempt to improve integrated resource planning. The context and culture determine how the planning framework is developed and whether it can adapt to changing conditions.

Adapting to cultural perspectives is especially challenging for IREM when cultures do not share similar worldviews, have a different way of understanding science and express knowledge in different ways. The integration of 'traditional ecological knowledge' (TEK) from indigenous communities, for example, into standard agency decision making provides a good example of the need for adapting different knowledge bases into an IREM framework.

As the name implies, the need for integration is an important consideration in the context of IREM. In some cases, however, there may be no need for integration or desire to pursue integrated strategies. It is important to stress then that IREM may not be a suitable decision-making framework for all resource management problems. It is also important to appreciate that integration does not happen without appropriate funding and commitment. Hooper *et al.* (1999) suggest that there are three vital questions in considering the need for IREM: (i) Is there is a defined problem? (ii) Is a coordinated and integrated approach appropriate to the problem under consideration? and (iii) Will an integrated approach give value to the solution of the problem? Thus, understanding the context of the problem is essential in deciding whether IREM is a suitable planning framework to pursue a workable solution.

Institutional arrangements

Institutional arrangements refer to processes, structures, policies, scale and other arrangements that influence decision making and affect the behaviour of individuals, groups or agencies. Mitchell (1989) identifies key institutional variables as:

- Legislation and regulations
- Policies and guidelines
- Administrative structures
- Economic and financial arrangements
- Political structures and processes
- Historical and traditional customs and values
- Key participants and stakeholders

The suitability of the institutional arrangements that are put in place is essential for the success of any integrated strategy for resource management. In the management of public lands and other public natural resources, the commitment of government and public agencies is reflected in the institutional arrangements that are funded and encouraged for integrated strategies. Nelson and Weschler (1998, p. 565) refer to the term 'institutional readiness' as the 'degree to which jurisdictions are aware of and primed for engaging each other in collaborative governance'. Both 'bottom-up' (decision making from local residents, stakeholders or community groups) and 'top-down' (decision making from senior bureaucrats or politicians) institutions are required for effective coordination within IREM regimes, both to get the grass-roots support at the local level and to gain regional (state or provincial) and federal support for administration, scientific support and funding. Special institutional arrangements are sometimes legislated to facilitate integration and to support coordinated decision making. The Fraser Basin Management Plan in British Columbia, the Tennessee Valley Authority, and the Hunter Valley Conservation Trust in New South Wales, Australia provide examples of institutions created to plan large-scale watersheds.

Bellamy *et al.* (1999) identify institutional structures as one of four critical components in the implementation of IREM. Criteria for defining the institutional conditions under which IREM programmes are likely to succeed include: (i) clear and consistent objectives to guide agency and staff; (ii) suitable funding and finance; (iii) coordination of agency activities and programmes at different levels of government; (iv) guidance to implementing agencies; (v) staff commitment; and (vi) provisions for access by outsiders such as community groups and stakeholders.

A final component associated with institutional arrangements is the concept of scale; the application of institutions must be matched to the scale at which the specific activity takes place. Gardiner *et al.* (1994), when making a comparison of watershed planning between selected Canadian and UK case studies, noted that the scale at which the watershed plans were prepared was important to the effectiveness of their implementation. Basin-scaled institutional structures, for example, that matched local knowledge and participation, were considered important for successful management strategies and outcomes in the planning process.

Overcoming boundaries

The concept of *boundaries* is important for understanding how to better integrate resource management at different institutional levels. Mitchell and Pigram (1989), for instance, have identified boundary problems in institutional arrangements in the Hunter Valley Conservation Trust in Australia. They define specific leverage or management intervention points to help overcome boundary issues with respect to facilitating and improving integrated resource management of land and water resources. Figure 9.2 outlines specific leverage points that identify key aspects to enhance integration. In summary, they have identified the following leverage areas as important: (i) *context* – the natural, social and economic environment; (ii) *legitimization* – defining the statute, policy and political commitment to ensure that IREM is successfully carried out; (iii) *functions* – the integration of management functions at different levels of government and between different agencies; (iv) *structures* – the organizational structures used to carry out IREM; (v) *processes and mechanisms* – levels of task forces, interagency committees and review procedures (regional planning, environmental assessment) used to encourage integration; and (vi) *organizational culture and attitudes* – the incentives or disincentives that are built into the process to encourage IREM, such as conflict resolution, interagency cooperation and bargaining strategies.

Interagency coordination

Born and Sonzongni (1995) recognize that organizational cultures, professional bias and traditional means of doing business are often barriers to innovative

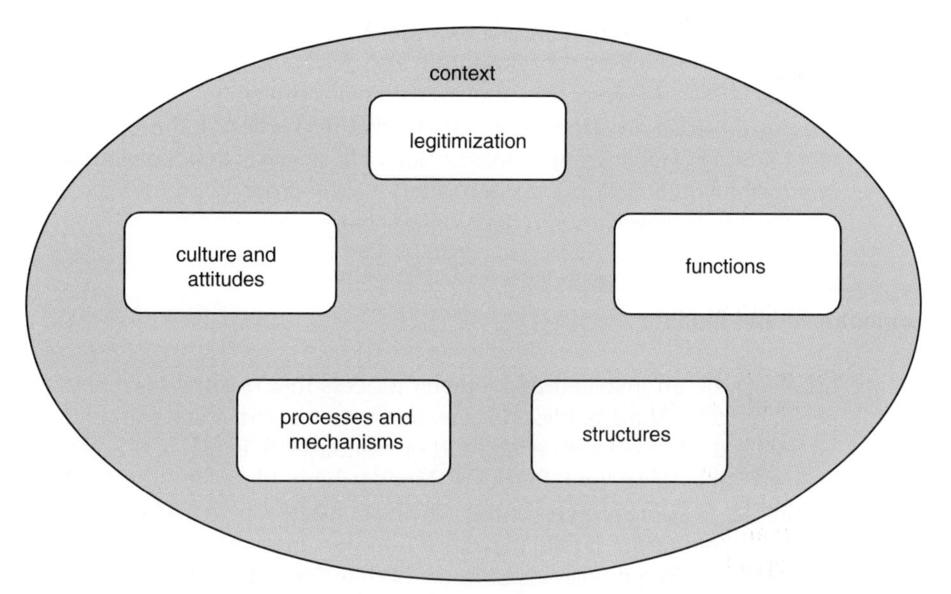

Fig. 9.2. Boundary problems in IREM. From Mitchell and Pigram (1989).

methods in the field. Strong agency cultures and defined mandates often provide impediments to cooperation, *interagency coordination* and mutual problem solving between agencies. Agencies that are mandated to build dams and water improvement projects, for example, regularly have quite different perspectives on a resource management problem as compared with agencies that protect fish or are concerned with environmental rehabilitation. Entrenched historical roles and traditional ways of 'doing business' sometimes provide distinct barriers to integrating interests between government departments, especially when the agencies have had considerable conflict with respect to their resource utilization interests.

Professional bias is determined by both the education and culture that define a profession. Ewert and Baker (2001) have evaluated various academic majors at a university level and their attitude to the environment. The results of the study suggest that individuals in different academic disciplines place fundamentally different levels of concern and express different beliefs regarding the environment. It is apparent from this study and elsewhere that different academic training programmes inculcate fundamentally distinct views of the environment that may be entrenched further by professional membership.

Methods for overcoming agency and professional bias, and integrating common goals are important strategies for sharing information, understanding different points of view and enhancing integration. A variety of strategies can be employed at different levels using joint committees as one method. For example, at the field and operational level, technical steering committees are a beneficial means to share scientific information and interagency data, and for individuals to get to know each other. Other methods may include management committees, issue-specific meetings (such as groundwater contamination workshops) or interagency planning committees. It is imperative to stress here that it is necessary that the structure of the committee should match the scale and issue being addressed and not exclude important stakeholders.

Conflict resolution processes can also facilitate interagency cooperation. Dorcey (1987) suggests that the inclusion of bargaining within the planning process provides an opportunity for creative interaction and a basis for agreement amongst agencies and stakeholders. He suggests that individuals need to improve their interaction and negotiation skills in order to improve the quality of interagency cooperation and decision making.

Commitment and funding

IREM can be an involved and lengthy process that extracts both physical and personal costs. At the public and stakeholder levels, individual commitment may be required for endless meetings and open houses to expedite this process. For example, the Land and Resource Management Planning process in British Columbia, in some regions of the province, lasted 4 years, where stakeholders and members of the public met once a month over that period. This entailed participants giving thousands of hours of their free time without compensation.

This is not an uncommon requirement for public involvement processes in the resource and environmental management arena.

Commitment is also an important variable at the agency level, where individuals need to be committed to IREM principles of communication, information sharing and finding common solutions. Agency allegiance and internal 'empire building' are often barriers to such a commitment. In addition, the management structure of different levels of government has to be committed to achieving integrated solutions through appropriate incentives and educational processes. Political commitment is also an essential component of the functioning of IREM (Walther, 1987). Short-term election horizons and political agendas are often impediments to IREM. IREM may take decades to implement and require continuous funding over that period.

Funding is strongly associated with commitment and is an essential element for IREM. Funding may be required for a variety of activities such as data collection, infrastructure, meetings, agency involvement, or regulation and enforcement. Many programmes or initiatives have disappeared with cuts in budgets or a withdrawal of funds. The financial arrangements for IREM are important at all levels of implementation and provide both the essential resources and validity for the application of the process.

Summary and Conclusions

IREM is a new and evolving field that is changing on a continual basis. The theoretical literature is based on limited case studies, most of which involve river basin or watershed management. From these case studies, the normative literature attempts to establish a standard model for both substantive and procedural aspects of the decision-making process. We have chosen the following criteria of: (i) *comprehensive/inclusive*; (ii) *interconnective*; (iii) *strategic/reductive*; (iv) *interactive/coordinative*; and (v) *holistic* as a basis from which to develop a theoretical model for IREM.

The descriptive literature gives direction as to how IREM should be implemented and focuses on barriers in the present resource planning process that impact the above criteria. The context and culture of *where* IREM can be applied is an important consideration for whether it is feasible to use this approach to solve resource allocation problems. As discussed in Chapter 4, 'driving forces' external to the resource management problem set a context that may inhibit internal solutions. Much of the research on IREM has been directed at improving institutional arrangements and identifying problems in decision-making processes that impede integration. Interagency coordination is recognized as another important factor to improve public agencies' communication and cooperation. Organizational cultures, professional bias and 'traditional ways' of doing business are all elements that can affect attempts to introduce more integrated ways of sharing decision-making power.

How well does theory inform practice? The following case study of the Cache Watershed in southern Illinois provides an excellent example of the complexities involved in watershed management. The challenges faced in the Cache Watershed involve interagency coordination amongst federal, state and local authorities; diverse stakeholders with different agendas; issues of legitimacy; and a complex regulatory environment. How do the boundary problems in Fig. 9.2 apply to the case study?

Case Study
Watershed Planning in Watersheds Dominated by Multiple, Largely Private Owners: The Cache River of Southern Illinois

STEVEN KRAFT, JANE ADAMS, TIM LOFTUS, CHRIS LANT, LESLIE DURAM AND J.B. RUHL

Preamble. The Cache River case study is characteristic of the challenges encountered in planning at a watershed level. Multiple uses within watersheds worldwide have led to polluted basins with little accountability amongst residents for clean-up. It is difficult to coordinate different stakeholders within a watershed to manage for ecological integrity and a wide range of social and natural values. A holistic approach that identifies interconnections within a basin is essential for successful planning. Yet, the history of a watershed, the private property owners and the traditional institutional arrangements may make changes difficult. The Cache River case study provides a good example of the need for an integrated approach to planning, but also highlights the considerable challenges in implementing changes within the basin.

Introduction

Driven by ongoing problems of non-point source pollution (such as chemicals used in agriculture) and decline of aquatic ecosystems, the 1990s witnessed a rapid development of watershed-scale planning initiatives. Variously called 'place-based', 'community led', 'locally led', 'integrated watershed management' or other similar terms, these initiatives now number over 1000 and are growing rapidly throughout the USA. Nevertheless, these initiatives face numerous obstacles, frequently more social than hydrological, in achieving improved water quality and aquatic ecosystems, or other natural resource goals that planning groups or state and/or federal agencies may identify. In particular, water resources and land use planning in watersheds, dominated by multiple, largely private owners, has been fragmented and subject to a variety of forces originating both within and outside the watershed.

Watersheds do not normally constitute formal, organized political jurisdictions; hence, resource planning groups, their planning processes and their plans face the challenge of acquiring political legitimacy and legal authority. Deyle (1995) observes that the fragmented decision making that is typical of watershed management constitutes an 'organized anarchy' where the involvement of stakeholders is fluid, while goals and the means of achieving them are poorly specified. Thus, too often the planning process produces the 'pet' solutions of agents who are only temporarily cooperating to address a particular watershed-based resource problem. Alternatively, planning groups organized around principles of decision making by consensus arrive at solutions representing the lowest common denominator that all participants can agree to but which do not necessarily have the potential to improve environmental conditions in the watershed.

The Cache River watershed in southern Illinois provides an instructive case study of the challenges of watershed planning to deal with multiple resource problems. Through the case study, we can observe how the socioeconomic driving forces external and internal to multiple-ownership watersheds influence and restrict the decision-making processes of landowners/users as well as agency personnel.

Study area and background

The Cache River watershed encompasses 1944 km^2 of southern Illinois near the confluence of the Mississippi and Ohio Rivers (Fig. 9.3). The lower reaches of the Cache River formed from an ancient bed of the Ohio River. During the massive floods of 1937, the Ohio River reclaimed its old channel for a number of weeks. The Cache and its tributaries descend from the Shawnee Hills, forming into a once-navigable stream through the confluence of its several branches. Originally, it emptied into the Ohio River about 5 miles north of the confluence with the Mississippi. Due to the coming together of multiple physiographic provinces, the watershed has diverse ecological resources and unique natural communities, including bald cypress (*Taxodium distichum* L. Rich.) and water tupelo (*Nyssa aquatica* L.) swamps at the northern edge of their range and other forested wetlands (see Fig. 9.4). At least 100 state-threatened or endangered plant and animal species are known within the watershed. The Cache River region also supports unique ecological communities and globally rare or endangered species. Because of the unique diversity of the region's ecosystems and its capacity to support migratory waterfowl, in 1990 the US Fish and Wildlife Service created the Cypress Creek Wildlife Refuge astride the most threatened portions of the bottomland reaches of the swamp. In 1994, the United Nations designated the wetlands as a Ramsar site, indicating their worldwide significance.

The ecological integrity of the Cache River ecosystem is threatened by: (i) loss and fragmentation of natural habitats as a result of agricultural activities and timber harvest; (ii) dramatically altered hydrological systems caused by drainage, channelization and other modifications; (iii) sediment deposition in wetlands causing deterioration of water quality and alteration of habitat conditions in areas of unique swamps; and (iv) land use and economic activities that are incompatible with long-term maintenance of ecological functions. Further, many of the land-based production activities contribute non-point source pollution not only to the Cache River but also to the river it empties into – the Mississippi. Moreover, the predominantly rural five-county area of the watershed has an impoverished economy with minimal infrastructure and weak linkages to the surrounding region, which make it sensitive to the cost and benefits of habitat restoration and protection in the Cache River region or changes in the economic viability of agriculture.

History

During the 19th century, with growing demand for railroad ties, mine timbers, pilings and lumber for construction and manufacturing, the extensive bottomland forests in the watershed

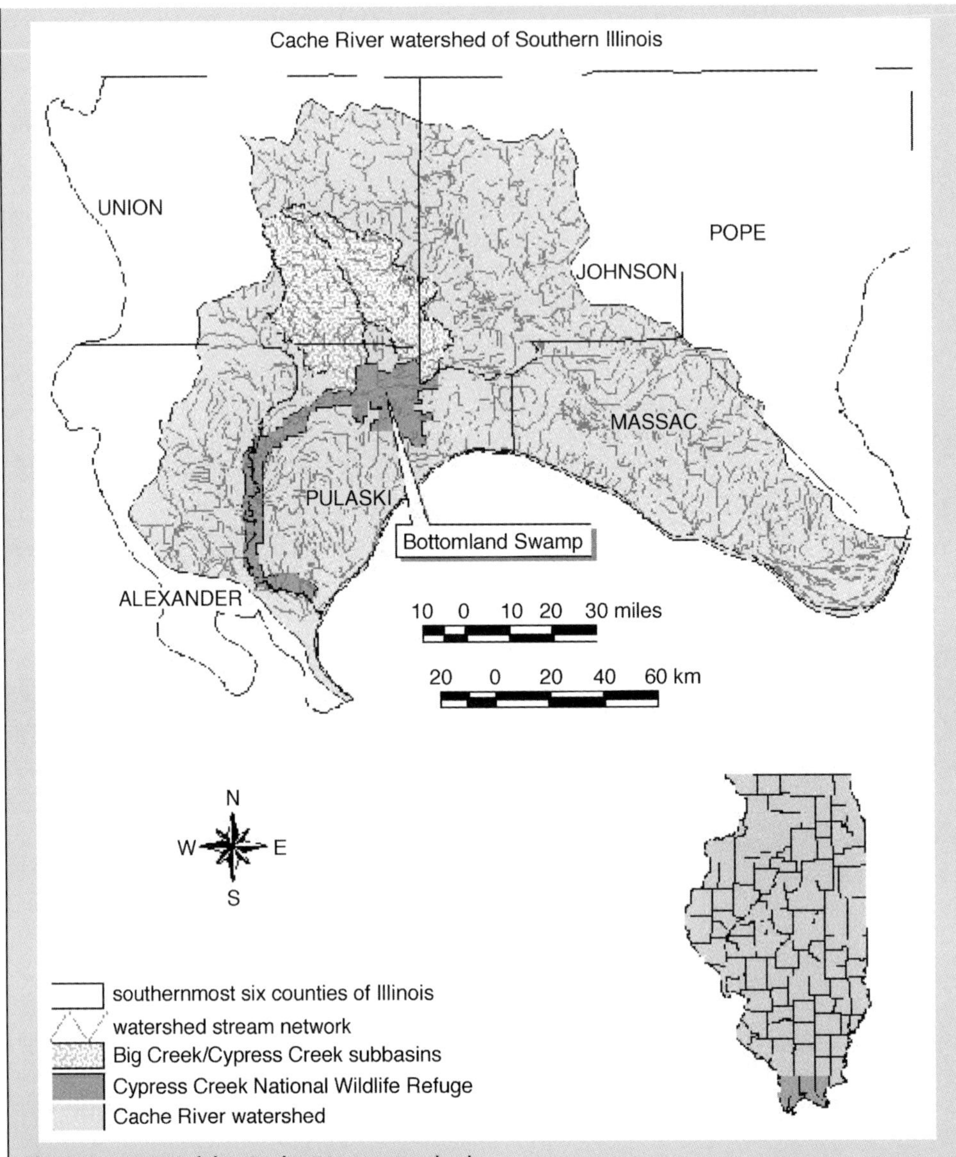

Fig. 9.3. Map of the Cache River watershed.

began to be timbered. The first significant drainage and ditching in the watershed appears to have been done by timber companies to facilitate the transport of logs from stump to mill. Farmers attempted to expand and improve tillable acreage through drainage. The clearing and drainage did not significantly shift the hydrology until the construction of the Post Creek Cutoff in 1913–1916. This ditch drained the Cache and its eastern tributaries into the Ohio River well above the Cache's natural mouth. The cutoff and the resulting drainage of extensive eastern wetlands opened large tracts of land to agriculture in the watershed. What became known

Fig. 9.4. Cache River wetlands. From Illinois Department of Natural Resources (http://dnr.state.il.us/lands/landmgt/parks/r5/cachervr.htm).

as the 'Lower Cache' – that area below the Post Creek Cutoff – remained largely swamp and forest. The Post Creek Cutoff essentially created two rivers where there had been one. For the next 60 years, reclamation work continued on both eastern and western tributaries. Starting in the late 1930s and early 1940s, the Army Corps of Engineers, working with locally constituted Drainage Districts, continued to straighten river segments to facilitate moving water out of the watershed and to drain the swamps. In 1950, these activities included a diversion canal that shifted the mouth of the lower portion of the Cache River from the Ohio River to the Mississippi; levees, railroads and highways further reconfigured the river basin.

This drainage was, apparently, uncontroversial for many years. However, in 1978, area landowners who made their living from fishing and/or seasonal use of their wetlands, and local hunters organized the Citizens Committee to Save the Cache and began to contest this drainage. Furthermore, the ecological significance of the region began to attract outside environmental groups, (e.g. The Nature Conservancy, Ducks Unlimited and the Sierra Club). These outside groups worked with local conservation/environmental groups as well as state and federal agencies. The combined efforts of these cooperating groups and agencies were also able to harness and direct state and federal resources as well as public policies to address the four factors mentioned above that threatened the ecological integrity of the area.

Building on international migratory waterfowl agreements among the USA, Canada and Mexico, the conservation provisions of a number of farm bills starting in 1985, a national concern with agricultural non-point source pollution, a changing Corps of Engineers reflecting greater environmental concerns, and funding to support planning to enhance water quality, the region experienced significant change. Such changes included a US wildlife refuge being established over strong local opposition, private environmental groups acquiring significant land resources while forming joint partnerships with state and federal agencies, the Corps studying how to restore ecological function to streams it once straightened and channelized as well as the land it once drained, extensive agricultural acreage entered the Conservation

Reserve Program, and Environmental Protection Agency (EPA) monies were made available to facilitate a watershed planning process.

Planning at the Watershed Level

Nationally, while agriculture had been exempted from much of the Clean Water Act, there seems to be an implicit assumption within the watershed planning community that participatory user-centred watershed planning structured along a consensus model would result in plans that will be readily adapted by landowners/users, resulting in significant environmental improvement with a minimum of economic cost. To this end, the Cache River Watershed Resource Planning Committee (RPC) was begun in 1992/1993 as a basis for developing an integrated resource management plan for the watershed. The RPC was an EPA-funded initiative, sponsored by TNC and the Natural Resources Conservation Service (NRCS), involving 25 citizens who developed a long-range plan for the use of land and water resources in the watershed. While many issues dominated the discussion, one of the reoccurring issues was the relationship among the planning process, the groups involved and the protection of the perceived property rights of local landowners/operators and other residents of the watershed. In addition to the citizen-based planning committee, there was a 20-member technical committee comprising representatives from public and private agencies [e.g. Illinois EPA, US Fish and Wildlife Service (FWS), US Forest Service, NRCS, Illinois Department of Natural Resources (IDNR), US Army Corps of Engineers and a local university] that functioned as a 'research arm' of the RPC.

To be successful, such a planning process requires information that merges ecological constraints with economic data in a framework relevant for farm level and regional analysis and decision making. Just as importantly, the watershed planning process, the resulting plan and management measures/practices therein, and anticipated outcomes from plan implementation at the parcel, farm and regional level must be seen as legitimate in the eyes of the various stakeholder groups in the watershed. Over a period of almost 24 months, the RPC identified a number of paramount natural resource concerns for the watershed, and the technical committee developed a range of alternatives for dealing with each concern. These alternatives were not without controversy, reflecting the diversity of the RPC membership. Nor were the process and the resulting plan necessarily seen as legitimate by all land users/operators or residents in the watershed.

On review of state and federal legislation, there are at least 25 laws that impact agriculture in the watershed; the multijurisdictional nature of the existing laws features a system of media-specific laws (e.g. water, soil and air) on the one hand, and action-specific laws (e.g. waste and chemical handling) on the other, resulting in a complex overlapping mosaic that is hard for citizens to understand and abide by. However, there are no laws that sanction and legitimate watershed planning as the mechanism for dealing with non-point source pollution coming from agricultural and non-agricultural sources. The consequence is a plan that has been created through a consensual planning process but whose legitimacy and ability to influence behaviour is problematic. However, the plan has been used extensively by agency personnel as a way to legitimize their requests to their managers for additional resources to help them in their work. Nevertheless, the plan's recommendations for the adoption of practices to reduce non-point source pollution remain very much a work in progress. In actuality, the adoption of practices to reduce non-point source pollution seems to depend much more on the availability of governmental payments of one form or another to induce actual changes on the landscape.

Summary: Current Issues and Questions

The policy environment for dealing with non-point source pollution is complex to say the least. However, if a sector such as agriculture, which is a known contributor to non-point source pollution of waterways, is exempted from major legislation designed to correct problems of water quality, achieving the policy ends of enhanced water quality will be difficult. In the absence of state or federal legislation authorizing watershed planning along with resources for monitoring and enforcement, the reliance on watershed planning to enhance water quality is problematic at best.

While the literature on legitimacy and natural resource use is quite varied, there are a number of themes that reoccur: legitimacy is the degree of social acceptance of an institution, rules, outcomes, etc.; legitimacy reflects the willingness of citizens to accept and follow sets of rules and processes; legitimacy reflects a perceived obligation to 'obey or abide by' rules or processes; and, in managerial situations, to be legitimate, measures must be designed based on existing law (see Tyler, 1990; Hatcher *et al.*, 2000; Jentoft, 2000). Results of survey research in the watershed, both while the watershed management plan was being developed and 7 years later, suggest that the legitimacy of at least the plan is questionable. For example, in both years, while overwhelming majorities of the non-farm residents indicated that recommendations of such a planning process should be followed by landowners/users, similar large majorities of farmers disagreed, i.e. for a significant group in the watershed who control the resources targeted by the plan, the legitimacy of the plan (i.e. there is a perceived obligation to follow the plan) is questionable. In-depth interviews with participants in the planning process revealed that the process lacked legitimacy for many of the farmers involved.

Our review of existing laws suggests there is not a strongly defined legal basis for the development and implementation of watershed management plans that are informed by law with an accompanying process for monitoring and enforcement. Consequently, if watershed planning is going to be the mechanism for dealing with non-point source pollution (see the EPA's Office of Oceans, Wetlands and Watersheds – http://www.epa.gov/OWOW/watershed – and the associated watershed-based websites), how is legitimacy conferred on the planning process and the plans such that they are actually implemented and sustained?

Embedded in watershed planning, as in the Cache River watershed, is the issue of property rights and how they are defined/discussed in popular culture versus a more formal/legal definition (see Byrne, 1995). In economics, for example, economists frequently talk about property rights as a bundle of entitlements that specifies one's rights, privileges and obligations in the use of some property. What these entitlements are, and their scope, is determined by society and they are not immutable. Rather, they can be changed in response to new social conditions, new knowledge, etc. Indeed, in the case of watershed planning in the Cache, we could argue that there are competing sets of entitlements: the landowners' perceived entitlement to use their land as they see fit in terms of production practices without regard for spillover effects on others, i.e. externalities, versus the perceived rights of residents in the watershed and further downstream to clean water and viable ecosystems. As the literature on ecosystem services has grown in recent years, it indicates (see de Groot, 1992; Daily, 1997; Daily and Ellison, 2002) that the conflicts between these competing sets of entitlements will only sharpen. The argument exists that historically resources such as clean water have been treated as 'open access', i.e. open to anyone to use in production, consumption or as a dump for waste. Until relatively recently, little was known about ecosystem services and their value to humanity and the natural ecosystems that sustained life. The services or the ecosystems that provide them were managed without regard for their long-term viability. However, in the case of water through legislation and for ecosystem services through current debate potentially leading to legislation, society has endeavoured to transform these open access resources to ones of a regulated commons with access and use socially specified, monitored and enforced. In watershed planning, we see the playing out of this shift in entitlements and property rights: while landowners are entitled to use their land, they are enjoined to use it in ways that protect the quality of water that the whole community is entitled to.

The history of watershed planning in the Cache River watershed strongly suggests that the process was 'jump-started' by outside groups that perceived significant ecological resources were being threatened and should be protected. These outside groups worked with local residents who shared their goals and concerns, as well as state and federal agencies to facilitate watershed planning. Local groups opposed to planning activities in the watershed worked with outside groups as well to bolster their positions. This complex interplay between local and outside groups can have significance for the perceived legitimacy of the planning process and the resulting plan. How does the presence of outside NGOs as well as public agencies with staff resources in the region shape and guide the planning process?

Nationally, there are a number of policy issues that could well be dealt with using watersheds and watershed-based planning processes. Among others, these include non-point source pollution remaining a problem in many watersheds. As a consequence of legislative and legal action, the US EPA has instructed states to develop total maximum daily load (TMDL) standards for nutrients on a stream segment basis across the USA. Once a TMDL is determined, the 'loading' is then allocated among potential contributors in the area draining through the stream segment. Another aspect of non-point source pollution that could well require watershed-based solutions is the contribution of landscape-level activities such as agriculture to the presence of hypoxia in the Gulf of Mexico. Additionally, there is a desire to protect the valuable role that riparian areas play in providing ecosystem services and the protection of endangered species. The effectiveness of watershed planning in dealing with these issues is closely tied to the questions of the legitimacy of the process and the plans. Yet, these are issues that have not been well addressed and need to be.

Discussion Questions

1. Differentiate between substantive and procedural aspects of IREM. How are the elements of comprehensive/inclusive, interconnective, strategic/reductive and interactive/coordinative used in the context of these two terms?

2. Of the five categories describing how IREM works (context and culture, institutional arrangements, overcoming boundaries, interagency coordination, and commitment and funding), which ones do you feel would present the greatest challenge in your community? Why?

3. Rank the components or 'leverage areas' listed in the text in order of importance. Would this ranking change within different boundary characteristics or stay relatively stable? If changeable, what factors would create this change?

4. Based on your experience, are there issues or areas involved in the theoretical framework of IREM that have not been discussed in this text? If yes, what would they be and describe these components.

Case Study Discussion Questions

5. Define the boundary problems that are prevalent and affect IREM in the Cache River watershed. What is the role of legitimization?
6. With respect to interaction and coordination in IREM, what is the relationship of outside groups with residents in the Cache Watershed basin?
7. Define non-point source pollution. What are the problems in the Cache Watershed with respect to non-point source pollution? How can this issue be solved in the basin? Answer this question with respect to the elements of IREM (Fig. 9.1).

References

Beanlands, G. and Duinker, P. (1983) *An Ecological Framework for Environmental Impact Assessment in Canada*. Institute for Resource and Environmental Studies, Dalhousie University and Federal Environmental Assessment Review Office, Halifax, Nova Scotia, Canada.

Bellamy, J., McDonald, G., Syme, G. and Butterworth, J. (1999) Evaluating integrated resource management. *Society and Natural Resources* 12(4), 337–353.

Born, S. and Sonzogni, W. (1995) Integrated environmental management: strengthening the conceptualization. *Environmental Management* 19(2), 167–181.

Byrne, J.P. (1995) Ten arguments for the abolition of the regulatory takings doctrine. *Ecological Law Quarterly* 22(1), 89–142.

Daily, G.C. (ed.) (1997) *Nature's Services: Societal Dependence on Natural Ecosystems*. Island Press, Washington, DC.

Daily, G.C. and Ellison, K. (2002) *The New Economy of Nature: the Quest to Make Conservation Profitable*. Island Press, Washington, DC.

de Groot, R.S. (1992) *Functions of Nature: Evaluation of Nature in Environmental Planning, Management, and Decision Making*. Wolters-Noordhoff, Amsterdam.

Deyle, R.E. (1995) Integrated water management: contending with garbage can decision-making in organized anarchies. *Water Resources Bulletin* 31(3), 387–398.

Dorcey, A.J. (1987) The myth of interagency cooperation in water resources management. *Canadian Water Resources Journal* 12(2), 17–26.

Ewert, A. and Baker, D. (2001) Standing for where you sit: an exploratory analysis of the relationships between academic major and environmental beliefs. *Environment and Behavior* 33(5), 687–707.

Friedmann, J. (1987) *Planning in the Public Domain: From Knowledge to Action*. Princeton University Press, Princeton, New Jersey.

Gardiner, J., Thompson, K. and Newson, M. (1994) Integrated watershed/river catchment planning and management: a comparison of selected Canadian and United Kingdom experiences. *Journal of Environmental Planning and Management* 37(1), 53–66.

Hatcher, A., Jaffry, S., Thebaud, O. and Bennett, E. (2000) Normative and social influences affecting compliance with fisheries regulation. *Land Economics* 76, 448–461.

Hooper, B., McDonald, G. and Mitchell, B. (1999) Facilitating integrated resource and environmental management: Australian and Canadian perspectives. *Journal of Environmental Planning and Management* 42(5), 747–766.

Jentoft, S. (2000) Legitimacy and disappointment in fisheries management. *Marine Policy* 24, 141–148.

Lang, R. (1986) Achieving integration in resource planning. In: Lang, R. (ed.) *Integrated Approaches to Resource Planning and Management*. University of Calgary Press, The Banff Centre School of Management, Calgary, Alberta, Canada, pp. 27–50.

Margerum, R.D. (1997) Integrated approaches to environmental planning and management. *Journal of Planning Literature* 11(4), 459–475.
Margerum, R.D. and Born, S. (1995) Integrated environmental management: moving from theory to practice. *Journal of Environmental Planning and Management* 38(3), 371–391.
Mitchell, B. (1986) The evolution of integrated resource management. In: Lang, R. (ed.) *Integrated Approaches to Resource Planning and Management*. University of Calgary Press, The Banff Centre School of Management, Calgary, Alberta, Canada, pp. 13–26.
Mitchell, B. (1989) *Geography and Resource Analysis*. Longman, London.
Mitchell, B. (1990) Integrated water management. In: Mitchell, B. (ed.) *Integrated Water Management: International Experiences and Perspectives*. Belhaven, London, pp. 1–21.
Mitchell, B. and Pigram, J. (1989) Integrated resource management and the Hunter Valley Conservation Trust, NSW, Australia. *Applied Geography* 9, 196–211.
Nelson, L. and Weschler, L. (1998) Institutional readiness for integrated watershed management: the case of the Maumee River. *Social Science Journal* 35(4), 565–577.
Tyler, T.R. (1990) *Why People Obey the Law*. Yale University Press, New Haven, Connecticut.
Walther, P. (1987) Against idealistic beliefs in the problem-solving capacities of integrated resource management. *Environmental Management* 11(4), 439–446.
Wondolleck, J.M. (1988) *Public Lands Conflict and Resolution: Managing National Forest Disputes*. Plenum Press, New York.

10 IREM in a Complex World

Introduction

This chapter examines the challenges of promoting conservation, which implies reducing natural resource demands in the present to ensure the security of future supplies, in landscapes involving many owners and managers who may have wide-ranging resource management goals as well as variable commitment to conservation ideals. This suggests the need for a high level of cooperation among natural resource managers. Coincidently, over the past couple of decades, the notion of partnership building has gained particular favour, especially as the public demand for conservation has increased and funding for environmental projects has been rationalized. This has forced many resource managers to rethink their attitudes towards resource management and consider various joint projects such as ecosystem management as opposed to estate management. Generally speaking, conservation practices such as IREM, when applied beyond a single natural resource manager's boundaries, require a high level of managerial cooperation to succeed. Be that as it may, this process is rarely as easy as it first appears.

To bring light to bear on this particular issue, this chapter addresses the following questions:

- How does the IREM process change or adapt when more than one owner or autonomous manager is involved in the IREM process?
- What are the fundamentally different types of multiagency management and do these different types substantially change the approach to IREM?
- How does the IREM process differ when individual managers have different resource management interests and objectives?
- How can we begin to understand the complex dynamics of the multiagency management situation to make us more effective and efficient managers?

IREM in single ownership situations

To understand the dynamics of multiagency, multi-interest management situations, it is first useful to look at the simpler single ownership resource management situation. Even in a less complicated managerial situation such as a family woodlot, the resource manager is faced with substantial technical and financial challenges where strategic and ground-level decision making are bounded by the competencies and interests of a single individual (see Chapter 7). As we saw in Chapter 7 concerning decision making and power, even the best intentioned line-worker can inadvertently distort policy efforts. What might appear to a senior manager as a straightforward instruction to a subordinate may be highly confusing to a fieldworker. For example, in a field visit by one of the authors to an exclusive lakeside cottage development, management frustration was noted when a contractor failed to follow prescribed environmentally sensitive tree stand thinning instructions.

Types of multiagency IREM

Despite the inherent complexity of implementing IREM strategies within a single organization, it is reasonable to expect that when comparable decisions cross organizational boundaries, compliance with intended IREM goals will be increasingly problematic. In attempting to make sense of this complexity, Mandell (1989) identified three distinct and progressively more convoluted levels of multiagency management (see Fig. 10.1).

1. The first multiagency management type has a dominant central authority with sufficient influence over its so-called partners, from either its legislative or economic power, to effectively control their decision-making behaviour. Bissix and Rees (2001) suggest that these partners are largely treated as externalities by

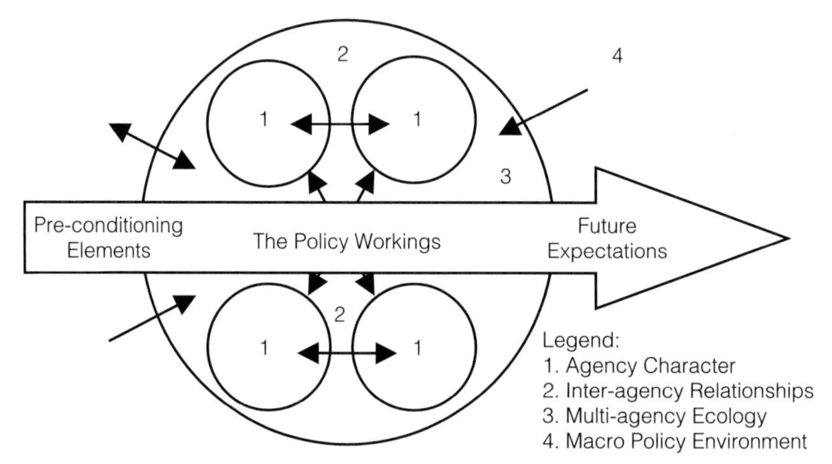

Fig. 10.1. The multiagency decision-making environment. Source: Bissix and Rees (2001).

the central authority to be reeled in as compliant agents whenever necessary. One example is the authority that manages the Great Barrier Reef Marine Park off the coast of Queensland, Australia (see Ottesen and Kenchington, 1995). A second example might be individual parks within the US National Park system. In this system, concessionaires, licensed guides and tour operators, for example, must comply with prescribed rules and regulations or face sanctions and serious business consequences. Generally speaking, these central agencies have not only sufficient legislative authority, but also, and just as importantly, the necessary credibility to implement its management objectives.

2. The second organizational type reflects a hub and wheel arrangement. Here a central agency holds notional legislative authority but lacks sufficient resources to execute its IREM plan without substantial cooperation from its partners. In this scenario, the agency is more dependent on the willing cooperation of external agencies that are not normally or easily coerced by command and control strategies. These partners tend to respond well, however, to incentives. The National Parks system in England and Wales operates very much in this way. In Dartmoor National Park, for example (see Bissix and Bissix, 1995), the park authority operates within a broad and guiding legislative framework. To reach its objectives successfully, however, the Park Authority must continuously negotiate with its partners such as farmers, tourism concerns, light industry and mine operators to encourage compliance with its conservation objectives.

3. The third typology relates to a complex interaction of many interests operating in what is typically a 'working landscape' rather than a protected area. While there may be an overlying legislative framework, most of the organizations operating within the landscape's boundaries largely act autonomously. They are not easily influenced by command and control measures nor are they predisposed to respond favourably to financial incentives – generally because so-called central authorities do not have the financial capacity to offer sufficiently large enticements to make conservation goals attractive enough. While many agencies and organizations may agree in principle to support a conservation initiative, they often compromise its objectives when the financial initiatives are insufficient. Any attempt to implement public policy through such a complex management environment is often analogous to herding cats! Without sufficient constraints and/or incentives, everyone appears to head in his/her own direction without regard to any over-arching conservation objectives.

Multi-interest, single owner resource management

In implementing a conservation management regime such as IREM, it is important to make a thorough analysis of how management actually functions rather than assume that written and agreed policy intentions are sufficient. While it was inferred earlier that promoting IREM in a single agency might be relatively simple, some singularly managed landscapes are in practice very challenging. For example, a large forest landscape managed by a national forest service or a large multinational company may on the surface seem uncomplicated, but its exterior veneer may hide a conundrum of opposing policy influences. In such situations, there are likely to be timber and pulpwood harvesting activities, recreation, tourism and scientific users, as well as mining and aggregate, housing, industry and water supply interests all vying for pre-eminence. Although any over-arching legislation may seem sufficient to rein in these interests, the reality of on-the-ground management may paint a much more complicated picture.

A Conceptual Framework for Multiagency, Multi-interest Analysis

Although the resource management literature concerning integrated resource management, landscape management and ecosystem management has grown significantly in the last 20 years, surprisingly little has been written on its particular decision-making dynamics and how these may vary over time. Given the organizational reality of multiagency IREM and its inherently complex decision-making environment, it is important to try to cut through this complexity so that resource managers can more effectively prescribe policy interventions. Bissix and Rees (2001) have proposed an analytical framework with four inter-related lenses that provide insights into this process (see Tables 10.1 and 10.2). The first lens focuses on the nature and character of the various agencies involved in the IREM process, while the second concentrates on the impacts of their various interagency relationships. The third lens is substantially more complex; it involves the outcomes of the various interagency relationships. This is referred to as a multiagency ecology or network. This third lens reflects the ways in which various agencies impact each other's effectiveness, how symbiotic or antagonistic

Table 10.1. The multiagency analytical framework.

Multiagency component	Component description
Agency lens	Concerns an agency's internal workings and its impact on the outside world. This includes autonomy, goal centrality, authority, internal decision style, resource control and managerial style. The multiagency context is concerned with the aggregate impact of constituent agencies and organizations.
Interagency lens	Involves the dynamics between two or more agencies. Associated variables include information management and communications, relationship propriety, delegation, cooption, corporatism, agency capture, relationship stage, resource flow and communication patterns.
Multiagency ecology lens	Concerned with the broader policy environment, especially how agencies inter-relate in a complex resource management sector. This includes field complexity – the quantity and the diversity of agencies involved; programme and policy complexity; and the sector's dominant ideology, traditions and policy inertia.
The macro policy environment lens	These influences include myriad pre-conditioning factors and *ex post* changes, especially political changes, paradigm shifts and natural resource transformations impacting policy workings.

Table 10.2. The multiagency framework components.

Agency lens	Interagency lens	Multiagency ecology lens	The macro policy environment lens
Agency autonomy	Relative relationship stage	Organizational and programme complexity	Biophysical transformations
Agency authority	Agency interdependence	Network baggage	Social and political economy
Agency permeability	Information management	Network goal congruency	Paradigm shifts
Goal congruency	Corporatism	Network resource flow	
Decision-making style	Receptivity to influence	Network coordination	
Management style	Level of delegation		

Adapted from Bissix (1999).

relationships form, and where the 'whole' may become greater than the 'sum of the parts' or where, more frequently, a system's outputs rarely reflect the sum of the inputs. The fourth lens places the system in the context of society. It provides insights concerning the impact of external factors impinging on the IREM process over which resource managers have little or no influence or control.

The agency lens

The agency lens focuses on a number of inter-related behaviours that help to explain and anticipate an agency's role in the multiagency IREM process. It also helps to identify how that role might vary over time. Agency analysis is largely concerned with internal factors that prescribe a more or less characteristic response to any problem or challenge that an agency faces. Each agency, for example, has its own special character that defines its potential and willingness to carry out IREM goals. Over time, this capacity will fluctuate according to variations in its internal structure such as staff composition and commitment, financial health, competing objectives and technical competence, as well as a whole host of changing external conditions. While the following list is not exhaustive, it provides a basic analytical framework as to how an agency or particular group of agencies might behave in any multiagency management situation.

Agency autonomy

This relates to an agency's ability to act on its own accord without fear of sanction from another group, agency or authority as well as to the degree an agency values and is willing to defend its independence. For a manager charged with influencing the behaviour of participating agencies within a landscape, the degree to which an agency protects its autonomy can be quite problematic. Such behaviour often translates into resistance by an agency to adopt and implement communal landscape management goals such as those incorporating IREM.

Agency authority

This relates in part to the level of influence an agency has over another agency, but can also be seen as the corollary of agency autonomy. It can, for example, relate to an agency's ability to pursue its own goals without undue influence or retribution from others.

Agency permeability

This concept has two fundamental components. The first refers to an agency's openness to new ideas and methods of going about its business, and the second, which is the converse of agency autonomy, refers to its susceptibility to external influences.

Goal congruency

This relates to the degree of overlap between established landscape management or ecosystem management goals – usually promoted by a government agency – and that of an involved agency. When there is considerable overlap between public policy and an agency's goals, there is likely to be substantial communal goal compliance. However, when its goals appear at loggerheads with broader conservation goals, policy compliance is problematic.

Decision-making style

Over the years, an agency is likely to have developed a number of standard ways of approaching problem solving which creates an inertia that can work against effective change. Various agencies make decisions differently, which may require different persuasive approaches. In some agencies, important decision making is centralized and very little discretion is given to field staff. In such cases, most effort of persuasion should be directed to central management. In other organizations, considerable discretion is devolved to fieldworkers; however, this can be a double-edged sword for policy influencers. While some field managers may be rather easy to influence, it may be impossible to persuade a critical mass, which may mean that a project falters. In some instances, a single agency may also present different views of itself to the outside world, which makes it quite difficult to decide on an appropriate approach to influence that agency.

Management style

This concept is closely associated with decision-making style but refers particularly to the overall ideology or philosophy that drives the way an agency conceptualizes and approaches key problems and issues. Over time, management style may fluctuate from one dominated by various professions such as ecologists, geologists, foresters and environmental chemists to one more closely attuned to accountants or publicists. Such shifts in management style can profoundly impact the way public policy is approached. When financial officers dominate policy, for example, there may be considerable focus on an agency's bottom line. This may mean cutting corners so that environmental management is compromised.

It is clear from this first part of the framework that even when only a single agency is involved, the application of IREM can be very complex and demanding. Typically, however, IREM involves the cooperation of at least two interacting agencies. It is for this reason that we must consider what happens when two or more agencies need to work together to implement policy.

The interagency lens

Agencies, no matter how powerful, cannot operate in isolation. Consequently, there is a need to examine the way that agencies interact with one another as they try to influence each other's behaviour. The interagency lens provides such insight and not only provides clues as to how attempted policy influence goes awry but also suggests ways to gain policy influence more effectively and efficiently. Six broad categories are highlighted here. They include: (i) the relative relationship stage; (ii) agency interdependence; (iii) information management; (iv) corporatism (including professionalism); (v) receptivity to influence; and (vi) levels of delegation.

Relative relationship stage

This refers to the ways in which relationships between agencies change over time. In the *formative* stage, for example, policy actors must make a special effort to find out who are the most influential or most appropriate decision makers in an agency to engage in policy development. This task might require considerable effort in the initial stages, but energy expended can often be reduced as a relationship matures. One frequent mistake is to assume that interagency relationships remain constant and that the focus of interagency relationship maintenance should remain the same. Unfortunately, depending on personality and circumstance, some interagency relationships degenerate into a process of game playing, while others that were initially strained or characterized by high levels of formality evolve into highly productive partnerships. Often, the key to success, especially with the most influential agencies, is constantly to monitor and nurture these relationships.

Agency interdependence

This refers to the mutual reliance one agency has on another to carry out their respective objectives. Thompson (1967 cited and refined by O'Toole *et al.*, 1984) developed a typology of interdependencies that explains the nature of these relationships and, from this, it is possible to better predict the power and influence of relationships between interacting agencies.

In a multiagency situation, it is possible for an agency to have a dominant and central role in policy application, usually through legislative power and/or fiscal influence. Often, several agencies are dependent upon a single dominant agency to accomplish their objectives. However, it is often the case that the central agency is not nearly as dependent on any single participating agency, as the production capacity of such a participating agency can easily be substituted whereas the central, more dominant agency's role cannot.

A more mutually dependent relationship exists when a chain of policy relationships complicates implementation. Such a chain is susceptible to breaks or congestion in the flow of resources from one intermediary to the next. For example, federal funding may need to move through state or provincial to regional or municipal governments before reaching the local target group. A lack of commitment or lowly assigned priority by any one intermediary can seriously disrupt policy implementation. This may create stress for that agency or it may actually confer a level of influence or control over policy success that is not normally afforded that agency.

A third situation is where there is a high level of mutual or reciprocal interdependence. For example, in the pulpwood industry, powerful multinational corporations often hold rein over many small woodlot suppliers. Should those suppliers remain unorganized in collective marketing, then their aggregate influence will probably remain small. However, if they successfully organize as a bargaining unit, they will probably exert considerable influence over what was a previously dominating agency.

Information management

As was seen in Chapter 7, the way that information is managed by one agency can be critical to the way that another responds and to the overall ecosystem management objective. Important information necessary for conservation management success and necessary to sway various agencies' resource management decision making can be freely circulated, suppressed, massaged or embellished, and each approach can have different impacts over the short, medium and long term. In promoting a new conservation management paradigm such as IREM, it is sometimes tempting to oversell short-term benefits that are difficult to achieve on the ground. This may move the process quickly forward in the short term but cause it to stumble over the longer term.

Corporatism

In the context of interagency relationships, corporatism refers to the special influence that certain agencies or groups have over public policy formulation and the way that it is implemented (Harrison, 1984). This phenomenon generally refers to large and powerful corporations and also to influential trade unions. Often, government officers and especially cabinet level politicians charged with, for example, overseeing natural resources, develop very close relationships with industry or top union officers. This results in certain agencies and people gaining special access to government's decision-making apparatus. This access is not usually available to others; for example, consultation time is often routinely provided to powerful industrialists, but rarely, if at all, to others such as environmental advocates or community groups.

Professionalism

Professionalism, which is a particular form of corporatism, suggests that certain professional groups, such as lawyers, doctors, foresters and planners, etc., develop relatively closed professional communication networks within these professions that work across agency boundaries. Here certain policy options find favour, usually those that bolster professional influence, while other, sometimes potentially more effective and efficient options go without proper consideration.

'Agency capture' refers to a particularly virulent form of corporatism where an agency that is supposed to be regulated by (usually) a government agency is in fact controlled by the so-called 'regulated' agency. This often occurs where one agency, such as a large natural resource company, makes a dominant contribution to the local economy. With such a dominant role, the natural resource company is often able to influence what rules and regulations are put in place, and when such manipulation has not worked at the design stage then it is able to control how these rules are implemented. Covert agreements or unspoken understandings emerge to 'go soft' on various rules and regulations that would, if implemented, incur costs for the so-called 'regulated'. A classic example is thought to be the relationship that US Steel historically developed in Gary, Indiana, regarding air pollution control. In Gary, because of US Steel's dominant economic role in this town, it was able to avoid costly pollution controls measures for many years, while a similar plant in Chicago, owned by the same company, underwent continuous environmental performance improvement. This difference in corporate behaviour was thought be the result of environmentalists' greater voice in Chicago, which led to the establishment and implementation of stronger environmental regulations.

Receptivity to influence

Resource flows: tracking the flow of management resources from one agency to another, can provide insights as to how certain agencies either exert influence or are influenced by other agencies.

Agency cooption also relates to how one organization is able to influence another's objectives. For example, an agency may agree that solid waste reduction is important for environmental health but be unwilling to voluntarily make changes to its production processes that might reduce pollution because of its added costs. In some cases, some combination of incentives or sanctions such as fines or bad publicity may sufficiently alter an agency's behaviour, which can be considered cooption – the adoption of behaviour that you would not otherwise carry out, in response to real or implied threat. If you as a government agency are charged with the responsibility of implementing a solid waste reduction policy, for example, you might consider cooption to be a good thing, but if you are a shareholder focusing on short-term investment gains, you may see this as costly and interfering.

Level of delegation

Some organizations are highly complex, with numerous divisions and levels. It is possible, therefore, to waste considerable effort developing strong relationships with certain personnel who might have little influence or authority to act on key management issues. Knowing when to deal with head office and when to do business with the local division can be critical to effective policy implementation. It is important not to necessarily assume that the higher authority is better; in some situations, those with local jurisdiction can make something work that would be impossible if directed from head office. Although developing partnerships is often championed as a positive initiative, it should be obvious from the discussion so far that it is not necessarily easy or efficient to do in the short term, and this can be quite problematic in the medium and longer terms. It should also be clear by now that not all smooth-running relationships are necessarily good for resource conservation in general and IREM in particular. It was seen, for instance, that corporatist relationships that symbolize close working relations between agencies, such as in Gary, Indiana, could spell harm to environmental objectives.

The multiagency ecology lens

This analytical lens focuses on the combined impacts of interagency relationships on resource and environmental management as they amass, accentuate and possibly cancel each other out. This level of analysis is then about interactions among various agencies and organizations rather than between them (Mandell, 1989). Some of the variables highlighted here have counterparts in previous parts of this framework. They are revisited here to emphasize their special significance in the organizational network or ecology. For example, goal congruency examined under the agency lens is also considered here as 'network goal congruency', and management style is examined under the category of 'network baggage'. In addition to these organizational dynamics, this lens considers: (i) organizational and programme complexity; (ii) network resource flow; and (iii) network coordination as key items for analysis.

Organizational and programme complexity

The sheer numbers of agencies and organizations – which is referred to as quantitative complexity – to be included in any IREM scheme can be quite daunting for anyone charged with policy oversight. Imagine, if you will, attempting to promote conservation management in an ecosystem with say six or seven landowners versus a situation involving 500 or 600 natural resource owner/managers. All other things being equal, it is obvious that the management challenge will increase as the quantitative complexity increases.

When qualitative complexity is added to the mix, however, the management challenge increases enormously. If some of these agencies are regional entities of large multinational corporations, for instance, and others are small local family businesses such as a mixed farm or a family-run sand and gravel operation, oversight management will involve weighing the appropriate influence of one agency against another. How should this be done? Should there be one multinational forestry company that controls 50% of the land base in an ecosystem and 300 small woodlot owners that manage a total of only 40%; how does the conservation policy overseer allot time and energy to each agency and what weighting will the policy manager attribute to each? Furthermore, if the ecosystem supports a broad variety of industries and one or two are particularly problematic as far as pollution is concerned, should the policy overseer focus more time and effort on trying to rein in the polluters or should the manager concentrate effort on maintaining relationships with those that are more supportive?

This aspect of multiagency management is referred to as programme complexity. Programme complexity does not just refer to the existence of different industries with diverse natural resource management objectives; it is also concerned with the broad spectrum of public policy which adds regulatory complexity. Each of the industries is probably driven by different market incentives and is constrained by different types of government legislation and regulation. One piece of legislation, for example, might be focused on maximizing production within an industry with all manner of financial incentives and subsidies, while another public policy may control the development of another industry so as to protect aesthetic qualities. Yet another may restrict location to minimize stream pollution. Given this broad range of policy possibilities, the application of an IREM paradigm may be superimposed on a simple policy framework or may be applied to a highly complex organizational environment where success requires overcoming the established and highly complex programme inertia.

Network baggage

This refers to all those human factors that pre-exist in a resource management situation including programme complexity, as explained above, that must either be built upon or dismantled for a new policy or paradigm to take hold. Here it is necessary to emphasize two key aspects of the policy process. The first is the influence of network ideology, and the second, as inferred above, is policy sector inertia. As was suggested with the first analytical lens, agency ideology relates to the dominant, pervasive and often not obvious driving forces that influence policy decision making.

At the network level, there is generally a dominant decision-making ideology that prevails, which is typically obscure and also poorly understood by those involved in the policy process. The prevailing network ideology examined here relates largely to the implicit rules used to make decisions about natural resources. For example, does allowing dairy farmers, who let their cattle graze up to the stream bank, allow them to drink directly from those streams and consequently defecate directly into the stream, causing water pollution, constitute an acceptance of a natural resource management ideology that supports water pollution? In situations such as this, the general public, the government and the farmers themselves often oppose actions that might directly impact milk production costs, as low-cost milk is implicitly seen as more important than stream pollution. No one directly admits to supporting such a policy, few speak of its existence and, perhaps most importantly, few in authority are willing to do anything about this. Behaviours such as this (which is no action) constitute policy sector inertia, which has its own momentum. Such behaviour is likely to strongly impede initiatives such as IREM.

Network goal congruency

Just how close a proposed policy goal is to prevailing practices or an emerging recognition of the need for change is also critical to the initial and the ongoing success of a policy initiative such as IREM. If the gap between established practices and proposed changes is too large or, alternatively, if there is little support or appreciation for change, then the acceptance of any new initiative will be problematic.

Network resource flow

As mentioned above, management resource flows, particularly regulatory authority and fiscal resources, but also the flow of natural resource products and even public opinion, are additional factors key to the efficient and effective implementation of conservation policy. Tracking the transfer and the receptivity of such resources helps in understanding who is communicating or working cooperatively with whom and why. From such analyses, especially if these are charted in an organizational map, important insights can be gained as to how resources are exchanged and what possible interagency influences these might have in superimposing a conservation policy such as IREM on a resource sector's everyday workings.

Network coordination

Given the complexity of the multiagency management setting, the IREM overseer must attempt to manage countless interacting forces impinging on the policy process. In most environmental management policy situations, a new environmental scheme is often framed in very loose legislation that relies heavily on goodwill rather than command and control mechanisms. The ability to effectively coordinate rather than control multiagency participants is then critical to the overall effectiveness of a new policy direction or the successful maintenance of an already well-established initiative. Unfortunately, the costs of effective coordination are too often underestimated and, if policy implementation seems to be progressing satisfactorily, then management resources are too readily reassigned elsewhere. As suggested above, interagency relationships are dynamic, and require constant monitoring and continued attention. All too often, network coordination is reduced to a process of 'putting out fires' rather than necessary pro-action, i.e. avoiding problems in the first place by observing trends and dealing with trouble before things reach boiling point. In the final analysis, however, as suggested in the discussion of network goals above, much of the success of implementing IREM will depend upon the goodwill and support of a critical mass of resource management participants (see also Chapter 8 and the free rider effect).

It is clear from examining this analytical lens that multiagency dynamics is not about the sum of the parts but is instead a highly complex state of affairs where the actions of one agency may neutralize the potential impact of another, where the seemingly simple action of one agency can generate momentum to create effective collective action or, alternatively, where considerable energy and management resources can be expended with all good intentions but attain very disappointing results. If the multiagency dynamic as already presented is not complicated enough, it should always be kept in mind that the application of IREM principles takes place in a broad and constantly changing political economy and evolving biophysical environment. To better appreciate this policy dynamic, the fourth analytical lens is outlined below.

The macro policy environment lens

This lens views the impact of all the outside influences on the management system, which include the broad biophysical transformations taking place within the ecosystem (this is where the human dimensions manager or social scientist integrates the work of the biophysical scientist and/or technician); the social, political and economic changes that impact the management system; and the longer term structural changes or paradigm shifts that fundamentally change the way we approach natural resource and environmental management.

Biophysical transformations

Often overlooked in the relatively short-term life span of human beings is the fact that ecosystems evolve and change naturally over time. In biological time frames, the biodiversity and structure of an ecosystem change to alter genetic structure, species composition and species distribution. In meteorological time, weather patterns change both in the relative short term (say over a period of 5–10 years) and in the longer term (centuries and millennia). In geological time, rivers change course, glaciers appear and disappear, and mountains form out of what was originally lake sedimentation. Landscapes are constantly changing whether we perceive those changes or not. In most circumstances, these natural processes are beyond the control of humans. By and large, natural resource managers must adapt to these changes as they occur. There are, nevertheless, many instances of human alteration of natural landscapes; many of these are inadvertent and some are deliberate, and often they occur with unintended and unanticipated consequences.

Perhaps the most controversial, in terms of clashing theories and abutting management prescriptions, is the issue of global warming. While over 100 Nobel Science Laureates agree that there is indeed global warming and it is induced by human behaviour, and that we should embark on a strict regime of restraint to reduce the ‘enhanced greenhouse effect’, several other quite reputable scientists suggest that this is quite probably a natural occurrence and that, in any event, the full implementation of initiatives such as the Kyoto Accord or any other human remediation effort can and will have no appreciable effect (Essex and McKitrick, 2002).

Regardless of the scientific and policy debate concerning the origins of and solutions for global warming, there is widespread agreement that human action can have quite devastating impacts on biophysical systems and that these systems have quite profound effects on our resource management efforts as well as our everyday lives. The degree to which we can adapt to or manage this is, however, situation-specific, usually requiring continuous adaptive management strategies. In this evolving biophysical context, it is essential, therefore, to have a clear understanding of what the biophysical goals are for an ecosystem, and how IREM or other conservation strategies might address these objectives or indeed whether the ecosystem left to its own devices might be more effective. Within the farming and forestry context, for instance, natural biological processes, such as so-called weed infestations and the broader physical shifts such as changes in weather patterns and river drainage configurations, can substantially alter the potential effectiveness of intended IREM goals. In some situations, such changes can render what once appeared as a sound sustainable development strategy rather obsolete. Such an event occurred recently in the Mountain Pine Ridge Forest Reserve in Belize, Central America. This forest was devastated by a bark beetle that caused almost total pine tree mortality. As a result, an established forest management plan was abandoned and replaced by one emphasizing pine log salvage.

The social and political economy

The earlier chapter on decision making and power (Chapter 7) outlined three basic theories of power in the macro political environment (pluralism, elitism and structuralism) that are useful to consider here in identifying the broad impacts of power and influence over the multiagency IREM process. Also useful in this analysis is the earlier discussion of globalization and the counter anti-globalization effort. While IREM typically requires long-range planning horizons, it is not unusual that short-term shifts in the political economy or indeed shifts in public attitudes can present both opportunities and constraints on the multi-agency management process. A pulp processing company, for example, may at one time be highly supportive and quite active in long-range conservation planning and management, but in a relatively short time frame find itself backing away from earlier commitments as market cycles – the result of shifting global demand – constrain managers' efforts to maintain investment in environmental management.

Paradigm shifts

Although the above discussion refers to the way in which different agencies may approach resource and environmental management as a reaction to shifts in the broad social and political economy, longer term shifts also occur in the way our society approaches various resource and environmental management problems. As suggested in earlier chapters, a paradigm shift results from new and profound ways of understanding public policy challenges and/or from substantially different circumstances that require new ways of dealing with such issues. Perhaps the most pervasive paradigm shift in resource and environmental management is the result of new thinking about sustainability. Such thinking requires managers to look beyond the maximization of production and profits to integrate the idea of environmental management early in the production planning process and, in the case of renewable natural resources, ensure that future generations have at least equal access to similar resources as the present generation enjoys.

Summary and Conclusions

It is clear that any prescriptive strategic planning models designed for multiagency ecosystem management environments must in practice remain very different from its single agency counterparts. It must also be emphasized that these more complex prescriptive frameworks must incorporate a clear and generally shared vision of acceptable strategies that are both motivating and effective. It should be noted that while the corporate planning literature documents numerous successes in corralling divergent intraorganizational interests to pursue a common vision, pursuing such a vision in the multiagency context appears much more challenging.

Without convincing evidence from elsewhere that command and control measures, which are often the backbone of internal corporate mechanisms, are particularly effective in the multiagency ecosystem management context, a high degree of consensual management behaviour appears essential for multiagency IREM success. Indeed, consensual management involves all interested and influential stakeholders and provides much of the philosophical and procedural foundation for IREM. Even with considerable support among participating agencies, organizations and civil society in general, it is likely to be advantageous and perhaps necessary to reinforce generally acceptable practices in contractual rules and/or public policy regulations. Such codification of generally acceptable practices helps avoid backward slides when external influences such as reduced budgets or sluggish markets encourage managers to make cutbacks which are often at the expense of longer term sustainability. As Bissix and Rees (2001) suggest, the widespread acceptance of sustainable ecosystem practices will not be easy with so many internal and external influences favouring short-term expediency. It seems that before a politically acceptable communal conservation management system can succeed, extraordinary efforts will be necessary to identify benefits, build and maintain trust, and provide mutual support.

The evidence from the Interagency Fire Management System in the USA, Dartmoor National Park in the UK, Whytecliff Marine Park in British Columbia, Canada, and the management of the Great Barrier Reef Marine Park in Australia shows, with varying degrees of success, that multiagency ecosystem management is indeed possible (Bissix and Rees, 2001). Unfortunately, we know of few prescriptive principles that can be applied routinely to complex ecosystem management systems. It seems imperative, nevertheless, that natural resource managers should not fall prey to the notion that IREM in complex ecosystems management situations will work just because we want them to. Success will require persistent and skilful management by both IREM overseers and a critical mass of the multiagency participants.

In Chapter 11, we draw on the theoretical and contextual underpinnings of IREM provided throughout this text to offer a basic procedural framework for analysing the human dimensions of IREM in a broad array of resource and environmental planning and management situations. While this framework is both strategic and comprehensive, it must be kept in mind that it is also reductive and as such may oversimplify the human aspects found in multiagency management processes. Because of this framework's deliberate simplicity, it behoves the user, when applying this framework in the more complex multiagency situation, to revisit this chapter frequently and, where appropriate, superimpose, in an iterative and pragmatic way, its essential lessons.

As with previous chapters, at the end of this chapter your understanding of the multiagency management context will be put to the test through a series of analytical study questions. A case study follows concerning the establishment of a new National Park in Scotland known as the Loch Lomond and the Trossachs National Park. To assist in this and future analyses, Appendix A provides a multiagency diagnostic that guides the user through the inherent complexity of the multiagency management process. This diagnostic is based on the multiagency analytical framework described in this chapter, but provides a systematic list of guiding questions that cuts through the system's inherent complexity. In essence, this diagnostic provides insights as to what multiagency dynamics are occurring in a particular situation and it also provides clues as to how the system might be influenced or managed.

As you will see, this chapter's case encapsulates the fundamental challenge of any multiagency, ecosystem management process, which is to tie together the multiple interests and the myriad resource management approaches of its organizational participants to pursue a common conservation goal. This national park creation initiative represents a conscious attempt to integrate resource and environmental management in what is a nationally and internationally significant landscape. It combines, among other uses, human settlement, wildlife conservation, farming, forestry, water management, water supply, outdoor recreation and tourism, and the military as well as various other industrial uses. In examining this case, you will be introduced to the drawn-out process of building a consensus for national park designation, and you will also be exposed to the many complementary and contradictory pressures that must be reconciled in a practical and politically acceptable way for this conservation initiative to succeed over the long term. As is usual with all the cases in this text, you are provided with a number of questions at its conclusion that focus your analysis and understanding of the multiagency IREM process. Such an appreciation will hopefully develop the skills necessary to apply IREM in the real world.

Case Study
Loch Lomond and the Trossachs: Scotland's First National Park

ROSS FIRTH

Preamble. This case study examines the establishment of a new National Park in Scotland known as the Loch Lomond and the Trossachs National Park. In this example, we look at some of the fundamental challenges involved in the multiagency, ecosystem management process, including the tying together of multiple interests with a variety of resource management goals and objectives held by the organizational participants as they pursue a common conservation goal.

Introduction

Situated 20 km outside Glasgow, Loch Lomond is Britain's largest freshwater loch (lake) with a surface area of 70 km^2. It runs for a length of 36 km and, at its widest point, is 8 km across. It reaches a maximum depth of 190 m and is the second largest loch in the UK by volume. Dotted throughout Loch Lomond are 33 islands. Some, like Inchmurrin (Island of St Murrin), are home to a hotel, houses and livestock grazing, while others, such as Ellan Vhow (Island of the Cow or Isle of Holy Women), hold the ruins of a castle constructed by the clan MacFarlane in the late 16th century.

Swans on Loch Lomond, Scotland. Notice the residential and tourism development encroaching on the water's edge. Photo by Glyn Bissix.

Conservation of the built environment in the village of Luss on Loch Lomond. Photo by Glyn Bissix.

Loch Lomond is located adjacent to Scotland's central belt, which is an area measuring roughly 80 km × 50 km between Edinburgh and Glasgow and is home to approximately 3.5 million inhabitants. As one of Britain's prime tourist destinations, it attracts visitors from all over the world in addition to large numbers of British holidaymakers, and experiences upwards of 5 million visitors per year passing through the area.

The breathtaking scenery that reflects a rich natural heritage attracts visitors to the area. This is complemented by a cultural history that stretches back for millennia. In order to understand the resource management pressures of Loch Lomond, it is necessary to place the Loch in a context that considers adjacent land uses. It is only by integrating both land and water issues that one can begin to comprehend the true nature of the management complexities of the area. One can then examine past and present management regimes and discuss future possibilities.

A mountainous fiord-like northern reach characterizes the natural landscape of the Loch Lomond area, while the southern end is broader and shallower. This is a result of geological forces where the Highland Boundary Fault cuts across Loch Lomond, leaving the loch at the southern edge of the Highlands and the northern edge of the Lowlands. Loch Lomond is subject to an oceanic climate that is damp with mild winters. Found in and around the loch are over 500 plant species and 31 native and introduced mammal species. Over 200 species of birds have been recorded in the area, and 18 species of fish reside in the loch, including the *Powan*, which is found in only one other Scottish loch.

There is evidence to suggest that Neolithic peoples had begun to settle in the area at some point during the 4th millennia BC – over 6000 years ago. Later peoples included those of the Bronze and Iron Age, Picts, Celts, Romans, Vikings and Anglo-Normans. The Vikings are known to have dragged their longboats across the 3 km isthmus from Arrochar to Tarbet and into the loch, where they raided shoreline settlements in the 13th century. More recently, families such as the Lennox, McGregor's, Colquhuon's and McFarlane's have played significant roles in the cultural heritage of the area. In 2002, the Scottish Parliament designated the Loch Lomond and the Trossachs National Park. It is Scotland's first National Park and is the culmination of a process that began in the 1920s. Within the National Park, a myriad of agencies and organizations operate to manage specific natural resources. They operate in an environment fraught with conflicting aims and a lack of an integrated and coordinated approach to resource management.

Background

In 1988, a 440 km^2 Regional Park was created that had Loch Lomond as its epicentre. The Loch Lomond Park Authority had, as its aims, a duty to: (i) conserve and enhance the natural beauty, heritage and natural history of the Loch Lomond Park; (ii) promote the enjoyment of the Loch Lomond Park by both residents and visitors; and (iii) promote the social and economic well-being of the people and communities within the Loch Lomond Park area.

The Park land base was overwhelmingly in private hands and was the focus of a wide range of land- and water-based issues. An Executive Committee comprising representatives of the three local Councils (local government) and non-elected members was responsible for administering the Park. In late 1999, the Loch Lomond Park Authority became the *Loch Lomond and the Trossachs Interim Committee* (LLTIC). The LLTIC was created in response to the developmental process to work towards the establishment of the Loch Lomond and the Trossachs National Park. The LLTIC was the temporary body that 'paved' the way for the National Park Authority and was composed of 24 members of whom 16 were local councillors and eight non-elected members. Employing approximately 60 full-time equivalent staff, the LLTIC had a wide remit that included issues of planning, visitor services and project management. As previously mentioned, the National Park was formally designated in 2002; the National Park Authority assumed control from the Interim Committee shortly afterwards. The National Park covers 1675 km^2 and contains a resident population of 14,000. The National Park is, according to IUCN definition, a Category V protected area where the primary management objectives include protection of specific natural/cultural features, tourism and recreation, and the maintenance of cultural/traditional attributes. This is in contrast to the Category II National Parks of Canada and the USA, where lands are set aside primarily for ecosystem protection and recreation.

Loch Lomond has, for many centuries, been the focus of a number of activities supported by a rich land and water resource base. These activities have been managed by a variety of public and private agencies and organizations and can be grouped under the following headings: tourism, recreation, conservation, agriculture, forestry, water abstraction and hydroelectric generation. Alongside these organizations can be found additional groups, such as academic and research institutions and non-profit interest groups. Each has its own set of aims and institutional values which are manifested in management strategies, policies and operating procedures.

Management may be contrasted not only by a public and private divide but also on a level of scale. Organizations such as the Friends of Loch Lomond and the Loch Lomond Angling Improvement Association are locally based groups interested specifically in the management of the Loch Lomond area and site-specific issues within that area. Forest Enterprise, the main government body responsible for forest management in Scotland, must consider management of its plantations as part of a wider regional and national context. Loch Lomond contains protected area designations that have been made in response to international initiatives such as The Convention on Wetlands of International Importance (Ramsar) as well as European Union directives (Natura 2000, Eurolakes Project).

It is useful to consider three of the aforementioned management headings as being illustrative of some of the ongoing management issues in the area. Tourism, conservation and recreation represent sectors that reflect the complexities of day-to-day management in the Loch Lomond area.

The tourism industry in Scotland generated £2.5 billion in 1999 and provided approximately 8% of Scotland's employment. Visitors spent £242 million within the local tourist board area in 1999. People travel to Loch Lomond primarily for its beautiful scenery. It is the aim of the local tourist board and the national body responsible for tourism to promote Loch Lomond as a visitor destination both nationally and internationally. In 1991, some 5 million visitors passed through the Loch Lomond area. In order to accommodate these numbers, adequate infrastructure and support services must be in place. Appropriate transportation networks (roads, railway, bus, airport), accommodation provision, additional built tourist attractions and access to natural and cultural landmarks must be considered, developed and managed.

Local residents depend in large part on the tourist trade for their livelihood in both a direct sense and in the spin-off effect. It is in the best interests of all those involved in tourism to ensure that tourism and its associated impacts do not jeopardize the cultural and natural resources upon which the industry is based. However, tourist developments do, on occasion, conflict with agricultural, recreation and conservation interests.

There is a plethora of protected areas designations in the Loch Lomond area. These include Sites of Special Scientific Interest (SSSI), National Scenic Area (NSA), National Nature Reserve (NNR), Special Areas for Conservation (SAC), Special Protection Area (SPA), Ramsar Wetland Site, Forest Parks, Regional Park, Country Park, Memorial Park and an Environmentally Sensitive Area (ESA). Scottish Natural Heritage (SNH) is the main government body responsible for safeguarding and enhancing Scotland's natural heritage and has a lead role in managing many of the natural heritage designations mentioned above. Other management bodies include local governments, other government agencies and individual landowners.

Two other organizations that own land in the Loch Lomond area who manage their land for conservation are worthy of mention. The Royal Society for the Protection of Birds (RSPB) owns a nature reserve, and The National Trust for Scotland own a 5400-acre property adjacent to the loch. Both are nationally constituted organizations that own and manage properties throughout the UK based on the general principle of conserving species and landscapes. The 'reserves' that these two organizations own and manage are already situated in a SSSI and/or an NSA. The Scottish government also operates an agri-environment scheme that provides incentives to landowners for the effective management of farmed environments that are important to biodiversity and protected areas. In addition, the loch-side village of Luss has been designated as an Outstanding Conservation Area.

A number of outdoor recreation activities take place in the Loch Lomond area that include such things as boating, hill walking, mountain biking, bird watching and fishing. On Loch Lomond alone, there are over 5200 registered boats. This particular activity brings with it issues such as hydrocarbon emissions into the loch, sewage from boat toilets, disturbance of wildlife, shoreline erosion, camping and associated impacts on the islands, and a decline in recreational quality and safety. Bylaws for Loch Lomond were instituted in 1995 but have been enforced in a very forgiving manner, with the vast majority of people stopped being given verbal warnings. Current management of boating issues is divided between the Interim Committee, SNH, Scottish Environmental Protection Agency, the Loch Lomond Association, various boating and cruising clubs, and Sport Scotland.

In order to accommodate such boating numbers, sufficient marinas and launching facilities must be made available with ancillary support services. The loch also supports a popular non-commercial fishery that is managed by the 1000-strong membership of the Loch Lomond Angling Improvement Association to ensure sustainable fish populations and prevent the introduction of non-native species. There is no shortage of resource management issues within the Loch Lomond area. At present, a variety of organizations have either a direct or indirect management remit for the loch and surrounding countryside. The logistics of attempting to bring together such diverse organizations in order to develop an integrated resource management strategy pose significant difficulties.

The Current Issue

The management of the Loch Lomond and the Trossachs National Park will be guided by four aims as set out in the National Parks (Scotland) Act 2000. These include: (i) to conserve and enhance the natural and cultural heritage of the area; (ii) to promote sustainable use of the natural resources of the area; (iii) to promote understanding and enjoyment (including enjoyment in the form of recreation) of the special qualities of the area by the public; and (iv) to promote sustainable economic and social development of the area's communities.

The mechanism by which these aims will be met will be through the development and implementation of a Park Plan. The Act states that a National Park Authority must prepare and submit a plan that sets out its policy for: (i) managing the National Park; and (ii) coordinating the exercise of: (a) the Authority's functions in relation to the National Park; and (b) the functions of other public bodies and office holders so far as affecting the National Park, with a view to accomplishing the purpose set out in Section 8(1). Section 8(1) states that the general purpose of a National Park Authority is to ensure that the National Park aims are collectively achieved in relation to the National Park in a coordinated way.

With the National Park Authority owning very little land and the remaining area in the hands of either other government agencies or private landowners, the task of 'collectively' achieving the National Park aims will be highly problematic. The National Park Plan will therefore be seen as the cornerstone document that sets out the vision for the area and articulates how its policies and management objectives will be met.

While it is hoped that through a lengthy and meaningful consultation process all government agencies that currently have a management remit in the area would support the National Park Plan, this is by no means assured. The National Park Act states that all public bodies in exercising functions so far as affecting the National Park must have 'regard to' the National Park Plan. The term 'regard to' has conveniently been left undefined in the Act. In the absence of a clear interpretation of this term, serious concerns must be raised when the time comes to coordinate the management goals of the Park with those of other government bodies currently operating in the area.

The Interim Committee consulted on the Loch Lomond and the Trossachs Local Forestry Framework in partnership with the Forestry Commission and Scottish Natural Heritage. It was hoped that this document would represent the type of coordinated joined-up thinking that the area needs. However, the extent to which Forest Enterprise's own Forest Design Plans will have 'regard to' this Framework and the Park Plan is not yet clear. Also, given the fact that some organizations are already showing signs of 'drawing a line in the sand' behind which they will operate to the exclusion of the National Park must raise significant concerns.

In addition, the European Community's Water Framework Directive will require member states to prevent deterioration and achieve 'good status' for all waters by 2016. This is to be attained through river basin management plans. The Scottish Environmental Protection Agency, Scottish Natural Heritage and East of Scotland Water have conducted public consultations toward the development of a catchment management plan for Loch Lomond. This catchment plan is intended to complement plans for the Loch Lomond and the Trossachs National Park. Interestingly, the proposed National Park boundary will not include the entire Loch Lomond water catchment area, thus compounding an already challenging management scenario.

The National Park's four aims, upon closer inspection, contain potential management anxieties with their call for the conservation of the natural and cultural heritage; promotion of the sustainable use of the natural resources; promotion of the understanding and enjoyment of the area; and, lastly, promotion of the sustainable economic and social development of the area's communities. The Park may be hard pressed at times to counter-balance conservation values successfully against what some might see as aims that may result in the further development and increased use of the area. However, in an attempt to address concerns regarding the primacy of the National Park aims, the Scottish Executive has included a section within the National Parks (Scotland) Act 2000 which states that where there is a conflict between one or more of the National Park aims, the Park Authority must give greater weight to the aim that seeks to conserve and enhance the natural and cultural heritage of the area. In addition, the National Park Authority replaced the four local councils as the statutory planning body for the Park. It is hoped that this will provide for a more coordinated approach to planning.

The Interim Committee saw the role of the public in the development of an integrated approach to resource management in the area as an essential component of any plan. While large government bodies own and manage land and associated resource values, there remain considerable holdings in the hands of local residents. This is particularly true with regard to the agricultural sector. With many residents remaining sceptical of the benefits of a National Park, the ability to involve them in the development and ongoing management of the natural resource base is crucial to any long-term success in integrated management.

Public consultations are one of the most favoured approaches to community involvement; however, there are other means available to residents that include the recently launched Community Futures project. There are five seats, or 20% of the total number, on the National Park Board that are locally elected. The National Park also operates a number of advisory groups, the membership of which is determined by the Park. It is anticipated that local residents will eventually comprise a large proportion of the membership of these groups.

Conclusion

Given the current multiagency approach to resource management in the Loch Lomond area, the National Park body is faced with the difficult task of both developing and implementing a Park Plan that reflects a much more integrated management system than is presently in place. As a newly incorporated statutory body, the National Park must work within a defined legislative framework that sets out its aims and allows for the means of achieving those aims. A very clear and powerful vision for the Park is required to enable resources to be focused and allocated effectively. The National Park is the subject of forces arising not only from a local level but also nationally and internationally. Under these circumstances, one of the challenges is the ability of the Park to take a strong leadership role in effectively integrating a wide spectrum of values and directives into a management regime that aspires to the long-term health of the landscape, including that of its residents. This must be done in a spirit of cooperation and a shared sense of vision amongst all those involved in the process.

Author's note. The views expressed in this case study are those of the author and do not necessarily reflect the views of the Loch Lomond and the Trossachs Interim Committee.

Discussion Questions

1. Give four key reasons why multiagency ecosystem management is more difficult than single agency management.
2. Provide your own example for each of the three different types of multiagency resource management, i.e. type 1, controlling partner interests; type 2, hub and wheel; and type 3, working among various partners. Explain why each example fits that particular type.
3. Think of a continuing relationship with a colleague that you developed a year or so ago. Recall the strategies you used to develop that relationship in the beginning and why that relationship was important to you. Now think about how that relationship has changed over time and what efforts you have made (if any) to maintain that relationship. Are there lessons that might be applied to working with a partnership agency? Whether yes or no, please explain your reasoning?
4. Consider the same relationship with a colleague that you reflected on in question 3. Think of a situation with that colleague where it was necessary to develop a consensus on a particular issue. What was the issue and what compromises or bargaining did you undertake to reach a consensus?
5. Identify a natural resource management situation that involved several different types of agencies to resolve an issue. Describe the different kinds of agencies involved and identify their interests or objectives in this matter. To what extent do you believe these objectives or interests are compatible?

Case Study Discussion Questions

6. What are some of the opportunities and constraints facing Park managers in the development of an integrated resource and environmental management plan for the Loch Lomond and the Trossachs National Park?
7. National Parks in the UK including Scotland represent an approach to protected areas management that includes areas with large resident populations and where landscapes often show signs of having been inhabited for thousands of years. This is in contrast to North American National Parks where residents, if any, are most often confined to relatively new town sites with tight planning controls. How might you meaningfully involve residents in the development of the Park Plan and continue to ensure that they have a role in its implementation? What things might you consider in the development of this strategy?
8. What approach might you take to successfully bring together resource management agencies and organizations so as to gain and maintain their support for a National Park Plan that represents an integrated approach to resource and environmental management?
9. Given the four aims of the National Park and taking into account that resource management agencies operate to different value and belief systems, prepare a draft vision statement for the Park Plan.

References

Bissix, G. (1999) Dimensions of power in forest resource decision-making: a case study of Nova Scotia's forest conservation legislation. Unpublished doctoral dissertation, Department of Geography and Environment, London School of Economics and Political Science, London.

Bissix, G. and Bissix, S. (1995) Dartmoor (U.K.) National Park's landscape management: lessons for North America's Eastern Seaboard. In: Herman, T.B., Bondrup-Neilsen, S., Willison, J.H.M. and Munro, N.W.P. (eds) *Ecosystem Monitoring and Protected Area*. Science and Management of Protected Areas Association, Wolfville, Nova Scotia, Canada, pp. 563–571.

Bissix, G. and Rees, J.A. (2001) Can strategic ecosystem management succeed in multiagency environments? *Ecological Applications: a Journal of the Ecological Society of America* 11(2), 570–583.

Essex, C. and McKitrick, R. (2002) *Taken by Storm: the Troubled Science, Policy and Politics of Global Warming*. Key Porter Books, Toronto, Ontario, Canada.

Harrison, M.L. (ed.) (1984) *Corporatism and the Welfare State*. Gower, Aldershot, UK.

Mandell, M.P. (1989) Organizational networking: collective organizational strategies. In: Rabin, J., Miller, G.J. and Hildreth, W.B. (eds) *Handbook of Strategic Management*. Marcel Dekker, New York, pp. 141–165.

O'Toole, L.J., Jr and Montjoy, R.S. (1984, November/December) Interorganizational policy implementation: a theoretical perspective. *Public Policy Review* 44(6), 491–503.

Ottesen, P. and Kenchington, R. (1995) Marine conservation and protected areas in Australia: what is the future? In: Shackell, N.L. and Willison, J.H.M. (eds) *Marine Protected Areas and Sustainable Fisheries*. Science and Management of Protected Areas Association, Wolfville, Nova Scotia, Canada, pp. 151–164.

Appendix A. Diagnostics for a Multiagency Strategic Management System

Lens A: agency character

Agency autonomy

To what extent does the agency control its own destiny?
How responsive is the agency to external influence?
How receptive is the agency to cooperative initiatives?

Agency authority

How much influence does this agency have over others in the network?
What are the mechanisms of its authority?
Where is the locus of decision-making control?

Network goal concurrence

How do this agency's goals match prevailing network goals?
How do this agency's goals match the network's intended vision?
Do this agency's values reflect the status quo or the new vision?

Decision-making style

PROCEDURAL RATIONALITY

To what extent does this agency follow a rational management process?
To what extent does this agency rationalize its goals by lowering 'the bar' to accommodate unrealized objectives?

ORGANIZATIONAL PROCESSES

To what extent do this agency's organizational subunits act in self, as opposed to agency, interests?

POLITICAL BARGAINING

To what extent does this agency bargain long-term goals against short-term expediencies?

Lens B: interagency relationships

Nature of relationship

What, if any, regulatory or contractual ties are there between agencies?
What is the nature of informal interagency relationships?
What are the political and economic ties between agencies?
Is this a developing or established relationship?

Delegation

To what extent does an agency delegate its statutory, moral, political or market authority to other agencies?
To what extent does an agency support and facilitate other agencies?
To what extent does an agency abdicate its authority and obligations?

Agency cooption

To what extent does an agency willingly conform its objectives to those of other agencies?
To what extent does an agency unwittingly conform to other agency influences?

Information management

How does an agency control or facilitate information management with another agency?
How does an agency conduct intelligence on another agency?
How does an agency use information to influence another agency?

Resource flow

How do management and market resources flow from one agency to another?
How does resource flow impact an agency's autonomy, authority and decision-making effectiveness?

Agency capture

What evidence exists that one agency controls the behaviour of a regulatory agency?
Is there evidence of tacit agreements that undermine legislative or policy objectives?
Is there evidence that the 'enforcement' of regulations and policies are handled behind closed doors?

Corporatism

Is there evidence of a statutory or lead agency bestowing special power on another?
Does an agency have systematic influence on a regulatory agency?
Does an agency enjoy special privileges not extended to another?

Professionalism

Are there strong professional ties between members of one agency and another?
Does this professional influence undermine or enhance sector goals?

Lens C: organizational ecology

Field complexity

QUALITATIVE COMPLEXITY

How diverse are the agencies, organizations and interests involved in the network?
How can they be categorized according to form and function?
Can their respective agency characters be usefully categorized under these conceptual categories?

QUANTITATIVE COMPLEXITY

How many units are involved in each of the above categories?
Is it possible to generalize about what management routines and other processes are used to maintain relationships within the network?

Programme complexity

What are the significant government and private sector policies and programmes impacting the network?
How compatible or incompatible are these programmes?
How many programmes must an agency regularly deal with?
How many organizational points of contact and influence are used to execute these programmes?
What is the management burden in maintaining programme integrity?
To what extent are regulations known and complied with?
What programme strategies appear to be most effective (financial incentive, regulation, education, etc.)?
What role do market forces have on programme effectiveness?

Goal congruency

What overlap and differences exist between agency purpose, missions, programme objectives, methods of operation, management style and ideology?
In what direction does the critical mass of agencies gravitate?
Are there significant minority groupings?

Coordination

Is there evidence of coordinated network action in this sector?
If so, who has the statutory authority to coordinate action?
What factors influence coordination efforts?
What role do various professional and sectoral associations play in coordinating efforts?
What management resources are expended on coordination efforts?
How effective are coordination efforts?
What gaps exist in coordination?
What evidence is there of discordant behaviour?
What are the limits of coordination effectiveness?

Historical baggage

What significant history has impacted this sector?
What impacts do these historical events have, or are likely to have, on coordination efforts?
What network inertia exists?

Management network ideology

What are the significant management styles of network agencies?
What is the dominant network paradigm?
To what extent does an *ecocentric* decision style pervade the network?
To what extent does a *technocentric* decision style pervade the network?

Interdependency

What interdependencies exist between and among agencies in the network?
How do they impact public policies and market processes?

Lens D: resource management external factors

Macro socio-political influences

What are the economic forces impacting the management system?
What are the social forces impacting the management system?
What are the political forces impacting the management system?
What traditions impact on the management system?
What are the technological forces impacting the management system?

Paradigm shifts

What social paradigm shifts or influences are evident?

Biophysical cycles and fluxes

What are the natural and human changes on the natural resource base?
What resource management techniques, practices and outcomes impact natural resource management?
What technologies are useful in managing this resource?

11 Planning Framework for IREM

Introduction

As no two resource management situations, natural resource managers or management teams are alike, it is difficult to imagine a simple or single prescriptive approach to IREM that can work effectively for all applications. The complexity and distinctiveness of each resource management situation, especially when attempting to accommodate a broad range of interests and values into the decision calculation, require management continually to adjust to the exacting demands of an IREM philosophy. In fact, when managers adopt an IREM strategy, they accept a considerably more challenging mission to ensure that their management practices are sustainable over the long term, and that a wide range of resource and environmental considerations are taken into account in the overall management process. The importance of IREM, its implied environmental management philosophy and its influence over management operations depend on a number of key factors such as an agency's financial health and aspirations, the potential of the natural resource itself, an agency's recognition and acceptance of its broader societal obligations, and a clear understanding of how it is able to apply IREM strategies effectively.

As made clear throughout this book, in order to implement IREM effectively, resource managers must carefully blend short-term economic and social goals with the longer term expectation of sustaining the natural resource base and protecting the natural environment over the long term. IREM, among other things, is the practical application of 'intergenerational equity' in natural resource management. As far as possible, it ensures that our generation hands down a reasonable quality of life-sustaining environmental assets to the next generation and a legacy of renewable natural resources that are in no overall worse condition than the stock of natural resources that our generation inherited (Brown, 2001, calls this the commonwealth of life). It also means, in the present, that we do not unduly externalize the environmental costs of our resource management business, for example creating downstream pollution in our mining operations.

As you will recall, earlier chapters laid out in some depth the historical and contemporary influences, as well as the driving forces that have given rise to the need for IREM. These earlier chapters provided a number of general strategies for viewing this natural resource management context; they also provided more detailed insights into the complex world of intra- and interorganizational processes that either expedite or impede the IREM process. While Chapter 9 outlined the theoretical underpinnings of IREM to provide management guidance on developing a planning framework, and Chapter 10 examined the complex world of multiagency ecosystem management, this chapter boils down the text's essential messages in order to provide a guiding planning and procedural template for managing a wide range of natural resource management situations and resource values. In this process, it is important for practitioners to ensure that they never lose sight of resource management's inherent complexity and, as a consequence, they must judiciously customize this template to meet their own planning and management requirements.

Developing a Planning Framework

Given the complexity regularly found in IREM situations, there is a very real danger that an overview chapter, such as this, will underestimate the inherent resource management challenge, leading practitioners to devise overly naïve prescriptions that are doomed to failure in the real world. The opposite danger is, of course, that practitioners are left with a perplexing assortment of contradictory theories that, when left unsorted, offer no reasonable way forward. To find the middle ground, this chapter attempts to cut through this inbuilt complexity and provide a number of conceptual and procedural principles for IREM that can be adapted for any planning and management situation.

To guide this process, a schematic planning framework is provided in Fig. 11.1. This shows a *comprehensive* decision support framework adapted from earlier work by Bissix (1995, 1999/2000). This IREM decision support system (DSS) concentrates on wide-ranging managerial decision processes. It comprises a broad five-step process that includes a number of inter-related and iterative planning operations. The first step develops a strategic direction, the second combines a review of IREM theory with the identification of broadly defined IREM opportunities, the third step involves a thorough analysis of the potential of the natural resource, while the fourth step concerns operational planning, design and implementation. The fifth broadly defined step considers assessment and evaluation.

The IREM Decision Support System

In its most simple form, IREM can be defined as managing a natural resource holding for several values at the same time (Bissix, 1995). This may, in the case

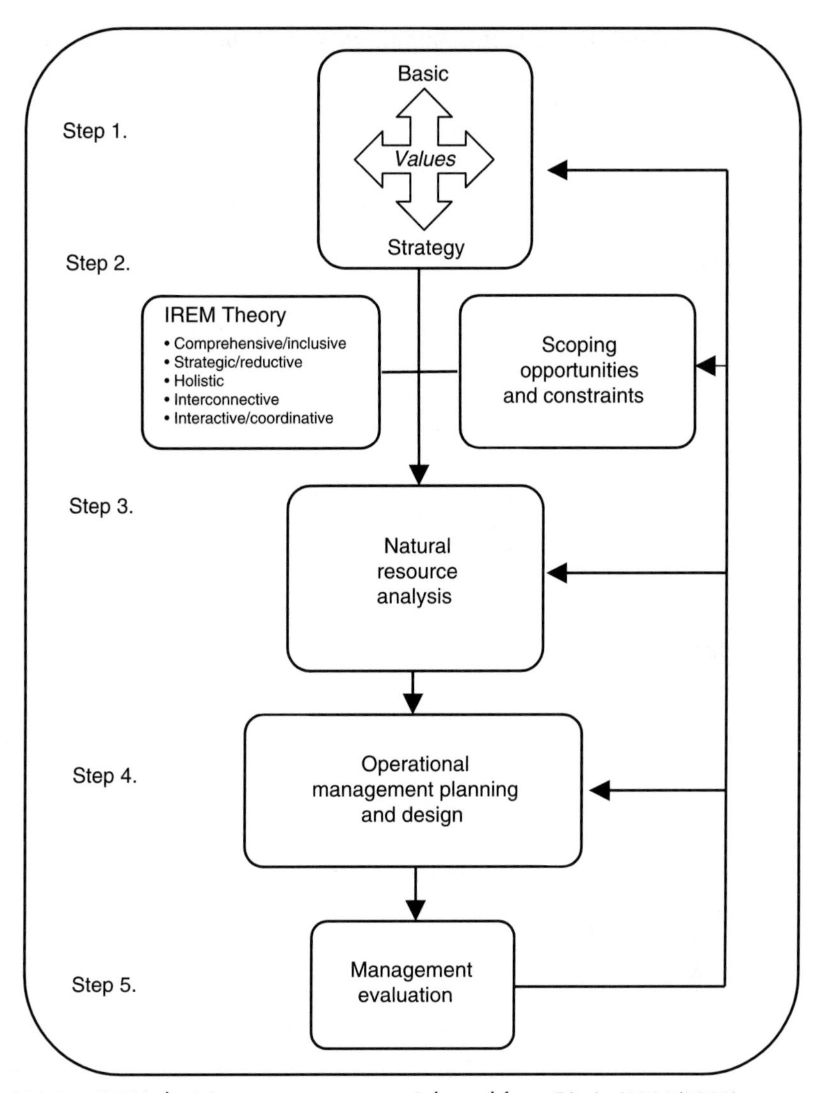

Fig. 11.1. IREM decision support system. Adapted from Bissix (1999/2000).

of forestry for example, include wood fibre production, soil and water conservation, wildlife management and outdoor recreation. IREM attempts to make the most of a resource holding's potential and, at the same time, provides a high level of environmental protection to ensure sustainable outputs over successive generations. When IREM is applied effectively, resource management strategies are carefully combined so that efforts to manage the holding for one resource entity enhance other resource values rather than diminishing or destroying them.

Initiating the IREM process can be as simple as making an inventory of the broad range of objectives you want to accomplish in a resource setting, deciding what is possible, what are the best methods and what are the major constraints.

A substantial challenge throughout the process is to think broadly enough about the options available, determine the necessary compromises and make practical choices, while never losing sight of the underlying purpose of IREM, which emphasizes environmental stewardship. This focus on fundamental management philosophy and underlying values is especially important when problems arise during implementation that might otherwise knock management goals off course. In this regard, two considerations help managers deal with this sometimes perplexing process. The first is that no matter how straightforward a resource management challenge seems to be, a manager will never have access to perfect information. This means that she/he will often have to make do with whatever is available or put off the decision until better data come along. Recognizing that IREM is an adaptive and changing process also means that there will often be future opportunities to fine-tune earlier decisions when better information becomes available.

The second point is that essential support for the implementation process is enhanced by broad participation in the ongoing IREM decision-making process. As a general rule, it is useful to involve senior resource management personnel and other key decision influencers throughout the IREM planning and implementation process. This not only avails management of the agency's or planning sector's full expertise, but a broadly based consensus on management direction also leads to enhanced commitment and shared responsibility as well as accountability during the implementation phase.

There is one other key concern. It is typical in most industries for strategic plans to reflect a 3–5 year time period. Such a planning horizon usually represents the outer limit that investment banks will wait for a return on their investment. Unfortunately, such a time frame is often insufficient for managers to accomplish reasonable natural resource management goals, such as making measurable gains in ecosystem rehabilitation or establishing a workable sustainable development plan. Given such investment constraints, managers must consider carefully what is absolutely essential to do in the short term in order to accomplish its long-term sustainability goals. Managers should also think realistically about their IREM objectives – what are its broad social and legal obligations; how these might affect future management processes; what it is reasonably possible to achieve; and what priorities must be set to ensure that core agency values are emphasized.

Step 1: Applying a Strategic Decision-making Process – the Strategic Diamond

When formulating a general strategic direction, it is useful to consider four broad questions that form a simplified but not necessarily simple agency values clarification process (see Fig. 11.1, step 1). This approach to strategic decision making is sometimes referred to as the 'strategic diamond' decision-making process (Bissix, 1999/2000; see also Fig. 11.2). These key questions are: What do managers want to do? What market opportunities exist? What should managers do? and What can managers do?

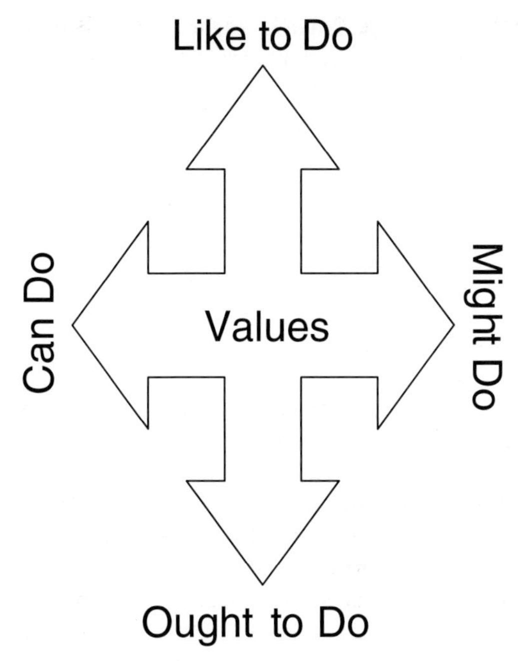

Fig. 11.2. The strategic diamond (source unknown).

What do managers want to do?

Answering this question requires managers to think about a wide range of possibilities for their resource holding; it allows managers to visualize a more or less idealized future for the natural resource holding that is relatively free of most everyday constraints. While such daydreaming, on the surface, may seem frivolous, this process makes an agency's values more explicit; it makes clear its preferred way of doing business if most everyday restrictions were lifted.

What market opportunities exist?

The general public often believes that marketing is simply another name for promotion and selling, but it is much more. In its broader context, marketing reflects an agency's commitment to satisfy customer or client needs rather than merely selling pre-conceived products or services. In this text's context, it is a process of systematically gauging customers' or clients' willingness to invest time, money and energy to support an agency's IREM goals. This is accomplished in a number of ways, such as purchasing its products or services, participating in its programmes, and by providing political and fiscal support. This support infers a mutually beneficial exchange between the resource agency and its customers, clients and various other stakeholders.

Marketing natural resources, whether that concerns products derived from the resource holding or is a service or experience enjoyed by users on site, creates a reciprocal exchange of agency and customer benefits. For example, if a woodlot manager contemplates constructing hiking trails for community use, the manager must be reasonably assured of sufficient interest to justify construction costs. If this manager hopes the venture will also stimulate the sale of its wood products, the manager must be confident that this investment will pay off.

What should managers do?

To be truly successful, a resource manager must also take account of an agency's social responsibilities. As previously stressed, IREM is about making sacrifices for future generations and making environmentally sound resource management choices that are likely to incur additional costs in the short term and may have uncertain corporate benefits in the long term. What managers should do then is concerned not only with legal obligations, such as environmental laws and legislative regulations, but also with best practices to enhance ecosystem health and safety, as well as the protection of the environment for future generations. This assessment process then draws upon and puts into practice Turner's (1993) typologies of sustainable development. It specifically considers the extent to which a resource manager is prepared to trade human and/or economic capital for ethical and ecological capital (see Chapter 8).

What can managers do?

Identifying market opportunities and responding to social pressures is one thing, but facing up to the practical realities and potential of an agency places the strategic decision-making process into a real-world context. Realism concentrates a manager's attention on the organization's capacity to manage its natural resources in the present as well as in the future – in less than ideal conditions. Agency and organizational limits are concerned with its motivations, its biophysical potential, its financial resources, its organizational and managerial expertise, and its internal and external political support. This practical focus relates to how an organization's values are operationalized and how this reality matches its declared IREM philosophy. This is about how everyday pressures drive organizational decision making, especially under conditions of duress.

Accounting for external constraints requires the constant monitoring and assessment of various natural forces, such as weather and endemic disease, and includes assessment of accessible technical expertise, management resources and available skilled labour. In addition, this analysis requires the manager to take account of various infrastructures, such as available transportation and adjacent industrial development; it must also account for evolving customer preferences. In addition, becoming aware of external factors requires assessment of various government programmes, including incentives and regulations, as well as various local traditions and conditions, such as pollution. As was made clear in Chapter 10, IREM in a multiagency ecosystem management context describes a harsh realism of substantial complexity where management consensus may be blurred, ambiguous and fleeting. Assessing the internal and external constraints as well as organizational potential to successfully implement and maintain a viable IREM management plan over the long term can become a management challenge of enormous proportions.

Strategic wrestling – determining the trade-offs

Formulating responses to each of these four strategic questions provides a sound basis for selecting an overall IREM direction; it is unlikely, however, that the response to each is philosophically and operationally compatible. For instance, the direction suggested by what an agency hopes to do may well clash with what an agency's key customers or clients want. To deal adequately with this tension, an agency must wrestle with these incongruities to establish – by making a value judgement – the most appropriate resource management direction for the agency. In the final analysis, the manager must wrestle with, and draw a conclusion from, opposing influences. While this process is often 'give and take', the strategic direction consciously taken – given all of the opposing influences – is the one that reflects an agency's actual operational values. This strategic wrestling procedure is fundamentally a practical, as well as a theoretical, process of values clarification.

From strategy to operational planning

To be successful with IREM, it should be taken into account that a resource manager must not only adopt a well-thought-out strategic direction, but she/he must also ensure that various subunit plans, as well as operational plans, nestle well together. It should be noted that few, if any, worthwhile resource and environmental management successes happen purely by chance. They most often materialize because resource managers carefully formulate a viable vision of where they want an agency to go, they monitor and adjust their plans to various internal and external changes, they commit and invest agency assets wisely, and work diligently to execute the plan as it evolves over time. Throughout the planning and implementation process, therefore, it is important for managers to carefully consider the driving forces impacting resource management (see Chapter 4) and prudently incorporate the best available guidance from the social and biophysical sciences so as to avoid, as far as possible, the mistakes of others (see Chapter 6).

Step 2: Reviewing IREM Theory and Assessing Preliminary Natural Resource Holding Opportunities

The second step in this planning process combines an examination of IREM theory (see Chapter 9) with a preliminary assessment of the resource holding's IREM potential. Before considering this operation in more depth, three key terms need to be defined: (i) resource holding; (ii) IREM opportunities; and (iii) the

IREM system. A *resource holding* is the combination of natural features and physical assets upon which the resource manager directly applies IREM strategies. A farm holding, for example, includes both natural and culturally shaped biophysical assets including farm buildings, farm roads and tracks, woodlots, various plantations, farm stock and wildlife, as well as natural and constructed water features. An *IREM opportunity* refers to the potential to apply IREM principles to a resource holding. An opportunity is the combination of the resource holding's natural and cultural features, the resource manager's desire to implement IREM strategies, the manager's IREM management skills, and the organizational and external constraints impinging on the management system. In this context, it is critical to understand that an opportunity is just that; it is not assurance that IREM will be put into practice or that it will necessarily be successful. An *IREM system* includes the combined activities of resource management that influence the natural resource holding and modify the future resource management opportunities. These activities generally include three sets of inter-related functions: resource management; user and customer/client management; and product/service management. In attempting to implement IREM at ground level, it is important to understand just how these management functions inter-relate.

This second step in the IREM process involves the resource holding manager making a preliminary assessment of the agency's expertise, desire and capacity to utilize applied biophysical science and technology to nurture or exploit the resource holding's natural assets (see Turner *et al.*, 2001). This is also where the agency assesses its accumulated managerial skill to mobilize its human (intellectual) capital in order to put its IREM goals into operation. As already emphasized, the application of IREM at ground level involves the blending of a complex set of biophysical and social scientific concepts and theories, which help create a practical management framework. What distinguishes a good manager from an ineffectual one is not necessarily an individual who considers every possible option, but rather one who is adept at selecting the most useful theories and applies these to optimize the resource holding's potential.

A useful step in this process is to develop a cross-impact analysis of potential and actual resource exploitation activities (see Fig. 11.3). A cross-impact analysis can be a relatively simple exercise, depending upon the sophistication of the manager's IREM goals for a resource holding, or it can be highly complex. One example recently established to streamline the application process and ensure appropriate stewardship of a valuable resource is that developed by the Great Barrier Reef Marine Park Authority (GBRMPA) in Queensland, Australia. According to the GBRMPA, 'the Protected Environment Management System (PEMS) improves the speed, accuracy and consistency of [its] permit application processing'. It streamlines the application process for any development proposal within the park's boundaries and integrates a wide range of information to assess potential cross-impacts before a project is approved.

The key to PEMS is its use of digital mapping and overlay techniques to identify permit impacts, removing the need to manually interpret a range of paper-based maps, plans and other documents to achieve the same result. According to GBRMPA project manager, Dr Adam Smith, the digital mapping and overlay techniques also allow the transparent identification and

	Commercial fishing	Scuba diving	Tourism infrastructure	Citrus plantations	Weather events	Forest clear-cutting	Urban development
Commercial fishing	CF CF						
Scuba diving							
Tourism infrastructure							
Citrus plantations		CP SD					
Weather events			WE TI				
Forest clear-cutting							
Urban development							

Fig. 11.3. Schematic for the cross-impact analysis of the Belizean Barrier Reef.

interrogation of critical information such as impacted plans of management and sensitive reefs to ensure that the appropriate conditions, definitions and permissions are attached to each permit. The process also automatically identifies any Native Title claims and, where necessary, generates customized notifications for each claimant (Henriss-Anderssen, 2002).

The process of cross-impact analysis ensures that a *comprehensive* assessment is initially completed and that *interconnected* biophysical and socioeconomic processes are identified. The initial objective of a cross-impact analysis is to identify how maximizing one resource management objective may impact the resource manager's ability to maximize a second or a third resource management objective (it might be useful for the reader to revisit the limitations identified for multiple-use resource management identified in Chapter 8, and the boundaries of decision making evaluated in Chapter 9). For example, in maintaining an IREM regime for the Belizean Barrier Reef off the coast of Central America, it might be necessary to consider the cross-impacts of several natural and cultural processes on the evolution of the reef and human activity, including commercial fishing, nature-based tourism activities, the indirect consequences of constructing and maintaining tourism support infrastructure, such as hotels and entertainment facilities, inland deforestation, citrus plantation and pesticide pollution, general urban development, and natural weather events such as hurricanes.

This hypothetical analysis of the Belizean Barrier Reef is proposed then at an extensive ecosystem scale that goes beyond the boundaries of the reef itself to stretch to the highest reaches of the inland watersheds whose waters eventually flow to the sea and across the reef. In this example, an IREM approach must be mindful of the potential damage, for example, of heavy pesticide use in the inland citrus plantations and their contaminated runoff that eventually finds its way to the sea. Such contamination may have the potential to poison the coral reefs that, in themselves, give rise to a vibrant ecosystem, and a spin-off fishing and tourism industry. An IREM approach would necessarily assess the cross-impacts and *interconnection* of both activities (citrus plantations and coral reef tourism), assess their respective markets and their contribution to social welfare, their environmental and sustainable management implications, and, as a consequence, provide workable management prescriptions. This would be done first to avoid any negative cross-impacts and secondly to mitigate environmental costs while also trying to build on, or capitalize on any positive cross-impacts, such as tourism providing a local market for the citrus industry.

Similar analyses can be made for other impacts such as the ones noted in Fig. 11.3. The cross-impacts highlighted there include major weather events such as hurricanes and their impact on tourism infrastructure, and the impact of one sector of commercial fishing on another. As one might readily appreciate, conducting a full array of cross-impacts (in the case of Fig. 11.3 this would amount to a minimum of 49 separate, two-way analyses of varying depth) is likely to be very involved and extremely costly. In addition, the assessment of the accumulative impacts of several variables one upon another is likely to be not only extensive but also extremely difficult to assess. As a consequence of this complexity, a reasonable cross-impact analysis is in practice *strategic* and *reductive*, and likely to be one where astute professional judgement is necessary in order to prioritize the various possible combined impacts. This reductive process focuses attention on the most pressing, rather than proceeding with all possible eventualities. A fully comprehensive analysis would probably produce large amounts of data that could not be usefully analysed and gainfully incorporated into the management process.

While the above example encompasses an ecosystem of substantial proportions (the coral reef itself is ~300 km long), a conceptually similar but much smaller scale analysis can be made to encompass, for instance, a single farm holding and its surroundings. Regardless of size, the overall goal is the same. It is to use the matrix formed by the cross-analysis process to raise key questions about the possible cross-impacts of one resource management activity on another and, as a consequence, assess the need for integrative management strategies. Such analyses can also begin answering those questions that raise serious concern and might be answered only by additional research or by long-term monitoring through the IREM implementation phase. This process then, begins to flesh out and put into practice the conceptual framework for IREM borrowed from Born and Sonzongi (1995) that was reviewed in Chapter 9. Cross-impact analyses thus begin to define just how *comprehensive* and *inclusive*, in terms of scale and complexity, IREM processes can and should be. Such an approach highlights the interconnections between biophysical variables and socioeconomic factors, and provides direction for what interconnections must be focused on in the IREM planning and management process.

This analytical process allows the management team to decide what the important management priorities are – once the IREM problem has been defined in its broadest terms and in a holistic way. It allows the management team to be more *strategic* – matching and aligning its management resources to its highest priorities. Thinking and acting strategically, in other words, focusing on what is considered to be an agency's most important priorities, should not be, however, a recipe for myopic thinking (narrow mindedness). The management team must be continually mindful of the broad scope of the IREM problem under examination and, as a consequence, be sufficiently broad-minded to ensure that key resource management values are identified and that they are not compromised by unforeseen events.

Beyond its value in aiding strategic thinking about the resource management problem, a cross-impact analysis also provides important insights for the IREM team to determine who the key stakeholders might be, and how their interests and expertise can best be represented and integrated into the management process. Two seemingly contradictory courses of action are considered here. The first cautions the manager to heed the concern outlined by Wondolleck (1988) cited in Chapter 9. He emphasized the need to *interact* with key stakeholders and involve them in formulating important decisions. The second, as explained in Chapter 10 (see Bissix and Rees, 2001), warns managers of the difficulty in maintaining a committed critical mass of stakeholders in a multiagency, ecosystem-based IREM process over time. Given this evolving challenge, it behoves the core IREM management team to consider carefully how stakeholder interests can be adequately represented and *coordinated* in any broad-based ecosystem management process over time.

Step 3: Resource Holding Analysis

The third step's main focus is to guide the resource manager in assessing the holding's IREM ground-level potential and determine, in more depth, appropriate uses as well as necessary management interventions and trade-offs. This step builds on the strategic assessment done in step 1 by reflecting key resource management values, and by validating the synergies and conflicts identified in step 2, especially in the cross-impact analysis. While step 3 focuses on the resource holding's physical attributes, this analysis still maintains a strong human dimension orientation by assessing managerial strengths and weaknesses to implement various IREM strategies effectively at ground level. An IREM resource holding analysis includes an identification and assessment of a holding's vegetation, wildlife and various other biophysical features such as lakes, marshes, bogs, rivers and streams, subterranean hydrology, cave systems, hills and valleys, and cliffs. The resource holding analysis is also concerned with variables that the manager has little control over, such as the weather and air pollution.

Assessing the IREM potential of the resource holding incorporates, among others, five interconnected analyses. Each of these analyses is layered to form a composite, multidimensional picture of the holding's attributes and potential. The first *comprehensive* assessment usually considers the resource holding's general natural and cultural assets. In the case of an estuary, for example, this might include a definition of the extent of a freshwater watershed feeding the estuary and some demarcation of saltwater boundaries, an overall assessment of biophysical assets (such as weather patterns, flora and fauna), and physiography and geology, and the identification of cultural assets, such as urban and rural settlements, and industry, such as aquaculture. Further assessments can be made of the holding's size and biophysical variation, various environmental factors that impact biological potential, and the resource holding's biophysical developmental phase. In the case of a small woodlot, this might include an assessment of its scale and its potential to accommodate various activities including wood fibre production, wildlife habitat provision, nature-based tourism and water conservation. An even-aged mature forest, for example, defines the woodlot's biophysical development stage and its potential for particular resource values. While a well-established and biologically productive estuary may offer one array of IREM possibilities, a freshly cut channel after a major flood would offer quite a different set of possibilities. The fifth consideration is the holding's location and proximity to supporting infrastructure and to potentially deleterious externalities. Proximity to a large city, for example, might limit the advisability of some IREM strategies and objectives while stimulating others. A coastal farm holding, for instance, may have limited agricultural potential because of poor soils, but substantial wildlife and nature-based tourism potential because of its proximity to the sea and to urban centres. Once a thorough analysis is completed, these previously untargeted resource values may be integrated into an IREM plan.

According to available resources, interests and priorities, various analyses, including more detailed cross-impact analyses of various scales, can be conducted to give the IREM manager an increasingly detailed picture of the consequences and the possible benefits of various physical management interventions. At all stages of this more detailed site analysis, the manager will be expected to capitalize on various beneficial opportunities that avail themselves, and make various informed trade-offs when it is obvious that action on one resource development front will compromise the potential of another resource management activity.

Step 4: Tactical and Operational Planning and Management

This next step in the comprehensive IREM planning process encourages resource managers to examine three concurrent operational management concerns before proceeding with IREM implementation. These are: (i) site planning, design and development; (ii) natural resource marketing; and (iii) operational management planning.

Planning, design and development

This aspect of the decision-making process helps the resource manager translate broad IREM objectives into actual physical, 'on-the-ground' developments. The outputs are usually a set of detailed maps, drawings and specifications that serve as a blueprint for future development. There are several key ingredients to successful project design including: technical (ground level) knowledge of the IREM design process; a clear understanding of management needs and capacity; and a clear picture of the character and potential of the resource holding concerning the maximization of certain resource potentials. It also includes an understanding of the general compromises necessary, including the sacrifice of short-term economic goals for longer term aspirations, and detailed knowledge of the users' and customers' needs; management creativity and innovation as well as organizational and public attitudes. It is important to keep in mind throughout the IREM adoption process that the manager should maintain both short-term pragmatism as well as a longer term, intergeneration equity orientation. For example, the manager should plan with the short-term needs of resource users and customers in mind, but always be mindful of the manager's broader responsibility for sustainable development.

Natural resource marketing

While it is inconceivable that every IREM activity, especially those that are specifically selected to enhance long-term environmental benefits, will have a positive impact on the organization's profit and loss accounts, it is nevertheless important to consider the impacts of the IREM process on the short- and medium-term viability of the agency. A common managerial oversight is to focus too much attention on present and proposed ground-level IREM initiatives that seem reasonable, from a resource management perspective, without adequately considering the implications of these resource management activities on the requirement for additional marketing initiatives, perhaps new techniques in operational management, and the need for different or enhanced approaches to project evaluation.

Natural resource marketing design requires appropriate market packaging (or integration) to provide the manager with a sense of what is required to identify key target customers or clients. In some instances, this might mean new political support constituencies. This may also require appropriate products or services to meet targeted customers needs – perhaps a new set of products and services produced in an environmentally friendly manner. Such a strategy will also require a suitable promotional programme – to inform potential customers and clients of the management's steps to integrate environmental management and sustainable development. Additionally, it will probably be necessary to develop an effective distribution strategy that connects potential customers with these products produced in an environmentally sensitive way, such as selling organically produced vegetables in specialty markets. Finally, the creation of a pricing strategy will certainly be necessary that reflects the customers' willingness to pay. This may involve premium-priced commodities or products that reflect the additional (internalized) costs of sustainable management, or it may reflect taxpayers' willingness to support a non-priced publicly available service.

It is important to note that 'packaging' is an integrative process that layers an additional conceptual dimension on the IREM process. This involves, essentially, a series of adjustments and accommodations in determining target customers, producing environmentally friendly natural resource products and services, developing an adequate promotion strategy, determining a reasonable pricing strategy, and working out a way to get products to the customers or, as in the case of nature-based tourism, attracting the client to the natural resource.

Operational management planning

'A 'B' quality plan fully implemented is better than an 'A' quality one that only gathers dust' (source unknown, cited in Bissix, 1995). Good planning can lead to fitting ground-level action and can also steer natural resource managers away from ill-advised activities. The transition process from planning activities to successful implementation is often wrought with difficulty. Such challenges are especially difficult when managers are faced with competing management demands, as is often the case in the IREM implementation process. Implementing IREM successfully not only requires a sound philosophical basis, good science, specialized resource management skills and meticulous application, it also requires persistence. Before any ground-level workforce is deployed, numerous operational issues must be addressed and weighed. This includes maintenance, budgeting and finance, partnership building, information management, access and risk management, legal administration, and legislative and regulatory compliance. While an in-depth explanation of these is beyond this book's scope, it is important to stress that each has its own significance.

Step 5: Planning and Management Assessment and Evaluation

Developing a well thought out evaluation plan before implementation begins helps to clarify IREM objectives and assists in making prudent decisions throughout the management process. A pre-established evaluation plan helps managers validate their IREM strategies and persuade others of their importance, whereas formative evaluation (that which takes place during the planning and management process) provides a means for *interactive* input. It provides timely opportunities to adjust key components of the plan as new information emerges and experience is gained. An appropriate evaluation process informs decision makers of whether IREM efforts are of value, effective, efficient and acceptable. An effective assessment and evaluation process guides managers as to whether they should keep the course, or whether they should change or even terminate a planned action. Carefully formulated evaluation strategies also aid in shaping appropriate implementation strategies by helping resource managers make wise choices among competing alternatives. Most importantly, systematic evaluation provides *accountability* to financiers, others that sanction management plans, and those that provide legal authorization. Since resource management practices are increasingly scrutinized by external entities, timely evaluations provide information for interested stakeholders so they can appropriately judge an IREM project's worth.

Implementing the IREM Planning Framework

It is inconceivable that the typical reader will be able to master the use of this planning framework in the field without considerable practice. For this reason, we provide an IREM case study at the end of this chapter that can be used to practise its various steps (if you need more practice, we suggest that you revisit one of the earlier cases in this book and apply the planning framework to it). Case studies provide a relatively safe, inexpensive and effective way to develop planning skills, whereas mistakes made in the field are costly and take considerable time and resources to remedy. To aid in your analysis of this case study and in future field applications, a detailed checklist is provided in Box 11.1 that will help guide you in your use of the planning framework. The case study is entitled, *Integrating Farming, Railway Safety and Recreation in Wolfville.* Unlike the other cases in this text, there are no specific questions for you to address. There is, however, an invitation for you to take on the role of the Chief Executive Officer (CEO) of the Windsor and Hantsport Railway to develop a strategy that would help resolve the issues identified. Once you are familiar with these techniques using this case, you will be better prepared to tackle field applications that are presented to you.

Summary and Conclusions

In applying this framework to a resource management situation, it is important to understand that arriving at the end of the IREM framework is merely the prelude to starting the process over again. It is very important to realize that the application process may have taken many detours from your originally intended direction. You may have travelled along numerous sidetracks, endured many interruptions and false starts, and undergone many partial departures of what initially seemed to be a rationally driven and linear planning process. To be successful in applying this IREM framework, it is important to understand that the natural resource manager must be comfortable with (or at least be tolerant of) the process of multitasking, i.e. dealing with several management tasks at the same time so as to continuously adjust various, inter-related management activities according to surfacing information, changing environmental conditions, evolving socioeconomic conditions, and emergent IREM goals, objectives and best practices.

In closing, it is important to stress that despite continually changing operational conditions, the resource manager must never lose sight of the fundamental principles and underlying purpose and philosophy of IREM, which is in essence

Box 11.1. IREM decision support system guiding questions.

Step one: developing a basic strategy

1. Who are the most important stakeholders in this proposed IREM planning process, what influence are they likely to have and what influence should they have?

2. What beliefs, attitudes, values and perceptions concerning this planning situation will be most influential in developing a satisfactory solution?

3. By wrestling with what you would like to do, what you might do in the marketplace, what you are capable of doing, and what you ought to do, narrow down the possible solutions to provide a direction for your subsequent analysis. This process of determining the trade-offs provides a working direction that can be modified or fine-tuned once the analysis begins in earnest.

4. Briefly state what your initial vision is for this IREM project (again this is likely to be revised as the process proceeds).

Step two: IREM theory and scoping

1. What are the physical boundaries of the case under scrutiny? This defines the resource holding and in more general terms establishes the boundaries of the IREM setting which may involve the broader ecosystem.

2. What concepts and theories from the IREM literature are most relevant and useful for analysing this situation?

3. What are the most important natural resource values of this holding (e.g. mining, forestry, transportation, farming, recreation and tourism, and air quality) and what are the values derived from the built environment such as roads, dykes, buildings and docks?

4. What are the most critical cross-impacts of each of those values (see Fig. 11.3)?

5. What are the most important external factors and key management opportunities available in this situation that will affect IREM implementation?

6. What are the most important factors affecting the natural resource management system?

7. What are the most critical external and organizational constraints that exist that will affect a reasonable solution in this situation?

8. Is this management situation resolvable or is it merely containable?

9. What are the main biophysical forces acting on this resource holding and what technologies are available to manage them in a sustainable and environmentally sensitive way?

10. Develop a concise report on the pertinent literature and also write an overview of the broad opportunities and constraints impacting this IREM situation.

Step three: natural resource analysis (resource holding analysis)

1. What detailed biophysical and social information concerning the resource holding will be useful in assessing IREM potential?

2. What information is available? Is it in a usable form? If not, how can it be transformed at a reasonable cost to make it useful?

3. Given available management resources, what information can be obtained in a reasonable time frame and at reasonable cost? What assumptions about information gaps will be necessary?

4. Given the available realistic options, what is the preferred IREM strategy that is to undergo further operational planning (this IREM strategy is more detailed than the basic strategy identified in question 2 – it will have been refined as a result of a much more detailed and iterative analysis)?

5. What management expertise is available to carry out this project?

6. Is there sufficient motivation within the management team to carry out this project?

7. Are there sufficient management resources available to complete this project successfully?

8. Complete a preliminary cross-impact analysis of competing values for the resource holding.
9. Complete a Natural Resource Analysis of the IREM holding.

Step four: operational management planning and design
1. In order to implement the preferred strategy, what additional planning, design and project development must be carried out to ensure project success?
2. A new or revised IREM strategy will require a marketing plan; what will be the essential elements of this plan?
3. Will this plan require adjustments to the agency's risk management strategy, its insurance coverage including liability, and its financial management plan?
4. Will this plan require additional or different operational management such as site supervision, maintenance and information management?
5. Produce a set of operational and implementation plans as well as any site (resource holding) maps and necessary infrastructure designs.

Step five: management evaluation
1. What assessment and evaluation processes will be used throughout the formative stages of this project?
2. What processes of assessment and evaluation will be used to provide a summative evaluation of this project? What will be considered an appropriate timeline to conduct a summative evaluation?
3. Who will be accountable for the success or failure of this process? How will they be accountable?
4. Who will decide whether the IREM process is a success or failure?
5. How will feedback from the evaluation process be integrated into the next iteration of the IREM management process?
6. Produce an evaluation plan.

Compiling an IREM plan
1. Using an acceptable report writing procedure, develop a final report that reflects your IREM plan.

to manage natural resources with a greater sense of environmental integrity and with a better sense of natural resource stewardship that can more appropriately serve both present and future generations.

Planning Framework Guidelines

The questions in Box 11.1 will guide you through the planning process. You should use these broadly based questions in conjunction with the planning framework figures (Figs 11.2 and 11.3), the detailed explanations provided in this chapter and, when appropriate, the more detailed explanations given throughout this book.

Keep in mind that planning in the real world is an iterative process. While the planning framework presented in Fig. 11.3 appears on paper as a linear step-by-step process, good planning and decision making requires management to go back over various steps as new insights are gained and as new directions are refined.

Always take into account that it is less expensive to revisit the planning process on paper than it is to try to make amends for planning shortcomings during implementation.

Also remember that during implementation, IREM is invariably an adaptive management process where on-the-ground experience always provides additional insights that are impossible to appreciate with desktop planning exercises. Some changes to IREM plans are inevitable, no matter how thorough your planning is, once implementation and formative evaluation begins.

Please note that the list of questions in Box 11.1 is only a representative list that managers should consider in developing an IREM plan. It does not replace the need for a manager to thoroughly analyse the situation her or himself in order to develop key questions that must be addressed to meet each situation's specific needs.

Case Study
Integrating Farming, Railway Safety and Recreation in Wolfville

GARY NESS

Preamble. In this case study, Gary Ness describes a situation involving a number of stakeholder groups who seek to exert influence and control over natural resource areas in Wolfville, Nova Scotia, Canada. How the issue is resolved represents an interesting and not uncommon example of the high level of concern various groups have toward a natural resource.

Introduction

The situation

Wolfville, Nova Scotia is located on Canada's east coast, on the southern shore of the Minas basin, which is an extension of the upper reaches of the Bay of Fundy. The Bay of Fundy is famous for its enormous tides, which can average 55 feet (16.8 m) at some points along the upper reaches of the estuary. These twice-daily fluctuations in the sea-level are associated with enormous water movements that shift much silt (and other nutrients) and have created extensive mudflats that are exposed at low tide. At mid-tide, more water flows into or out of the Bay of Fundy than all the freshwater flows throughout the world.

In the first half of the 1700s, early French settlers recognized the agricultural potential of these areas and began a process of building dykes to restrain the sea and to allow for drainage of these potentially arable lands. This practice has continued to this day, and large expanses of dyked farmlands now lie around the Minas basin shoreline and run for miles inland along its river estuaries. These fields and tidal flats offer much more than fertile fields for farming; the Minas basin – because of nutrient availability due to tidal surge – produces an abundance of crustaceans and other foods for migrating and nesting birds, and other wildlife. Such is the fecundity of this area that the Southern Bight of Minas basin and its surrounding shoreline and

The Wolfville dykelands before harvesting. The photo was taken from the sea dykes. Note in the foreground the farm road used regularly by recreationists and the line of the railway in the mid-ground. The railway separates the town, which is above sea level, from the farmland that is below sea level. The farmland is protected from saltwater incursion by the dykes. Photo by Sue Bissix.

The railway adjacent to the Wolfville Waterfront Park and the Wolfville Harbour. Note the sea ice at high tide in winter, the ski tracks alongside the rails, and the tracks of walkers between them. Photo by Glyn Bissix.

related estuaries was declared a wetland site of international significance under the Ramsar Convention, and the area has become part of the Western Hemisphere Shorebird Reserve network that aims to protect migratory shorebirds between the Arctic and southern South America.

The dykes extend for many kilometres, forming a barrier between the reclaimed lands and the sea. Each dyke has an associated access road along the base of the dyke or along the top of the dyke. Because of the natural setting, and the fact that these roads provide accessible flat, easily traversed trails, they have been a very popular natural recreational area that has been used for generations by those who walk, jog, cycle or simply wish to enjoy and experience or observe nature. In fact, the use of these paths has been so widespread for so long that many users are of the opinion that the public has the legal right to use these trails, when in fact they are trespassing on privately owned lands! Marsh Bodies, each of which comprises owners of the fields within the respective dyke walls, govern the lands bounded by the dykes. The Marsh Bodies are jointly responsible with the Nova Scotia government under provincial statute, the Land Reclamation Act, for the maintenance of the dykes, roads, drainage and general governance of their properties.

The dilemma

While provincial statute designated these dyked lands for agricultural purposes only, from time to time other incursions and development have infringed on them. The Windsor and Hantsport Railway (WHRC), for example, runs on an east–west route through the northern edge of the university town of Wolfville, Nova Scotia (Fig. 11.4). The railway corridor borders the dykelands to the north and, for the most part, it separates the town from the dykelands. However, over decades, a number of homes and businesses have located north of the railway line. This had largely gone unremarked, but in the 1980s significant legal controversy erupted when the town of Wolfville placed their sewage treatment facility on the farmlands, and then again in the 1990s, the town announced plans to allow 'industrial/business' development to encroach further on to dykeland space that the town had previously purchased. The plan was to allow the town's new maintenance offices and garages plus several businesses to relocate. This controversy was finally resolved, but only after litigation that produced considerable ill will between the Marsh Bodies and town officials.

Railway trespass

For decades, the general public has crossed the railway at many points including public and private crossings and many other undesignated spots as a means to access the dykelands. However, the most contentious invasion of the railway right of way has been to use the tracks as a trail. The railway, which is flat, semi-isolated and has become 'wild' after the railway abandoned line-side herbicide spraying, is an easy scenically attractive 'nature trail' across the northern boundary of the town. In summer, the 't(rail)' is used by some as a means to by-pass the town core, and in winter the railway has even been used by a small number for cross-country skiing!

In 2000, the Railway Association of Canada, in cooperation with Transport Canada, announced the *Direction 2006* plan to reduce railway accidents, injuries and trespass throughout Canada by over 50% by the year 2006. This directive was relayed to the Nova Scotia Department of Transportation and Public Works, which is responsible for implementing such policies at the provincial level. Simultaneously, railway insurance companies were informed of this initiative. The WHRC was told that non-compliance with these initiatives to halt trespass would lead to cancellation of their liability insurance policy and their operating licence would not be renewed.

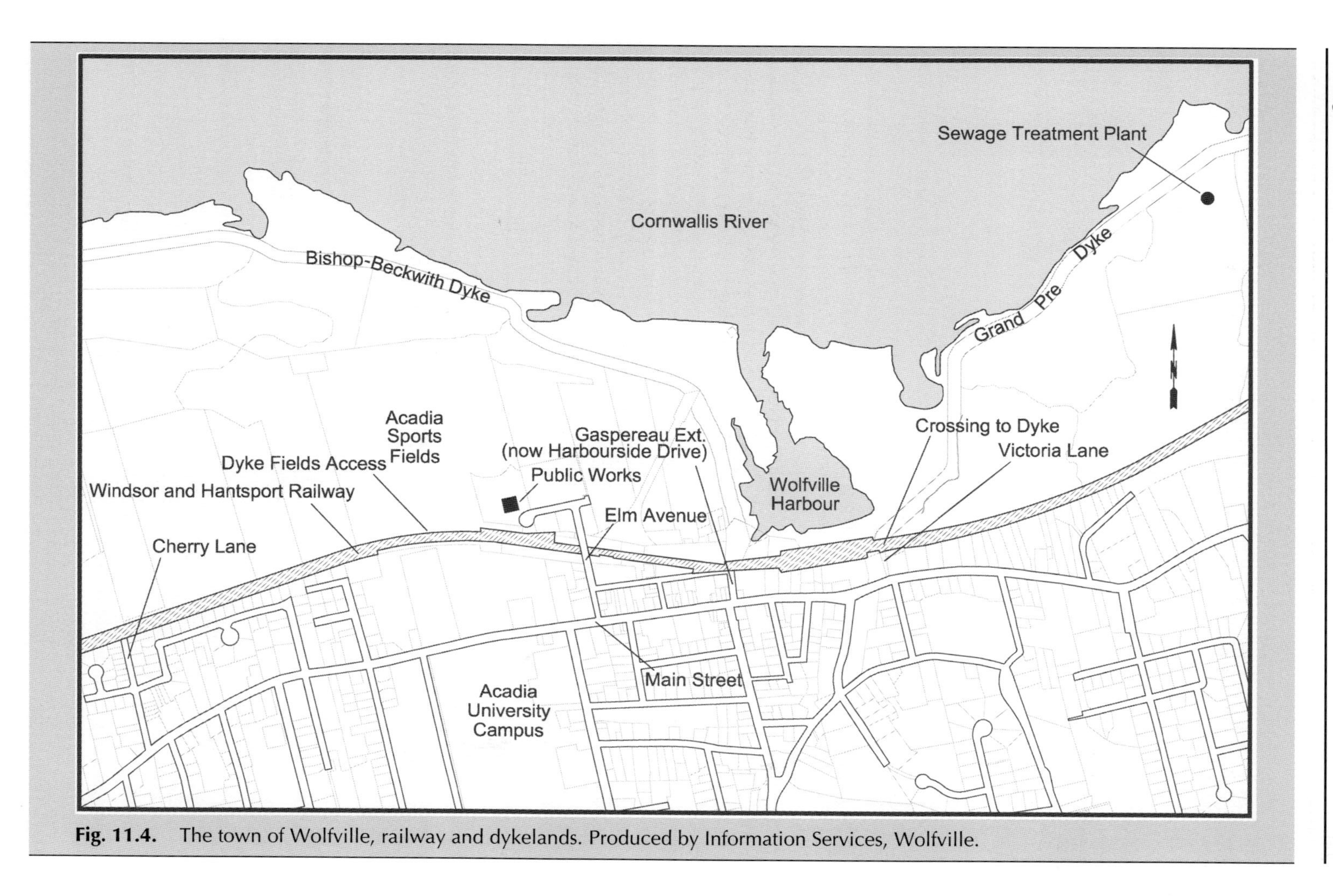

Fig. 11.4. The town of Wolfville, railway and dykelands. Produced by Information Services, Wolfville.

The WHRC immediately initiated a consultative process; a meeting was called which included various stakeholders in this matter. The first meeting of the 'Railway Safety Committee' was held in Wolfville during the late summer of 2000, and included representatives from the WHRC, Transport Canada, the NS Department of Transportation and Public Works, Operation Lifesaver (a public education programme of the Railway Association of Canada), the town of Wolfville, Acadia University's School of Recreation Management and Kinesiology (SRMK), and the Royal Canadian Mounted Police (RCMP). The meeting served as a means to acquaint the participants with the trespassing problem. The ensuing round-table discussion of the matter raised many issues, particularly with regard to the role of the participants in finding solutions, the actual extent and nature of the problem, and the challenges to be addressed.

The Role of the Participants

The WHRC's role was transparent from the outset, as were the roles of Transport Canada, and NS Transport and Public Works representatives. The presence of a representative from Operation Lifesaver, a national agency with the task of educating the public about railway safety, was also easily understood. The RCMP was represented because of the legal implications and its role in enforcing the laws of trespass on private property. The presence of the town representatives, including the Mayor, the Health and Safety Officer, the Chief Administrative Officer, the Public Works Director and the Recreation Director, was somewhat less obvious, and the presence of the Director of Acadia University's SRMK was also unclear, but the need for their participation became clearer as the discussions unfolded.

Problems, challenges and finding a solution

The simplified version of the problem was that citizens of the town and members of the Acadia University community used the railway illegally as a trail, or crossed the railway at many unauthorized points along the railway right of way. While both of these actions may constitute trespass, a clear distinction between the use of the tracks as a trail and crossing the line at various points was drawn. Of particular concern was the very large number of people using the railway as a trail between the public crossing at the Gaspereau Avenue Extension eastward to the entry to the Grand Pre dykelands beginning at Victoria Lane. Other forms of trespass using the railway as a trail were noted, particularly that the railway tends to be treated as a nature/walking path at many points within the town limits, but the greatest intrusions are along the short segment noted previously.

The distinctions about designated crossings of the railway were discussed at some length. There exist two *public crossings* in the town, one at the Gaspereau Avenue extension (the 'Rafuse' crossing) and one at the Elm Avenue extension (to the town's Public Works facility). By law, these are the only points where the public may cross the railway tracks; at these sites, there are specific rules governing railway safety practices, including signage and warning signals (use of train whistle signals and the like), and there is public liability insurance. The railway and the provincial government (NS Transport and Public Works Department) maintain these public crossings to a high safety standard. There also exist three *private crossings* of the railway within the town. These include: the crossing at Victoria Lane, which is used by the town to access the waste treatment plant on the Grand Pre dykelands; the crossing at Cherry Lane to access the Bishop-Beckwith dykelands; and the crossing to the north of Acadia University to access the Acadia soccer fields. These private crossings are not authorized or intended for general public use, and railway safety and liability insurance are primarily the responsibility of the 'owner' of the crossings. The town is the sole owner of the Victoria Lane and Cherry Lane crossings and is a partner with Acadia University in the soccer field access.

While it was clear that the WHRC's concerns were mostly about the use of the railway as a trail, these broader discussions forced town officials to focus on their liability related to their private crossings, which were used extensively for public recreational purposes, particularly to gain access to the dykelands. Clearly, the problems of allowing unauthorized use of the town's privately owned crossings to access the dykelands was a further complication; the dykelands are private property, owned, maintained and operated as farmlands by the Grand Pre Marsh Body and the Bishop-Beckwith Marsh Body. While the WHRC felt that the continued public use of these private crossings was acceptable, provided that the sight lines and warnings were adequate, the town realized that they had to deal with their legal liability related to allowing continued public use of these points and the problems of authorizing the public to use town-owned crossings to gain access to property owned by the Marsh Bodies. The previously existing ill-will between the town and Marsh Bodies was a potential hindrance to discussions, and thus the potential for a loss of public recreational access to the dykelands loomed.

It was noted that the use of the dykelands for informal recreational purposes has been a long-standing and very popular tradition, that citizens viewed access to the dykelands as a right, not a privilege, and that efforts to curtail that access would have a considerable public opinion backlash. Clearly, any efforts to deal with trespass, safety and liability concerns would need to be sensitive to these life-style issues and to the need to ensure sustained recreational use while protecting the owners of the crossings, the railway and the dykelands. A related and equally important issue was the process for implementing changes in public policy and behaviour related to railway trespass. The RCMP, the WHRC and the government agencies proposed that a period of public education, accompanied by a grace period during which warnings would be issued, was preferred to the immediate issuance of trespassing fines of Can$500.00. A general need to 'get the public onside' was noted.

Solutions and Actions

Educational sessions were conducted at the local public schools. Educational handouts and mail-outs were circulated with town bills. New, more obvious signage was erected to educate the public about trespass and the associated penalties (this appears to have had very little impact). Greater RCMP visibility was initiated to re-educate and to warn offenders. This has been a challenge, given the time that must be dedicated to other policing priorities; to date, no tickets have been issued (early 2003).

The SRMK was asked and agreed to undertake a consultative and support role in the proceedings, and volunteered to provide leadership and consultative expertise, and to engage students in a practical project to assist in dealing with these matters.

The town convened a well-attended public meeting to educate the public, to outline the concerns of the various stakeholders (including the WHRC, the RCMP and Operation Lifesaver) and to hear the public's views. The Director of the SRMK facilitated the session. Of particular note was the public concern about continued informal recreational access to the dykelands. The public was very supportive of the railway company's need to stop the use of the railway as a trail and for the need to find a solution to the Gaspereau Avenue extension–Victoria Lane short cut on railway property.

This input spurred further activities and meetings during 2001. The two Marsh Body executive committees were contacted – by townspeople who had long-standing relationships and 'established credibility' with members of the local, rural community – to ask if they would authorize the continued use of their private lands for recreational purposes. In a display of magnanimity, the Marsh Bodies were agreeable provided that it was understood that specific conditions were made clear. The following is an excerpt from the letter to the Marsh Bodies:

> It is clear that some guidelines are needed to ensure continuing recreational access to the dykelands. We note that:

- The dykelands are private lands owned and operated by your members. Additionally, the primary purpose of the area is for farming operations and this takes precedence over any other use. Recreational visitors to the dykelands must give way to farm machinery.
- Recreational access to the dykes is a privilege not a right, and users are expected to treat the dykelands with appropriate respect.
- Recreational use has been informal and should remain so. To be specific, the dykelands should not be advertised or promoted formally as recreational places.[1]
- Informal use of the dikes may include walking and cycling, but motorized vehicular traffic (cars, motorcycles, ATVs, etc.) is discouraged, except for the town of Wolfville Public Works vehicles that are en route to the treatment plant on the Grand Pre Dykelands. The locked, vehicle-control gate at the end of Victoria Lane will remain in service, as will the barricade at the Wolfville entry to the Bishop-Beckwith dyke. Recreational users of the dykes do so at their own risk.

[1] Note: this particular statement relates to several brochures and magazine articles that have encouraged organized public use of the railway as a trail and the dykelands as a nature walk; these publications and other efforts to conduct guided tours on the dykes have infuriated the members of the Marsh Bodies. This informal agreement paved the way for the town to allow access to the dykelands at its private crossings. This use of private crossings necessitated, however, a thorough examination of the town's liability and of the safety issues related to those crossings.

Discussion Questions

1. Refer to the strategic diamond assessment process and imagine that you have just purchased a homestead and an adjoining small woodlot in the countryside. Decide in general terms what you would like to do with your woodlot; determine what would be the most important market opportunity for you in developing the woodlot's resources, what major public constraint would you expect to find concerning your woodlot, and what set of management skills or limitations do you have that would affect your woodlot management. To what extent are the answers to these questions compatible and what are their inherent contradictions?

2. What are the type of considerations you would make in an IREM site analysis for a river basin compared with that of an upland plateau.

3. Consider yourself as the manager of an extensive coastal zone that stretches along 200 km of coast and extends out to the continental shelf. You have just recently negotiated a tentative agreement with all licensed commercial fishers in this region; they have voluntarily agreed to a substantial reduction in fish harvesting to sustainable levels. This agreement also includes halting dragging (a fishing method that tends to destroy the seabed) as well as establishing a no commercial fishery zone. Overall, this agreement is thought to have no significant impact on the total amount or the kind of species caught. The agreement will be ratified once an acceptable marketing plan is developed that aims to encourage the highest possible economic returns for each fish harvested. In developing a marketing plan, to what type of customers will you target these fish? How will you distribute these fish? What pricing strategy will you use? What promotional strategies will you recommend?

4. Develop a cross-impact matrix for a mixed farm operation (livestock and crops) that is contemplating integrating *agritourism* (farm-based tourism) as a value added activity that considers five key social and environmental objectives. Select three of the possible cross-impacts (select three cells in the matrix) and provide an example of the likely cross-impacts to be found.

5. Given that Wondolleck warns against leaving important stakeholders out of the IREM planning and management process and Bissix and Rees' caution that various power relationships will serve to undermine any apparent consensus in developing multiagency IREM processes over time, what steps might you take early in the planning process as well as later in the implementation process to try to reconcile this contradiction?

Case Study Discussion Question

6. After numerous external and internal discussions, the Windsor and Hantsport Railway developed a corporate strategy to integrate its need to run a regional railway at a profit and develop a policy that could best serve the needs and desires of the various interests represented in this case as well as integrate its desire to promote safety and environmental stewardship. Taking on the role of the CEO of the Windsor and Hantsport Railway and applying the planning framework provided in this chapter, select, develop and justify a strategy that you believe would best meet these stated long-term objectives.

References

Bissix, G. (1995) *Woodlot Recreation*. Nova Scotia Department of Natural Resources, Wolfville, Nova Scotia, Canada.

Bissix, G. (1999/2000) A forest recreation decision support system: the woodlot outdoor Recreation Opportunity Spectrum (woROS). *Leisure/Loisir: the Journal of the Canadian Association of Leisure Research* 24(3/4), 299–319.

Bissix, G. and Rees, J.A. (2001) Can strategic ecosystem management succeed in multi-agency environments? *Ecological Applications: a Journal of the Ecological Society of America* 11(2), 570–583.

Born, S. and Sonzogni, W. (1995) Integrated environmental management: strengthening the conceptualization. *Environmental Management* 19(2), 167–181.

Brown, P.G. (2001) *The Commonwealth of Life: a Treatise on Stewardship Economics*. Black Rose Books, Montreal, Quebec, Canada.

Henriss-Anderssen, C. (2002) *Newsletter – New GBRMPA Permit Processing System Cuts the Paper Chase*. Great Barrier Reef Marine Park Authority (GBRMPA), Queensland, Australia.

Turner, M.G., Gardner, R.H. and O'Neill, R.V. (2001) *Landscape Ecology in Theory and Practice: Pattern and Process*. Springer-Verlag, New York.

Turner, R.K. (ed.) (1993) *Sustainable Environmental Economics and Management: Principles and Practice*. Belhaven Press, London.

Wondolleck, J.M. (1988) *Public Lands Conflict and Resolution: Managing National Forest Disputes*. Plenum Press, New York.

12 Summary and Conclusions: Where Are We Going?

IREM continues to evolve and change. It is not a static process. Decisions facing resource managers will continue to be complex, involving distribution issues of fairness, inclusion, costs and benefits, and sustainability. 'Who gets what?' and 'How much do they get?' are universal questions for the distribution of resources. In order for IREM to be an effective method for the planning and allocation of resources in the 21st century, it must be transparent and fair. Yet, the contexts in which these decisions are made involve uncertainty, multiple truths, varying scales, and a range of institutional and public forces that add to the complexity.

This concluding chapter provides an overview of the main points of the book and the many questions still surrounding those points. How has IREM evolved as a method for making decisions on natural resource distribution? What are the driving forces underlying IREM? How has theory helped develop the concept of IREM? What are the analytical, conceptual and procedural models that provide a framework for the process? The challenges and future of application of the IREM process to natural resource decision making is also discussed.

A Synopsis of the Book

The history of natural resource management has strongly influenced the development of IREM and continues to have an impact in many countries regarding how IREM is (or is not) applied. The historical record of natural resource management can be traced to four broad phases: (i) early Upper Palaeolithic and Mesolithic; (ii) Neolithic; (iii) Historic; and (iv) Modern. The historic ties to IREM in North America can be traced to Powell's 1888 treatise on arid lands, and in the early 20th century to Pinchot and Roosevelt's Inland Waterways Commission. One of the first applied strategies of IREM occurred

Integrated Resource and Environmental Management (A.W. Ewert *et al.*)

with the Muskingum Watershed Conservancy District and the Tennessee Valley Authority in 1933. IREM continued to evolve in North America as a means of addressing the impacts of large-scale development and to deal with increasing environmental conflict.

The emerging use of IREM provides a wealth of case studies in the USA, Canada, Europe and Australia. IREM practice has been applied in watershed/catchment management, forestry, parks and protected areas, as well as mega-project development. Case studies throughout this book provide examples of the varied applications of IREM both in practice and with respect to larger issues being managed.

The application of IREM is contingent upon driving forces that determine how resources are harvested and distributed – driving forces are the agents of change. Primary **driving forces** discussed in this book include: *demographics*, *trade and markets*, *public perception of natural resources*, *environmental injustice*, *international conventions*, *personal choice* and *indigenous peoples*. These forces are completely interconnected, and when individual components change they may often have unintended consequences. Both the social and natural environments are highly unpredictable and complex, and, as a result, require IREM to operate within a 'stormy' and often controversial context. Environmental conflict and public disputes over natural resources have changed traditional methods of resource management and facilitated more integrated perspectives. Competing demands for individual and public access to natural resources are founded on changing views of property rights and differing environmental values. This has given rise to alternative dispute resolution (ADR) approaches which deal with conflict and alternative methods that include public input in decision making.

The role of social science is increasingly important in IREM because of trends that reflect a growing diversity of values that people place on natural resources, the political power of science and the ever-increasing impact of humans on the earth's surface. People are an important link in the management of natural resources, and an emerging challenge is to incorporate the social sciences into the policy and decision-making agenda. Despite this growing importance, Field (1996) points out that the social sciences are still struggling to be viewed with greater legitimacy by the land management agencies and have not always been successful in being accepted by the other scientific disciplines.

A normative theoretical model that attempts to explain decision making in IREM is the 'rational comprehensive model', which posits that humans act in a manner consistent with their perceived goals and needs. However, there are other descriptive models including the procedural rationality, organizational and political bargaining models that reflect the 'real world' more closely. A global perspective includes power as an agent in the decision-making process and is explained by three fundamental groups of theories: the pluralist, the elitist and structural approaches. The dynamics of the public policy process, and how power and decision making are viewed in resource management, may further our understanding of IREM in the context of institutional development and change.

The Framework of IREM

The development of IREM as a concept and practice comes from natural resource management. The theoretical foundations of natural resource management are broad and include a wide range of disciplines. Five **theoretical themes** in natural resource management include: (i) *conservation management*; (ii) *environmentalism*; (iii) *multiple use*; (iv) *sustainability and ecological modernization*; and (v) *the impact of global trade policies*. All of these themes have contributed to the development of IREM over the last several decades and continue to be reflected in how IREM 'should' be practised.

The composition of IREM is divided into normative research that outlines the essential elements required for IREM and prescriptive literature that defines criteria required for IREM to work as a planning process. **Normative research characteristics** discussed in this book consist of: *comprehensive/inclusive, interconnective, strategic/reductive, interactive/coordinative* and *holistic*. All of these elements are required in order for IREM to be an effective decision-making model. The prescriptive literature gives direction as to how IREM should be implemented and the barriers to successful integration. Much of this work has focused on institutional arrangements and interagency cooperation.

Challenges in the application of IREM are examined in detail with respect to multiagency and multi-interest ecosystem management. Complex institutional arrangements, a wide range of stakeholders and complex ecosystems make it difficult to implement IREM. Three types of multiagency resource management models that define agency roles are examined in this chapter. An analytical framework from which to better understand how various agencies and organizations react to ecosystem management initiatives, over time, consists of defining: agency character; interagency relationships; multiagency networks; and macro forces acting on the ecosystem. This framework recognizes the complexity of ecosystem management and how challenging general prescriptive remedies and/or solutions are for all contexts.

The planning framework proposed in this book offers a set of general guiding principles to implement IREM. It is intended to be of service to both the student and the practitioner. We recognize the limitations of attempting to offer a framework such as IREM as the solution to a very complex process that changes according to the situational context and within a wide range of variables such as culture, scale, time frames or institutional arrangements.

Concluding Remarks

22 April 2004 marked the 34th anniversary of Earth Day. That day represented a world that was warmer, more complex and crowded, and less rich in biodiversity than it was 34 years ago. While celebrated in numerous ways since it was first founded in 1970, perhaps the most important function that Earth Day serves is as a benchmark. It has become a convenient 'yardstick' in determining the direction the earth's environment is taking. In one sense, the founding of Earth Day represents the starting point for the modern environmental

movement. Also, there have been signs of improvement in some areas, such as reduction in waste production; in certain processes, such as industrial materials, forest replanting in specific locations; rehabilitation in certain fisheries; and the reduction in certain areas of specific environmental pollutants, and toxic agents such as water-borne pathogens.

However, taken in total, by most accounts, the state of the world's environment has deteriorated over the past 30 years and continues to do so (Worldwatch Institute's, *State of the World*; World Resource's Institutes, *World Resources*; and the UNEP's *Environmental Data Report*). Contributing to this deterioration is a host of factors and trends (O'Riordan *et al.*, 1995). Among these trends are the following:

- Continued growth in world population
- Lack of access to safe drinking water and adequate food for millions of people
- Dramatically increased resource development and consumption
- Environmental contamination from pesticides and other forms of pollution
- Loss of various 'systems' and resource integrity, such as various fisheries, forested lands, animal and plant species, and biodiversity indicators.

What seems increasingly clear about this growing list is that there is not one central cause of these environmental problems, and any long-term solutions will be costly and complex. Many of the factors affecting the environment have little to do with the environment itself. For example, increased meat consumption in one part of the world can cause habitat fragmentation and loss of biodiversity in another. Any attempts at developing a sustainable relationship with the natural resources base must account for these outside influences (e.g. external forces) ultimately to be effective.

Likewise, for IREM to be effective, a number of issues and questions must be resolved. A sampling of these issues and questions includes the following.

1. Will an IREM approach be politically accepted? Wilbanks (1994) points out that 'big' natural resource issues are, by definition, complicated, and often contain inter-related variables, such as the natural environment, social issues (e.g. poverty, business concerns, trade issues and debt relief), cultural beliefs, normative values and community wellness. The extent to which IREM processes can be linked to these other components will be indicative of its effectiveness and, ultimately, its level of acceptability.
2. Will IREM be effective in situations or areas where a wide spectrum of 'stakeholders' is unwelcome? This question is particularly important in situations where there are one or two relatively powerful parties involved, often in collaboration, followed by a number of weaker, less affluent, or less powerful interested parties.
3. Can IREM be implemented in circumstances where there are competing interests, differing values and/or a high level of institutional inertia? For example, Dale (2001) talks about the inability of large organizations and agencies: (i) to see how their culture influences their decision making; and (ii) to better understand and adapt to changing realities.

4. By its very nature, IREM represents a 'messy' decision-making process. A greater number of stakeholders are involved in the decision-making process, more outcomes from any decision need to be examined, and more information needs to be considered. In addition, implementing and assessing a decision will be more complex and involve issues of equity. Can IREM work in circumstances that demand quick decisions or expedited actions? Is IREM more appropriate for smaller scale and less complex questions, or for situations that pose larger and more complex decision-making challenges?

The Problem of Consensus

Whatever the answers to the previous questions may be, reaching societal consensus about how to manage our natural resources will remain problematic in the foreseeable future. There is a growing reluctance on the part of society and individual citizens to treat natural resource management issues as something beyond their control or controlled exclusively by the professional manager or politician. Wondolleck (1991) points to this issue by suggesting that natural resource management possesses neither broad public support nor a clear sense of mission and future vision, both characteristics that had distinguished it for more than a half a century.

Indeed, as Fiske (1990) pointed out in the early 1990s, natural resource management is increasingly viewed as 'human management', i.e. any long-term resolution of a natural resource problem will ultimately hinge on effectively reaching some form of consensus and acceptance among various sectors of society. As has been discussed in this book, however, reaching consensus is often difficult and, in part, represents the 'real' challenge in contemporary and future natural resource management.

This difficulty is, in part, a result of differences in worldviews, particularly in dealing with uncertain outcomes or incomplete data. For example, Bretherton (1994) writes that some people view the world as a stable place where change is often a positive attribute; others view the world as an unstable entity where any change often seems negative and threatening. Still others view the world as unpredictable and place a high value on – what they believe to be – the certainty of data and science. In each case, the way in which an individual views the world is often a determining factor in what type of natural resource management is most desirable and needed.

Demographic diversity and cultural background can also play important roles in deciding whether a society can reach consensus regarding natural resource issues. A number of authors including Floyd (1999) and Virden and Walker (1999) remind us that variables such as ethnicity, gender, age and racial diversity can influence how individuals attach meaning and value to natural resource environments in addition to guiding their behaviours and actions in, and toward, those environments.

Another area of discord in reaching social consensus regarding natural resource management involves how society actually accumulates resource information and makes decisions regarding those resources. Historically, natural resource management was considered the responsibility of the resource professional. Increasingly, however, natural resource professionals are viewed with some suspicion as being too much under the influence of politicians' or government control. While few would argue that sound natural resource decision making is usually assisted by reliable and accurate scientifically generated information (Ecological Society of America, 1995), some professionals now question making too strong a connection between natural resource science and natural resource management. For example, Hutchings *et al.* (1997) point out that, in the case of the Atlantic cod (*Gadus morhua*) and Pacific salmon (*Oncorhynchus* spp.) fisheries in Canada, a premature linking of science with management helped suppress scientific uncertainty, and legitimate differences in scientific opinion. Moreover, Wolosoff and Endreny (2002) found that scientists and policy makers/managers are often faced with fundamentally different decision-making arenas and, as such, have differing perspectives regarding the desired type of information needed. For example, policy makers are often faced with developing a policy that can be applied across entire counties, provinces or states. Scientists, on the other hand, often look for information that examines the complexities among specific variables and their research often requires longer time frames, as well as more specific or abstract applications. Thus, it is not surprising that the desired information and interpretation of that information differ widely between the two groups.

In addition, while acknowledging that scientifically derived information has value in certain situations and settings, others would readily point to the importance, but diminishing visibility, of 'traditional environmental knowledge' (TEK). Sherry and Myers (2002) indicate that while TEK can be of limited value in many settings and circumstances, it can be a useful tool, particularly in specific situations requiring a co-management approach involving aboriginal groups.

Consensus and agreement, regarding how to manage our natural resources, often become a matter of both values and world opinions. For many, the natural resources represent an economic resource that should be safeguarded but utilized. Pearce (2001) contends that economic valuation of forests must move from a simple accounting of timber values to also include other 'direct use values' such as fuelwood and charcoal, biodiversity and genetic information, and tourism and recreation. Further, other values such as 'indirect values' (e.g. watershed protection and carbon sequestration), and 'option and existence values' must also be considered in determining the full value of a forested or other natural environment. Blahna and Yonts-Shepard (1989) would argue that understanding the natural resource values held by the public is increasingly difficult for land management agencies because there is often little direction on *how* to collect the information and make sense of it. For example, the standard use of paper and pencil questionnaires to record information regarding psychological issues, such as values and attitudes, often belies the multidimensions and 'richness' underlying these types of complex constructs (Satterfield and Gregory, 1998).

In addition, other values held by individuals and society are focused on the link between ecosystem health and human health (Ewert, 2002; Rapport *et al.*, 2002). The common belief is that, if the natural ecosystem is unhealthy, the individual health of a person living in or near that ecosystem will also be adversely affected. Extending this belief, authors such as Rapport *et al.* (2001) and McMichael (2001) posit that many of the attributes associated with human health, such as levels of asthma, can be linked to environmental degradation.

Finally, other members of our society examine how natural resources are managed as a point of personal responsibility. For them, attitudes, beliefs and behaviours toward natural resources are bound-up and interconnected by feelings of personal commitment or duty (Kaiser *et al.*, 1999). Thus, one's actions and the actions of others toward a natural resource should be guided by a sense of responsibility and stewardship, and strongly influenced by social and moral values. For people with similar values, who view natural resources as purely an extension of the economic or political base, reaching consensus is difficult. More importantly, this lack of consensus is amplified by what some feel are the cumulative effects of environmental abuse that appear all powerful and ever-advancing, or what Kahaner (1988) refers to as 'creeping degradation'.

In closing, perhaps what IREM represents most of all is a change in thinking regarding leadership. As Sirmon (1991) surmised over 10 years ago, resource management was traditionally built upon a hierarchical model where decisions were left to the experts. Later, in response to public concern over how natural resource decisions were being made, public input was sought through structured forums, focus groups, etc., as a way to hear from other 'stakeholders'.

A number of authors now suggest a new form of leadership, namely a 'community of interests' type of leadership, where decision making is shared by both the agency and members of the community who have an interest in the issue (Sirmon, 1991; Shands, 1999). Not surprisingly, this approach fits both the philosophy and intent of IREM, and essentially means moving from an *authority-based* model to a *partnership* model (Eisler, 1987).

Although finding common ground will remain an elusive goal, both now and in the future, it is perhaps the ultimate goal of IREM to institute a process where reaching consensus about how our natural resources should be used and managed, through a partnership, where many groups and entities play meaningful roles in the decision making, leads to greater long-term acceptance and ultimately stability. For the reality is that without an IREM-type process, winners and losers are created without a sense of fairness or understanding, ultimately causing the losers to 'regroup' their political resources to wait for a time when they can turn the decisions around, thus creating a never-ending cycle of resource management conflicts and short-lived decisions that continue to fuel social bitterness. Like democracy, IREM may not be a perfect solution to natural resource management decision making, but it is probably better than any of the alternative processes currently available.

Discussion Questions

1. Select a large natural resource management agency and describe: (i) its culture; (ii) how this culture influences the decisions made in this agency; and (iii) how this agency responds to emergent issues and challenges.
2. What emerging trends in society will serve to profoundly impact natural resource use? In turn, how will these trends influence how individuals 'value' and 'use' natural resources? Will there be a dichotomy between what values society places on natural resources and how natural resources are actually used?
3. In what ways has the liberalization of global trading policies impacted natural resource management and use? Have policies such as NAFTA served to promote sound natural resource management policy making or disrupted it?
4. Examine the political structure and make-up of your local community. Given this situation, will your community and political structure be more or less likely to support an IREM process? What do you believe the long-term trend will be?
5. This text has promoted IREM as one of the most sustainable ways for society to make decisions regarding the sustainable use and management of natural resources. Do you believe this to be true? If yes, what are the key elements of the IREM model that makes it so useful? If not, what other approaches should be considered to manage our natural resources more sustainably?

References

Blahna, D.J. and Yonts-Shepard, S. (1989) Public involvement in resource planning: toward bridging the gap between policy and implementation. *Society and Natural Resources* 1, 209–227.

Bretherton, F. (1994) Perspectives on policy. *Ambio* 23(1), 96–97.

Dale, A. (2001) *At the Edge: Sustainable Development in the 21st Century*. UBC Press, Vancouver, British Columbia, Canada.

Ecological Society of America (1995) *The Scientific Basis for Ecosystem Management*. Ecological Society of America, Washington, DC.

Eisler, R.T. (1987) *The Chalice and the Blade: Our History, Our Future*. Harper and Row, Cambridge, Massachusetts.

Ewert, A.W. (2002) Quality of life, recreation, and natural environments: exploring the connection. In: Rapport, D.J., Lasley, W., Rolston, D., Nielson, O., Qualset, C. and Damania, A. (eds) *Managing for Healthy Ecosystems*. CRC Press, Boca Raton, Florida, pp. 199–205.

Field, D.R. (1996) Social science: a lesson in legitimacy, power and politics in land management agencies. In: Ewert, A. (ed.) *Natural Resource Management*. Westview Press, Boulder, Colorado, pp. 249–255.

Fiske, S.J. (1990, Winter) Resource management as people management: anthropology and renewable resources. *Renewable Resources Journal* 16–20.

Floyd, M. (1999) Race, ethnicity, and use of the National Park System. *Social Science Review* 1(2), 1–23.

Hutchings, J.A., Walters, C. and Haedrich, R.L. (1997) Is scientific inquiry incompatible with government information control? *Canandian Journal of Fisheries and Aquatic Sciences* 54, 1198–1210.

Kahaner, L. (1988) Something in the air. *Wilderness* 52(183), 18–27.

Kaiser, F.G., Ranney, M., Hartig, T. and Bowler, P.A. (1999) Ecological behavior, environmental attitude, and feelings of responsibility for the environment. *European Psychologist* 4(2), 59–74.

McMichael, A.J. (2001) *Human Frontiers, Environments and Disease: Past Patterns, Uncertain Futures*. Cambridge University Press, Cambridge, UK.

O'Riordan, T., Clark, W.C., Kates, R.W. and McGowan, A. (1995) The legacy of Earth Day: reflections at the turning point. *Environment* 37(3), 7–15, 37–42.

Pearce, D.W. (2001) The economic value of forest ecosystems. *Ecosystem Health* 7(4), 284–295.

Rapport, D.J., Howard, J.M., Lannigan, R., Anjema, C.M. and McCauley, W. (2001) Strange bed fellows: ecosystem health in the medical curriculum. *Ecosystem Health* 7(3), 155–162.

Rapport, D.J., Lasley, W., Rolston, D., Nielson, O., Qualset, C. and Damania, A. (2002) *Managing for Healthy Ecosystems*. CRC Press, Boca Raton, Florida.

Satterfield, T. and Gregory, R. (1998) Reconciling environmental values and pragmatic choices. *Society and Natural Resources* 11, 629–647.

Shands, W.E. (1999) Leadership in a community. In: Aley, J., Burch, W.R., Conover, B. and Field, D.R. (eds) *Ecosystem Management: Adaptive Strategies for Natural Resources Organizations in the Twenty-first Century*. Taylor and Francis, Philadelphia, pp. 117–123.

Sherry, E. and Myers, H. (2002) Traditional environmental knowledge in practice. *Society and Natural Resources* 15, 345–358.

Sirmon, J.M. (1991) Evolving concepts of leadership: towards a sustainable future. *Forest Perspectives* 1(2), 8–9.

Virden, R.J. and Walker, G.J. (1999) Ethnic/racial and gender variations among meanings given to, and preferences for, the natural environment. *Leisure Sciences* 21, 219–239.

Wilbanks, T.J. (1994, December) Sustainable development in geographic perspective. *Annals of the Association of American Geographers* 541.

Wolosoff, S.E. and Endreny, T.A. (2002) Scientist and policy-maker response types and times in suburban watersheds. *Environmental Management* 29(6), 729–735.

Wondolleck, J.M. (1991, December) *Natural Resource Management in the 1990s and Beyond: Problems and Opportunities*. Prepared for the Pinchot Institute's Project on Leadership in Natural Resources, Milford, Pennsylvania.

Glossary

Anthropocentric: a human-oriented perspective of the environment that serves to provide a distinction between humans and non-humans.
Biodiversity: diversity, complexity and quantity of the biota in an area. Areas thought to be high in biodiversity generally have a broad range of organisms from micro to macro levels, varied and abundant integrated systems, and high degrees of resilience as a result of this complexity and 'richness'.
Biological carrying capacity: the maximum population of a species or organism that can be sustained within the biological parameters (e.g. habitat) provided.
Biomass: the weight of organisms (producers, consumers, decomposers, other organic material, etc.) present at a specific time. Often called 'living weight' and sometimes inaccurately used as a surrogate measure for biodiversity.
Carrying capacity: maximum level of specific populations of organisms that can be sustained in a given location. Often divided into levels (optimum or 'safe' and maximum) or type (biophysical or social).
Conservation: a land management philosophy that is generally associated with the 'wise use' of natural resources and promotes the belief that resources should be managed to provide a long-term flow of products and uses.
Cumulative effects: the sum total of a series of stresses acting on a particular organism, population, species, ecosystem, physical feature or landscape.
Decoupling: the deliberate or inadvertent separation of one social or biophysical trend from another. For example, the attempt to decouple an expected increase in environmental degradation from socioeconomic growth.
Demographic entrapment: the effect of current or projected population numbers will exceed the carrying capacity of a specific environment and create conditions for famine, disease and instability within the society.
Dominant use: a situation in which the use of one or more natural resources is considered of primary importance and given consideration over and above other resources or uses.

Economic values: values associated with the use of a natural resource or a product derived from a natural resource that is usually calculated in monetary units such as dollars. Often referred to as a 'market value or good' as opposed to those uses or products that are difficult to quantify in monetary units such as a beautiful vista. Such an assessment is often referred to as a 'non-market value'.

Ecosystem: perceived as a holistic collection of biological units operating within the physical parameters of a given area. Usually thought of as an area where the processes of primary production, consumption, decomposition and material cycling are largely self-contained.

Ecosystems management: a management strategy that focuses on the long-term health of an ecosystem as opposed to looking primarily at the goods and services it produces. Ecosystem management is often thought of as being more concerned with the functioning and stability of the entire ecosystem rather than specific parts.

Environmental uncertainty: unpredictable factors that create changes in the environment. Some examples would include weather, human demographic changes that were unexpected, or the presence of a new disease or pest.

Externality problems: unintentional issues and problems that occur as a result of other decisions. These externalities often produce negative consequences on the environment, community or economy. For example, improperly conducted timber harvesting can lead to the externality of increased siltation in local watersheds.

Habitat: physical locations where an organism lives and travels, most often associated with wildlife issues.

Human (social) capital: practical knowledge, learned skills and abilities that allow individuals to effect change in the environment and be productive.

I = PAT: an equation describing the impact humans have on natural systems where impact (I) is a function of population size (P), level of affluence (A) and level of technology (T).

Instrumental value: as opposed to intrinsic value, instrumental value refers to the worth or utility that an entity or natural resource component has to humans.

Integrated resource and ecosystem management: a management process and philosophy that takes into account the values of multiple resources of an area, considers the linkages between these resources and other organisms, and views these resources within a long-term, sustainable perspective.

Intrinsic value: the worth of an entity or natural resource component independent of utility or value to humans. See Instrumental value.

Land management: the control and management of natural resources including water, air and below-ground natural resources. How these resources are managed (e.g. developed, preserved, made available for use, etc.) is often a very complex and political process fraught with controversy and disagreement.

Land use regulation: laws, ordinances and other legal instruments that serve to enforce the land management philosophy and decisions made by a governing body overseeing the use of a natural resource or area.

Multiple objective resource management: a forerunner of integrated resource and environmental management and is often used synonymously with multiple use. This management process attempts to optimize the outputs of selected resource management targets such as water quality and quantity, timber production and outdoor recreation.

Multiple use: the managed or unmanaged simultaneous use of a natural resource such as public lands by different stakeholder and interest groups. Often results in conflicting demands and incomplete protection of the resource. As a management strategy, it promotes the use and protection of varying resources in a given area. The USDA Forest Service has long been a proponent of multiple use and stresses that in a particular area water, wildlife, timber, grazing and recreation may all be important management targets.

Nature: the non-human environment including both the physical (e.g. mountains, rivers, etc.) and the biological (e.g. plants, wildlife, microorganisms).

Non-market values: those aspects of the world that have worth and importance but are not normally assigned (or are incapable of being assigned) an economic value (usually in monetary terms) and traded in typical markets. Such values may include moral values, spiritual values and intrinsic values.

Non-point source pollution: pollution that occurs over a wide area and enters the ecosystem or area in question via a variety of sources and locations. One example of non-point source pollution is fertilizer and pesticide runoff into local watersheds from agricultural lands.

Old growth: a term implying that an area has not been previously harvested or was harvested very early in the settlement of an area. Often referred to as Ancient Forests or Virgin Forests. The harvesting of old-growth forests has become controversial for a variety of political and biological reasons.

Option value: assigning a value (usually economic) to a resource for future use.

Point source pollution: often referred to as 'tailpipe' pollution; refers to pollution that can be traced to a single source or point such as a sewage or an industrial drainage pipe.

Public good: an activity, project or area deemed to be useful and promoting the general welfare of the public. Commons areas or undeveloped open spaces near urban developments are often viewed as a public good because of the positive benefits they are generally associated with (e.g. catharsis, experience of nature, etc.).

Regulatory authority: the legal and governing foundation that describes the areas of responsibility and authority of an agency or organization. State Fish and Wildlife agencies often have regulatory authority for the management of fish and wildlife populations and species.

Resilience: the ability of a system to maintain its integrity and structure when faced with stresses or disturbance.

Social carrying capacity: maximum number of humans that a specified area can hold without upsetting the physical/biological parameters or creating a significant detrimental change in the 'experience'.

Species diversity: level of species 'richness' or proportional distribution.

Species richness: the number of species in an area or specific location.

Sustainable: the concept that promotes the use and development of a specific natural resource in ways and at levels that can be continued for an extended time. This use and development takes into account the renewal capacities of the organism(s) in question. For example, in order to be sustainable, the harvesting of specific fisheries must be at a level that does not exceed the reproduction rate and associated variables of the species in question.

Sustainable development: the process of managing the natural resources in a manner that will provide for the overall well-being of the resources, people and communities, today and in the future, while (i) maintaining productivity; (ii) safeguarding the capacity of the ecosystem; and (iii) minimizing adverse impacts (Barichello *et al.*, 1996).

Systematic effects: the ongoing and usually predictable impacts of various stresses or variables on a natural resource often occurring over large areas. Global warming is an example of a systematic effect.

Threatened: a term implying a species is in danger of becoming extinct. From a legal perspective, when a species becomes 'threatened', certain countries have in place specific remedies and management actions that are enacted to provide heightened protection for that species.

Traditional environmental knowledge (TEK): that knowledge that has been accumulated over generations about a particular place, physical feature or ecological system that is not widely known or appreciated in formal academic or professional circles, that is often passed on using oral traditions, and is sometimes critical in understanding the potential of a particular natural resource.

Tragedy of the commons: a theoretical proposition that resources owned by the public will eventually degrade or become depleted because there is no individual incentive to safeguard their use and levels of exploitation.

Watershed: the entire land area that delivers water, sediment and dissolved substances via surface runoff and small streams to a major stream (river) and ultimately to the sea.

Utilitarian: a term used to describe any activity that is deemed beneficial to humans.

Zoning: land protection and use policy instrument that designates certain uses and precludes other uses for a specific area or location.

References

Barichello, R., Porter, R.M. and van Kooten, G.C. (1996) Institutions, economic incentives, and sustainable rural land use in British Columbia. In: Scott, A., Robinson, J. and Cohen, D. (eds) *Managing Natural Resources in British Columbia: Markets, Regulations, and Sustainable Development*. UBC Press, Vancouver, British Columbia, Canada, pp. 7–53.

Miller, G. (1992) *An Introduction to Environmental Science: Living in the Environment*. Wadsworth Publishing Company, Belmont, California, p. 139.

Bibliography

Ajzen, I. and Fishbein, M. (1980) *Understanding Attitudes and Predicting Social Behavior*. Prentice-Hall, Englewood Cliffs, New Jersey.

Alberta Forestry Resource Evaluation and Planning Division (1986) *Kananaskis Country Sub-regional Integrated Resource Plan*. Queen's Printer, Edmonton, Alberta, Canada.

Alberta Forestry, Lands and Wildlife Resource Planning Branch (1991) *Integrated Resource Planning in Alberta*. Queen's Printer, Edmonton, Alberta, Canada.

Albrecht, D., Bultena, G., Hoiberg, E. and Nowak, P. (1982) The new environmental paradigm scale. *Journal of Environmental Education* 13, 39–43.

Allison, M.T. (1993) Access and boundary maintenance: serving culturally diverse populations. In: Ewert, A., Chavez, D. and Magill, A. (eds) *Culture, Conflict, and Communication in the Wildland Urban Interface*. Westview Press, Boulder, Colorado, pp. 99–107.

Anderson, M.S. (1994) *Governance by Green Taxes: Making Pollution Prevention Pay*. Saint-Martin's Press, New York.

Atkinson, N. (1991) *Dartmoor National Park: Second Review 1991*. Dartmoor National Park Authority, Bovey Tracey, UK.

Atkinson, N. (1992) *Dartmoor National Park Local Plan: Including Minerals and Waste Policies – Consultation Draft 1992*. Dartmoor National Park Authority, Bovey Tracey, UK.

Australia's Continental Odyssey (1988, February) *National Geographic* (map supplement).

Babbie, E.R. (1979) *The Practice of Social Research*, 2nd edn. Wadsworth, Belmont, California.

Baird-Olson, K. (2000) Recovery and resistance: the renewal of traditional spirituality among American Indian women. *American Indian Culture and Research Journal* 24(4), 1–35.

Baker, D.C., Young, J. and Arocena, J. (2000) An integrated approach to reservoir management: the Williston Reservoir case study. *Environmental Management* 25(5), 565–578.

Barichello, R., Porter, R.M. and van Kooten, G.C. (1996) Institutions, economic incentives, and sustainable rural land use in British Columbia. In: Scott, A., Robinson, J. and Cohen, D. (eds) *Managing Natural Resources in British Columbia: Markets, Regulations, and Sustainable Development*. UBC Press, Vancouver, British Columbia, Canada, pp. 7–53.

Bateman, R.B. (1996) Talking with the plow: agricultural policy and Indian farming in the Canadian and U.S. Prairies. *Canadian Journal of Native Studies* 16(2), 211–228.

Beanlands, G. and Duinker, P. (1983) *An Ecological Framework for Environmental Impact Assessment in Canada*. Institute for Resource and Environmental Studies, Dalhousie University and Federal Environmental Assessment Review Office, Halifax, Nova Scotia, Canada.

Beck, B. (2000, 9 September) Survey – Australia: a sorry tale. *Economist* 356, 8187, p. S12.

Becker, L. (1977) *Property Rights – Philosophic Foundations*. Routledge and Kegan Paul, London.

Behan, R.W. (1990) Multiresource forest management: a paradigmatic challenge to professional forestry. *Journal of Forestry* 88(4), 12–18.

Bellamy, J., McDonald, G., Syme, G. and Butterworth, J. (1999) Evaluating integrated resource management. *Society and Natural Resources* 12(4), 337–353.

Beltrame, J. (1999, 6 December) Land claims by Canadian tribes gain court ruling and a treaty increase the pressure: non-natives are nervous. *Wall Street Journal*, p. A27.

Bennett, J.W. (1976) *The Ecological Transition*. Pergamon Press, New York.

Berner, R.A. (1990) Atmospheric carbon dioxide over Phanerozoic time. *Science* 249, 1382–1386.

Bettinger, P. and Boston, K. (2001) A conceptual model for describing decision-making situations in integrated natural resource planning and modeling projects. *Environmental Management* 28(1), 1–7.

Bevan, A. (1992) *Alternative Dispute Resolution*. Sweet and Maxwell, London.

Bingham, J. (1986) *Resolving Environmental Disputes: a Decade of Experience*. The Conservation Foundation, Washington, DC.

Bissix, G. (1995) *Woodlot Recreation*. Nova Scotia Department of Natural Resources, Halifax, Nova Scotia, Canada.

Bissix, G. (1999) Dimensions of power in forest resource decision-making: a case study of Nova Scotia's forest conservation legislation. Unpublished doctoral dissertation, Department of Geography and Environment, London School of Economics and Political Science, London.

Bissix, G. (1999/2000) A forest recreation decision support system: the woodlot outdoor Recreation Opportunity Spectrum (woROS). *Leisure/Loisir: the Journal of the Canadian Association of Leisure Research* 24(3/4), 299–319.

Bissix, G. (2002/2003) Residual recreation and sustainable forestry: historic and contemporary perspectives in Nova Scotia. *Leisure/Loisir: the Journal of the Canadian Association of Leisure Research* 27(1–2), 31–50.

Bissix, G. and Bissix, S. (1995) Dartmoor (U.K.) National Park's landscape management: lessons for North America's Eastern Seaboard. In: Herman, T.B., Bondrup-Neilsen, S., Willison, J.H.M. and Munro, N.W.P. (eds) *Ecosystem Monitoring and Protected Area*. Science and Management of Protected Areas Association, Wolfville, Nova Scotia, Canada, pp. 563–571.

Bissix, G. and Rees, J.A. (2001) Can strategic ecosystem management succeed in multi-agency environments? *Ecological Applications: a Journal of the Ecological Society of America* 11(2), 570–583.

Bissix, G., Levac, L. and Horvath, P. (2002) The political economy of the wilderness designation in Nova Scotia. *Proceedings of the 2001 Northeastern Recreation Research Symposium* (Northeastern research station – general technical report NE-289). US Department of Agriculture. Washington, DC, pp. 377–382.

Blahna, D.J. and Yonts-Shepard, S. (1989) Public involvement in resource planning: toward bridging the gap between policy and implementation. *Society and Natural Resources* 1, 209–227.

Blowers, A. (1984) *Something in the Air: Corporate Power and the Environment*. Harper and Row, London.

Blunden, J. and Curry, N. (1988) *A Future for Our Countryside*. Basil Blackwell, Oxford, UK.

Booth, A.L. and Kessler, W.B. (1996) Understanding *linkages* of people, natural resources, and ecosystem health. In: Ewert, A. (ed.) *Natural Resource Management: the Human Dimension*. Westview Press, Boulder, Colorado, pp. 231–248.

Bormann, B. (1993) Is there a social basis for biological measures of ecosystem sustainability? *Natural Resource News* 3, 1–2.

Born, S. and Sonzogni, W. (1995) Integrated environmental management: strengthening the conceptualization. *Environmental Management* 19(2), 167–181.

Botkin, D.B. (1990) *Discordant Harmonies: a New Ecology for the Twenty-first Century*. Oxford University Press, New York.

Boule, L. and Nesic, M. (2001) *Mediation: Principles, Process, and Practice*. Butterworths, London.

Boyack, S. (2000) *National Parks for Scotland: Consultation on the National Parks (Scotland) Bill*. Scottish Executive, Edinburgh, UK.

Bozeman, B. and Straussman, J.D. (1991) *Public Management Strategies: Guidelines for Managerial Effectiveness*. Jossey-Bass, San Francisco.

Bradley, G.A. and Bare, B.B. (1993) Issues and opportunities on the urban forest interface. In: Ewert, A., Chavez, D. and Magill, A. (eds) *Culture, Conflict, and Communication in the Wildland Urban Interface*. Westview Press, Boulder, Colorado, pp. 17–31.

Bretherton, F. (1994) Perspectives on policy. *Ambio* 23(1), 96–97.

Bromley, D.W. (1989) *Economic Interests and Institutions: the Conceptual Foundations of Public Policy*. Basil Blackwell, Oxford, UK.

Brown, G. and Harris, C. (1998) Professional foresters and the land ethic, revisited. *Journal of Forestry* 96(1), 4–12.

Brown, H. and Marriot, A. (1993) *Alternative Dispute Resolution Principles and Practice*. Sweet and Maxwell, London.

Brown, J.R. and MacLeod, N.D. (1996) Integrating ecology into natural resource management policy. *Environmental Management* 20(3), 289–296.

Brown, P.G. (2001) *The Commonwealth of Life: a Treatise on Stewardship Economics*. Black Rose Books, Montréal, Québec, Canada.

Buck, S. (1989) Multi-jurisdictional resources: testing a typology for problem-structuring. In: Berkes, F. (ed.) *Common Property Resources: Ecology and Community-based Sustainable Development*. Belhaven Press, London, pp. 127–147.

Bumpass, L. and Sweet, J. (1989) National estimates of cohabitation. *Demography* 26, 615–625.

Burton, L.T. (1972) *Natural Resources Policy in Canada: Issues and Perspectives*. McClelland and Stewart, Toronto, Canada.

Buttel, F. and Taylor, P. (1992) Environmental sociology and global environmental change: a critical assessment. *Society and Natural Resources* 5, 211–230.

Byrne, J.P. (1995) Ten arguments for the abolition of the regulatory takings doctrine. *Ecological Law Quarterly* 22 (1), 89–142.

Canadian Forests (2002) *Canada–US Softwood Lumber Dispute*. Retrieved June 20, 2002, from http://www/canadian-forests.com/

Carley, M. (1980) *Rational Techniques in Policy Analysis*. Heinemann, London.

Carson, R.L. (1962) *Silent Spring*. Houghton Mifflin, Boston, Massachusetts.

Charmaz, K. (1983) The grounded theory method: an explication and interpretation. In: Emerson, R.M. (ed.) *Contemporary Field Research: a Collection of Readings*. Waveland Press, Prospect Heights, Illinois, pp. 109–126.

Chase, R.A. (1993) Protecting people and resources from wildfire: conflict in the interface. In: Ewert, A., Chavez, D. and Magill, A. (eds) *Culture, Conflict, and Communication in the Wildland Urban Interface*. Westview Press, Boulder, Colorado, pp. 349–356.

Chatwin, B. (1987) *The Songlines*. Viking Press, New York.

Christensen, N.L. (chair) (1996) The report of the Ecological Society of America Committee on the scientific basis for ecosystem management. *Ecological Applications* 6(3), 665–691.

City of Chicago Department of Planning and Development (2002) *Calumet Land Use Plan*. City of Chicago, Chicago.

Cobb, C., Halstead, T. and Rowe, J. (1995) If the GDP is up, why is America down? *The Atlantic Online*. Available at: http://www.theatlantic.com/politics/ecbig/gdp.htm Retrieved 18 December, 2002.

Cockfield, G. (1999) The political economy of the Australian farm landscape: the application of social science to resource management in the Pacific region. *Abstracts of the International Symposium on Society and Resource Management*. University of Queensland, Brisbane, Queensland, Australia.

Copp, D. (1986) Some positions and issues in environmental ethics. In: Hanson, P. (ed.) *Environmental Ethics: Philosophical and Policy Perspectives*. Institute for the Humanities, Simon Fraser University, Burnaby, British Columbia, Canada, pp. 181–195.

Cordell, H.K. and Bergstrom, J.C. (1999) *Integrating Social Sciences with Ecosystem Management*. Sagamore, Champaign, Illinois.

Cordell, H.K. and Overdevest, C. (2001) *Footprints on the Land: an Assessment of Demographic Trends and the Future of Natural Lands in the United States*. Sagamore, Champaign, Illinois.

Cordell, H.K. and Tarrant, M.A. (2002) Changing demographics, values, and attitudes. *Journal of Forestry* 100(7), 28–33.

Cormick, G. (1982, September) The myth, the reality and the future of environmental mediation. *Environment* 24, 15–39.

Cortner, H.J. (1996) Public involvement and interaction. In: Ewert, A. (ed.) *Natural Resource Management: the Human Dimension*. Westview Press, Boulder, Colorado, pp. 167–179.

Cortner, H.J. and Shannon, M.A. (1993) Embedding public participation in its political context. *Journal of Forestry* 91(7), 14–16.

Coser, L.A. (1956) *The Functions of Social Conflict*. Free Press, New York.

Crane, J.A. (1982) *The Evaluation of Social Policies*. Kluwer-Nijhoff, Boston, Massachusetts.

Cronin-Fisk, M. (2001, October 22) 200-year-old land dispute nets $247.9M. *National Law Journal* 24(9), p. A6.

Crowley, T.J. (1996) Remembrance of things past: greenhouse lessons from the Geologic Record. *Consequences* 2(1), 3–12.

Dahl, R.A. (1984) *Modern Political Analysis*, 4th edn. Prentice-Hall, Englewood Cliffs, New Jersey.

Daily, G.C. (ed.) (1997) *Nature's Services: Societal Dependence on Natural Ecosystems*. Island Press, Washington, DC.

Daily, G.C. and Ellison, K. (2002) *The New Economy of Nature: the Quest to Make Conservation Profitable*. Island Press, Washington, DC.

Dale, A. (2001) *At the Edge: Sustainable Development in the 21st Century*. UBC Press, Vancouver, British Columbia, Canada.

de Groot, R.S. (1992) *Functions of Nature: Evaluation of Nature in Environmental Planning, Management, and Decision Making*. Wolters-Noordhoff, Amsterdam.

Demsetz, H. (1967) Toward a theory of property rights. *American Economic Review* 57, 347–359.

De-Shalit, A. (2000) *The Environment: Between Theory and Practice*. Oxford University Press, New York.

Deutsch, M. (1987) A theoretical perspective on conflict and conflict resolution. In: Sandole, D. and Sandole-Staroste, I. (eds) *Conflict Management and Problem Solving: Interpersonal to International Applications*. Frances Pinter, London, pp. 38–49.

Devall, W. and Sessions, G. (1985) *Deep Ecology*. Peregrine Smith, Layton, Utah.

Deyle, R.E. (1995) Integrated water management: contending with garbage can decision-making in organized anarchies. *Water Resources Bulletin* 31(3), 387–398.

Dietrich, W. (1992) *The Final Forest*. Simon and Schuster, New York.

Dixon, J.A. and Fallon, L.A. (1989) The concept of sustainability: origins, extensions, and usefulness for policy. *Society and Natural Resources* 2, 73–84.

Doka, K. (1992) When gray is golden: business in an aging America. *Futurist* 26(4), 16–20.

Dorcey, A.J. (1986) *Bargaining in the Governance of Pacific Coastal Resources: Research and Reform*. Westwater Research Centre, University of British Columbia, Vancouver, British Columbia, Canada.

Dorcey, A.J. (1987) The myth of interagency cooperation in water resources management. *Canadian Water Resources Journal* 12(2), 17–26.

Dorcey, A.J. and Riek, C. (1989) Negotiation-based approaches to the settlement of environmental disputes in Canada. *The Place of Negotiation in Environmental Assessment*. Canadian Environmental Assessment Research Council, Hull, Québec, Canada.

Drake, W. (2001, July/August) Green Britain. *Environment* 43(6), 7.

Dudley, N., Jeanrenaud, J.P. and Sullivan, F. (1995) *Bad Harvest? The Timber Trade and the Degradation of the World's Forests*. Earthscan, London.

Dukes, E.F. (1996) *Resolving Public Conflict*. Manchester University Press, Manchester, UK.

Dunlap, R.E. and Saad, L. (2001) Only one in four Americans are anxious about the environment. *Gallup Poll Monthly* 427, 6.

Dunlap, R.E. and Van Liere, K. (1978) The 'new environmental paradigm': a proposed instrument and preliminary results. *Journal of Environmental Education* 9, 10–19.

Dunlap, R.E. and Van Liere, K. (1978) The new environmental paradigm. *Journal of Environmental Education* 9, 10–19.

Ecological Society of America (1995) *The Scientific Basis for Ecosystem Management*. The Ecological Society of America, Washington, DC.

Eisler, R.T. (1987) *The Chalice and the Blade: Our History, Our Future*. Harper and Row, Cambridge, Massachusetts.

Elridge, A.F. (1979) *Images of Conflict*. St Martin's Press, New York.

El-Swaify, S.A. and Yakowitz, D.S. (eds) (1997) Multiple objective decision-making for land, water, and environmental management. *Proceedings of the First International Conference on Multiple Objective Decision Support Systems (MODSS) for Land, Water, and Environmental Management: Concepts, Approaches, and Applications.* Honolulu, Hawaii.

Ely, R.T. (1914) *Property and Contract.* Macmillan, New York.

Emel, J. and Brooks, E. (1988) Changes in form and function of property rights institutions under threatened resource scarcity. *Annals of the Association of American Geographers* 78(2), 241–252.

Essex, C. and McKitrick, R. (2002) *Taken by Storm: the Troubled Science, Policy and Politics of Global Warming.* Key Porter Books, Toronto, Ontario, Canada.

Estrin, D. and Swaigen, J. (1978) *Environment on Trial.* Canadian Environmental Law Research Foundation, Toronto, Ontario, Canada.

Ewert, A. (1990) Wildland resource values: a struggle for balance. *Society and Natural Resources* 3, 385–393.

Ewert, A. (1991) Outdoor recreation and global climate change: resource management implications for behaviors, planning, and management. *Society and Natural Resources* 4, 365–377.

Ewert, A. (1996) Gateways to adventure tourism: the economic impacts of mountaineering on one portal community. *Tourism Analysis* 1, 59–63.

Ewert, A. (ed.) (1996) *Natural Resource Management: the Human Dimension.* Westview Press, Boulder, Colorado.

Ewert, A. (2002) Quality of life, recreation, and natural environments: exploring the connection. In: Rapport, D.J., Lasley, W., Rolston, D., Nielson, O., Qualset, C. and Damania, A. (eds) *Managing for Healthy Ecosystems.* CRC Press, Boca Raton, Florida, pp. 199–205.

Ewert, A. and Baker, D. (2001) Standing for where you sit: an exploratory analysis of the relationships between academic major and environmental beliefs. *Environment and Behavior* 33(5), 687–707.

Ewert, A. and Pfister, R. (1991) Cross-cultural land ethics: motivations, appealing attributes and problems. *Transactions of the North American and National Resource Conference* 56, 146–151.

Ewert, A. and Williams, G.W. (1994, September) *Getting Alice Through the Door: Social Science and Natural Resources Management.* Presented at the Social Dimensions session at the meeting of the Society of American Foresters National Convention, Anchorage, Alaska.

Ewert, A., Chavez, D. and Magill, A. (1993) *Culture, Conflict, and Communication in the Wildland Urban Interface.* Westview Press, Boulder, Colorado.

Farley, R. and Allen, W. (1987) *The Color Line and the Quality of Life in America.* Russell Sage Foundation, New York.

Feldman, D. (1991) International decision-making for global climate change. *Society and Natural Resources* 4, 379–396.

Field, D.R. (1996) Social science: a lesson in legitimacy, power and politics in land management agencies. In: Ewert, A. (ed.) *Natural Resource Management.* Westview Press, Boulder, Colorado, pp. 249–255.

Field, D.R. and Burch, W.R. (1990) Social science and forestry. *Society and Natural Resources* 3, 187–191.

Filley, A. (1975) *Interpersonal Conflict Resolution.* Scott, Foresman and Company, Glenview, Illinois.

Fine Jenkins, A. (1997) Forest health: a crisis of human proportions. *Journal of Forestry* 95(9), 11–14.

Fiske, S.J. (1990, Winter) Resource management as people management: anthropology and renewable resources. *Renewable Resources Journal* 16–20.

Flanders, N.E. (1998) Native American sovereignty and natural resource management. *Human Ecology* 26(3), 425–449.

Floyd, M. (1999) Race, ethnicity, and use of the National Park System. *Social Science Review* 1(2), 1–23.

Force, J.E. and Machlis, G.E. (1997) The human ecosystem. Part II: social indicators in ecosystem management. *Society and Natural Resources* 10, 369–382.

Fosler, R., Alonso, W., Meyer, J. and Kern, R. (1990) *Demographic Change and the American Future*. University of Pittsburgh Press, Pittsburgh, Pennsylvania.

Friedmann, J. (1987) *Planning in the Public Domain: From Knowledge to Action*. Princeton University Press, Princeton, New Jersey.

Frissell, C.A. and Bayles, D. (1996) Ecosystem management and the conservation of aquatic biodiversity and ecological integrity. *Water Resources Bulletin* 32(2), 229–240.

Gamble, D.J. (1978, March) The Berger Inquiry: an impact assessment process. *Science* 19, 946–952.

Gardiner, J., Thompson, K. and Newson, M. (1994) Integrated watershed/river catchment planning and management: a comparison of selected Canadian and United Kingdom experiences. *Journal of Environmental Planning and Management* 37(1), 53–66.

Garreau, J. (1991) *Edge City: Life on the New Frontier*. Doubleday, New York.

Garrett, W.E. (1989, October) La Ruta Maya. *National Geographic* 176(4), 424–479.

Geller, J.M. and Lasley, P. (1985) The new environmental paradigm scale: a reexamination. *Journal of Environmental Education* 17, 9–12.

Ginther, K., Denters, E. and De Waart, P.J.I.M. (eds) (1995) *Sustainable Development and Good Governance*. Martinus Nijhoff, Dordrecht, The Netherlands.

Glasbergen, P. (1995) *Managing Environmental Disputes*. Kluwer Academic Publishers, Dordrecht, The Netherlands.

Glaser, B.G. (1978) *Theoretical Sensitivity*. Sociology Press, Mill Valley, California.

Glaser, B.G. and Strauss, A.L. (1967) *The Discovery of Grounded Theory: Strategies for Qualitative Research*. Aldine, New York.

Glasscock, R. (ed.) (1992) *Historic Landscapes of Britain from the Air*. Cambridge University Press, Cambridge, UK.

Glavovic, B., Dukes, E. and Lynott, J. (1997) Training and educating environmental mediators: lessons from experience in the United States. *Mediation Quarterly* 14(4), 269–291.

Goodwin, R.F. (1999, April–September) Redeveloping deteriorated urban waterfronts: the effectiveness of US coastal management programs. *Coastal Management* 27(2–3), 239–269.

Hall, A.J. (1975, December) A traveller's tale of ancient Tikal. *National Geographic* 148(6), 799–811.

Hall, P., Land, H., Parker, R. and Webb, A. (1972) *Change, Choice and Conflict in Social Policy*. Heinemann, London.

Hammitt, W.E. (2000) The relation between being away and privacy in urban forest recreation environments. *Environment and Behavior* 32(4), 521–540.

Hardin, G. (1968) The tragedy of the Commons. *Science* 162, 1243–1248.

Harriman, J. and Baker, D. (2003) Applying integrated resource and environmental management to transmission right of way maintenance. *Journal of Environmental Planning and Management* 46(2), 199–217.

Harrison, M.L. (ed.) (1984) *Corporatism and the Welfare State*. Gower, Aldershot, UK.

Harrison, M.L. (1987) Property rights, philosophies, and the justification of planning control. In: Harrison, M.L. and Mordey, R. (eds) *Planning Control: Philosophies, Prospects, and Practice*. Croom Helm, London, pp. 32–58.

Hatcher, A., Jaffry, S., Thebaud, O. and Bennett, E. (2000) Normative and social influences affecting compliance with fisheries regulation. *Land Economics* 76, 448–461.

Heberlein, T.A. (1972) The land ethic realized: some social psychological explanations for changing environmental attitudes. *Journal of Social Issues* 28, 79–87.

Heberlein, T.A. (1988) Improving interdisciplinary research: integrating the social and natural sciences. *Society and Natural Resources* 1, 5–16.

Heinrichs, J. (1991, March/April) The future of fun. *American Forests* 21–24, 73–74.

Hetherington, J., Daniel, T.C. and Brown, T.C. (1994) Anything goes means everything stays: the perils of uncritical pluralism in the study of ecosystem values. *Society and Natural Resources* 7(6), 535–546.

Himes, J.S. (1980) *Conflict and Conflict Management*. University of Georgia Press, Athens, Georgia.

Hodgkinson, S.P. and Innes, J.M. (2001) The attitudinal influence of career orientation in 1st-year university students: environmental attitudes as a function of degree choice. *Journal of Environmental Education* 32(3), 37–40.

Holden, C. (1988) The ecosystem and human behavior. *Science* 242, 663.

Holling, C.S. (1986) The resilience of terrestrial ecosystems: local surprise and global change. In: Clark, W.C. and Munn, R.E. (eds) *Sustainable Development of the Biosphere*. Cambridge University Press, Cambridge, UK, pp. 292–317.

Hollingshead, D. (1992) 'White'gaze, 'red' people – shadow visions: the dis-identification of 'Indians' in cultural tourism. *Leisure Studies* 11, 43–64.

Hooper, B., McDonald, G. and Mitchell, B. (1999) Facilitating integrated resource and environmental management: Australian and Canadian perspectives. *Journal of Environmental Planning and Management* 42(5), 747–766.

Horn, B., Agpaoa, L., Bailey, J., Chambers, V., Kissinger, J., McMenus, K., Morris, G., Smith, R. and Zwang, C. (1993) *Strengthening Public Involvement: a National Model for Building Long-term Relationships with the Public*. USDA Forest Service, Washington, DC.

Howlett, M. (2002) Policy instruments and implementation styles: the evolution of instrument choice in Canadian environmental policy. In: Van Nijnatten, D.L. and Boardman, R. (eds) *Canadian Environmental Policy: Context and Cases*, 2nd edn. Oxford University Press, Don Mills, Ontario, Canada, pp. 25–45

Humphrey, S., Burbridge, P. and Blatch, C. (2000) US lessons for coastal management in the European Union. *Marine Policy* 24(4), 275–286.

Hutchings, J.A., Walters, C. and Haedrich, R.L. (1997) Is scientific inquiry incompatible with government information control? *Canandian Journal of Fisheries and Aquatic Sciences* 54, 1198–1210.

Hutton, W. (2002, April 28) Log cabin to White House? Not any more. *Observer*. Retrieved 28 April, 2002, from http://www.observer.co.uk/comment/story/0,6903,706484,00.html

Janicke, M. (1990) *State Failure*. Pennsylvania State Press, University Park, Pennsylvania.

Jentoft, S. (2000) Legitimacy and disappointment in fisheries management. *Marine Policy* 24, 141–148.

Johnson, D.W. and Johnson, F.P. (2000) *Joining Together: Group Theory and Group Skills*, 7th edn. Prentice Hall, Englewood Cliffs, New Jersey.

Johnson, R.S. (1986) *Forests of Nova Scotia: a History*. Halifax Department of Lands and Forests/Four East Publications Halifax, Nova Scotia, Canada.

Jones, E.L. (1998) From steel town to 'ghosttown': a qualitative study of community change in southeast Chicago. Unpublished master's thesis, Loyola University, Chicago.

Kahaner, L. (1988) Something in the air. *Wilderness* 52(183), 18–27.

Kaiser, F.G., Ranney, M., Hartig, T. and Bowler, P.A. (1999) Ecological behavior, environmental attitude, and feelings of responsibility for the environment. *European Psychologist* 4(2), 59–74.

Kane, H.K. (1974, December) The pathfinders. *National Geographic* 146(6), 758–759.

Kaplan, R. and Kaplan, S. (1989) *The Experience of Nature: a Psychological Perspective*. Cambridge University Press, New York.

Kates, R.W. and Clark, W.C. (1996) Expecting the environmental surprise. *Environment* 38(2), 28–34.

Keenan, S.P., Krannich, R.S. and Walker, M.S. (1999) Public perceptions of water transfers and markets: describing differences in water use communities. *Society and Natural Resources* 12, 279–292.

Kelly, J. (1989) Leisure behaviors and styles: social, economic, and cultural factors. In: Jackson, E. and Burton, T. (eds) *Understanding Leisure and Recreation: Mapping the Past, Charting the Future*. Venture, State College, Pennsylvania, pp. 89–112.

Kelsey, E., Nightingale, J. and Solin, M. (1995) The role of partnerships in implementing a new marine protected area: a case study of Whytecliff Park. In: Shakell, N.L. and Willison, J.H.M. (eds) *Marine Protected Areas and Sustainable Fisheries*. Science and Protected Areas Association, Wolfville, Nova Scotia, Canada, pp.151–164.

Kempton, W. (1997) How the public views climate change. *Environment* 39(9), 12–21.

Kessler, W.B., Salwasser, H., Cartwright, C.W. and Caplan, J.A. (1992) New perspectives for sustainable natural resources management. *Ecological Applications* 2(3), 221–225.

Ketchington, R. and Crawford, D. (1993) On the meaning of integration in coastal zone management. *Ocean and Coastal Management* 21(2), 109–127.

Kimmins, H. (1995, March/April) Clear-cutting: a long history to controversial practice. *BC Professional Forester* 2, 21–22.

King, M., Elliott, C., Hellberg, H., Lilford, R., Martin, J., Rock, E. and Mwenda, J. (1995) Does demographic entrapment challenge the two-child paradigm? *Health Policy and Planning* 10, 376–383.

Kline, J.D. and Armstrong, C. (2001) Autopsy of a forestry ballot initiative. *Journal of Forestry* 99(5), 20–27.

Knopp, T. (1972) Environmental determinants of recreation behavior. *Journal of Leisure Research* 4, 129–138.

Kolb, W.E., Wagner, M.R. and Covington, W.W. (1994) Concepts of forest health: utilitarian and ecosystem perspectives. *Journal of Forestry* 91(9), 32–37.

Kretzmann, J.P. and McKnight, J.L. (1993) *Building Communities from the Inside Out: a Path Toward Finding and Mobilizing a Community's Assets*. Asset-based Community Development Institute, Northwestern University, Evanston, Illinois.

Lang, R. (1986) Achieving integration in resource planning. In: Lang, R. (ed.) *Integrated Approaches to Resource Planning and Management*. University of Calgary Press, The Banff Centre School of Management, Calgary, Alberta, Canada, pp. 27–50.

Laue, J. (1987) The emergence and institutionalization of third party roles. In: Sandole, D. and Sandole-Staroste, I. (eds) *Conflict Management and Problem Solving: Interpersonal to International Applications*. Frances Pinter. London, pp. 17–29.

Lean, J. and Rind, D. (1996) The sun and climate. *Consequences* 2(1), 27–36.

Lee, K. (1994) *Compass and Gyroscope: Integrating Science and Politics for the Environment*. Island Press, Washington, DC.

Leopold, A. (1949) *A Sand County Almanac*. Ballantine Books, New York.

Lessinger, J. (1987) The emerging region of opportunity. *American Demographics* 9(6), 33–37, 66–68.

Lester, J.P., Allen, D.W. and Hill, K.M. (2001) *Environmental Injustice in the United States: Myths and Realities*. Westview Press, Boulder, Colorado.

Lindblom, C.E. (1980) *The Policy Making Process*, 2nd edn. Prentice-Hall, New York.

London, J. (1908) To build a fire. *Century Magazine* 76, 525–534.

Long, L. (1988) *Migration and Residential Mobility in the United States*. Russell Sage Foundation, New York.

Mabry, M. (2002, March 11) South Africa is not Zimbabwe: the two countries have similar pasts, but crucial differences. *Newsweek*, p. 13.

Machlis, G. (1992) The contribution of sociology to biodiversity research and management. *Biological Conservation* 62, 161–170.

Machlis, G. (1999) New forestry, neopolitics, and voodoo economics: research needs for biodiversity management. In: Aley, J., Burch, W., Conover, B. and Field, D. (eds) *Ecosystem Management: Adaptive Strategies for Natural Resources Organizations in the 21st Century*. Taylor and Francis, Philadelphia, Pennsylvania, pp. 5–16.

Macpherson, C.B. (1978) *Property: Mainstream and Critical Positions*. University of Toronto Press, Toronto, Ontario, Canada.

Macridis, R.C. (1986) *Contemporary Political Ideologies: Movements and Regimes*, 3rd edn. Little, Brown and Company, Canada.

Maloney, M. and Ward, M. (1973) Ecology: let's hear from the people. *American Psychologist* 28, 583–586.

Mandell, M.P. (1989) Organizational networking: collective organizational strategies. In: Rabin, J., Miller, G.J. and Hildreth, W.B. (eds) *Handbook of Strategic Management*. Marcel Dekker, New York, pp. 141–165.

Mannion, A.M. (1991) *Global Environmental Change: a Natural and Cultural Environmental History*. Longman Scientific and Technical, London.

Marchak, P., Guppy, N. and McMullan, J. (eds) (1989) *Uncommon Property: the Fishing and Fish-processing Industries in British Columbia*. Methuen, Toronto, Ontario, Canada.

Marcin, T. (1993) Demographic change: implications for forest management. *Journal of Forestry* 91(11), 39–45.

Margerum, R.D. (1997) Integrated approaches to environmental planning and management. *Journal of Planning Literature* 11(4), 459–475.

Margerum, R.D. and Born, S. (1995) Integrated environmental management: moving from theory to practice. *Journal of Environmental Planning and Management* 38(3), 371–391.

Marin, G. and Marin, B. (1991) *Research with Hispanic Populations*. Sage, Newbury Park, California.

Maslund, T. and Newton, K. (2002a, March 11) The grievance of all grievances. *Newsweek* 139(10), p. 36.

Maslund, T. and Newton K.N. (2002b, March 25) Betting the farms: Mugabe stole Zimbabwe's election. Now his supporters are out to prove that his land grabs can yield economic success. *Newsweek* 53.

May, E. (1998) *At the Cutting Edge: the Crisis in Canada's Forests*. Sierra Club of Canada.

Mayntz, R. (1983) The conditions of effective public policy: a new challenge for policy analysis. *Policy and Politics* 11(2), 123–143.

Mazmanian, D.A. and Kraft, M.E. (1999) The three epochs of the environmental movement. In: Mazmanaian, D.A. and Kraft, M.E. (eds) *Toward Sustainable*

Communities: Transition and Transformations in Environmental Policy. MIT Press, Cambridge, Massachusetts, pp. 3–43.

McDowell, B. (1980, December) The Aztecs. *National Geographic* 158(6), 704–751.

McEnery, J.H. (1985) Toward a new concept of conflict evaluation. *Conflict* 6(1), 37–72.

McFarland, A.S. (1969) *Power and Leadership in Pluralist Systems*. Stanford University Press, Stanford, California.

McGrew, A.G. and Wilson, M.J. (eds) (1982) *Decision-making: Approaches and Analysis*. Manchester University Press, Manchester, UK.

McKibben, B. (1989) *The End of Nature*. Random House, New York.

McMichael, A.J. (1997) Global environmental change and human health: impact assessment, population vulnerability, and research priorities. *Ecosystem Health* 3(4), 200–210.

McMichael, A.J. (2001) *Human Frontiers, Environments and Disease: Past Patterns, Uncertain Futures*. Cambridge University Press, Cambridge, UK.

Meyer, J.L. and Swank, W.T. (1996) Ecosystem management challenges ecologists. *Ecological Applications* 6(3), 738–740.

Miller, G. (1992) *An Introduction to Environmental Science: Living in the Environment*. Wadsworth Publishing Company, Belmont, California.

Mills, C.W. (1959) *The Power Elite*. Oxford University Press, New York.

Minty, C.D., Sutton, D.A. and Rogers, A.D.F. (2001) *A Wildlife Impact Assessment for the Proposed Macal River Upper Storage Facility*. Natural History Museum, London.

Mitchell, B. (1979) *Geography and Resource Analysis*. Longman, London.

Mitchell, B. (1986) The evolution of integrated resource management. In: Lang, R. (ed.) *Integrated Approaches to Resource Planning and Management*. University of Calgary Press, The Banff Centre School of Management, Calgary, Alberta, Canada, pp. 13–26.

Mitchell, B. (1990) Integrated water management. In: Mitchell, B. (ed.) *Integrated Water Management: International Experiences and Perspectives*. Belhaven, London, pp. 1–21.

Mitchell, B. and Hollick, M. (1993) Integrated catchment management in Western Australia: transition from concept to implementation. *Environmental Management* 17(6), 735–743.

Mitchell, B. and Pigram, J. (1989) Integrated resource management and the Hunter Valley Conservation Trust, NSW, Australia. *Applied Geography* 9, 196–211.

Mitchell, B. and Sewell, D. (eds) (1981) *Canadian Resource Policies: Problems and Prospects*. Methuen, Toronto, Ontario, Canada.

Mitchell River Watershed Management Group (2000) *Mitchell River Watershed Management Plan*. Mitchell River Watershed Management Group, Mareeba, Queensland, Australia.

Mol, P.J. (2001) *Globalization and Environmental Reform: the Ecological Modernization of the Global Economy*. MIT Press, Cambridge, Massachusetts.

Mulvihill, P.R. and Baker, D.C. (2001) Ambitious and restrictive scoping: case studies from Northern Canada. *Environmental Impact Assessment Review* 21, 363–384.

Mungall, C. and McLaren, D.J. (eds) (1990) *Planet Under Stress: the Challenge of Global Change*. Oxford University Press, Toronto, Ontario, Canada.

Murdock, S. and Ellis, D. (1991) *Applied Demography: an Introduction to Basic Concepts, Methods, and Data*. Westview Press, Boulder, Colorado.

Nash, R.F. (1989) *The Rights of Nature*. University of Wisconsin Press, Madison, Wisconsin.

Nassauer, J.I. (1997) Cultural sustainability: aligning aesthetics and ecology. In: Nassauer, J.I. (ed.) *Placing Nature: Culture and Landscape Ecology*. Island Press, Washington, DC, pp. 65–84

Nelson, L. and Weschler, L. (1998) Institutional readiness for integrated watershed management: the case of the Maumee River. *Social Science Journal* 35(4), 565–577.

Neumann, R. (2001) *Resolving Environmental Disputes: Principles and Practice Study Guide*. Australian School of Environmental Studies, Griffith University, Brisbane, Queensland, Australia.

Nord, M., Luloff, A.E. and Bridger, J.C. (1998) The association of forest recreation with environmentalism. *Environment and Behavior* 30(2), 235–246.

O'Connor, D.R. (2002) *Report of the Walkerton Inquiry: the Events of May 2000 and Related Issues*. Publications Ontario, Toronto, Ontario, Canada.

O'Keefe, D. (1990) *Persuasion: Theory and Research*. Sage, Newbury Park, California.

Oltmann, R. (1997) *My Valley: the Kananaskis*. Rocky Mountain Books, Calgary, Alberta, Canada.

Orians, G.H. (1995) Cumulative threats to the environment. *Environment* 37(7), 6–36.

O'Riordan, T. (1981) *Environmentalism*, 2nd edn. Pion, London.

O'Riordan, T. (1989) The challenge for environmentalism. In: Peet, R. and Thrift, N. (eds) *New Models in Geography*. Unwin Hyman, London, p. 1.

O'Riordan, T., Clark, W.C., Kates, R.W. and McGowan, A. (1995) The legacy of Earth Day: reflections at the turning point. *Environment* 37(3), 7–15, 37–42.

O'Toole, L.J., Jr and Montjoy, R.S. (1984, November/December) Interorganizational policy implementation: a theoretical perspective. *Public Policy Review* 44(6), 491–503.

Ottesen, P. and Kenchington, R. (1995) Marine conservation and protected areas in Australia: what is the future? In: Shackell, N.L. and Willison, J.H.M. (eds) *Marine Protected Areas and Sustainable Fisheries*. Science and Management of Protected Areas Association, Wolfville, Nova Scotia, Canada, pp. 151–164.

Ovington, J.D. (1965) *The Role of Forestry: an Inaugural Lecture*. The Australian National University, Canberra, New South Wales, Australia.

Palacin, P.C. (1992) Spain. In: Wibe, S. and Jones, T. (eds) *Forests – Market and Intervention Failures: Five Case Studies*. Earthscan, London, pp. 165–200.

Parkes, M. and Panelli, R. (2001) Integrating catchment ecosystems and community health: the value of participatory action research. *Ecosystem Health* 7(2), 85–106.

Patterson, M.E. and Williams, D.R. (1998) Paradigms and problems: the practice of social science in natural resource management. *Society and Natural Resources* 11, 279–295.

Pearce, D.W. (2001) The economic value of forest ecosystems. *Ecosystem Health* 7(4), 284–295.

Petty, R., McMichael, S. and Brannon, L. (1992) The elaboration likelihood model of persuasion: applications in recreation and tourism. In: Mandredo, M. (ed.) *Influencing Human Behavior*. Sagamore, Champaign, Illinois, pp. 77–102.

Pfeil, R.W. and Ellis, J.W. (1995, April 30) Evaluating GIS for establishing and monitoring environmental conditions of oil fields. *American Association of Petroleum Geologists Bulletin* 79(4), 595.

Pirages, D.C. and Ehrlich, P.R. (1974) *Ark 2: Social Response to Environmental Imperatives*. W.H. Freeman, San Francisco.

Place, G.S. (2000) The impact of early life outdoor experiences on an individual's environmental attitudes. Unpublished doctoral dissertation, Indiana University, Bloomington, Indiana.

Polsby, N. (1980) *Community Power and Political Theory*. Yale University Press, New Haven, Connecticut.

Portney, K.E. (1992) *Controversial Issues in Environmental Policy: Science vs. Economics vs. Politics*. Sage, Newbury Park, California.

Postel, S. (1992) Denial in the decisive decade. In: Brown, L. (ed.) *State of the World*. W.W. Norton and Company, New York, pp. 3–8.

Rabe, B.G. (1999) Sustainability in a regional context: the case of the Great Lakes Basin. In: Mazmanian, D.A. and Kraft, M.E. (eds) *Toward Sustainable Communities: Transition and Transformations in Environmental Policy*. MIT Press, Cambridge, Massachusetts, pp. 247–281.

Radeloff, V.C. (2000) Exploring the spatial relationship between census and land-cover data. *Society and Natural Resources* 13(6), 599–612.

Rafson, H.J. and Rafson, R.N. (eds) (1999) *Brownfields: Redeveloping Environmentally Distressed Properties*. McGraw-Hill, New York.

Randall, A. (1987) *Resource Economics: an Economic Approach to Natural Resource and Environmental Policy*, 2nd edn. John Wiley & Sons, New York.

Rapport, D.J., Howard, J.M., Lannigan, R., Anjema, C.M. and McCauley, W. (2001) Strange bed fellows: ecosystem health in the medical curriculum. *Ecosystem Health* 7(3), 155–162.

Rapport, D.J., Lasley, W., Rolston, D., Nielson, O., Qualset, C. and Damania, A. (2002) *Managing for Healthy Ecosystems*. CRC Press, Boca Raton, Florida.

Raven, P.H., Berg, L.R. and Johnson, G.B. (1993) *Environment*. Saunders College, New York.

Rees, J.A. (1985) *Natural Resources: Allocation, Economics and Policy*. Methuen, London.

Rees, J.A. (1990) *Natural Resources: Allocation, Economics and Policy*, 2nd edn. Methuen, London.

Reeves, H. (2002, 13 March) Metro briefing New York: Syracuse: Indian payment stopped for now. *New York Times*, p. B8.

Rego, F.C. and Coelho-Silva, J.L. (2002) Rural change and resource management in Mediterranean mountain areas: the case of Serra da Malcata, Central East of Portugal. In: Ewert, A., Voight, A., McLean, D., Hronek, B. and Beilfuss, G. (eds) *Proceedings of the 9th International Symposium on Society and Resource Management*. Indiana University, Bloomington, Indiana, pp. 217–218.

Reidel, C. (1992) Asking the right questions. *Journal of Forestry* 90(10), 14–19.

Reynolds, K.M. (2001) Using a logic framework to assess forest ecosystem sustainability. *Journal of Forestry* 99(6), 26–30.

Richardson, E. (1973) *Dams, Parks and Politics*. University of Kentucky Press, Lexington, Kentucky.

Ross, M. and Bissix, G. (2000) Extending Turner's spectrum of sustainable development typologies: application to global and Canadian forest management practices. *Proceedings of the 8th International Symposium on Society and Resource Management*. University of Western Washington, Bellingham, Washington.

Roston, H. (1985) Valuing wildlands. *Environmental Ethics* 7, 23–48.

Rowley, T., Gallopin, G., Waltner-Toews, D. and Raez-Luna, E. (1997) Development and application of an integrated conceptual framework to tropical agroecosystems based on complex systems theories. *Ecosystem Health* 3(3), 154–161.

Ryan, A. (1984) *Property and Political Theory*. Basil Blackwell, Oxford, UK.

Ryan, A. (1987) *Property*. Open University Press, Milton Keynes, UK.

Saegert, S. and Winkel, G.H. (1990) Environmental psychology. *Annual Review of Psychology* 41, 441–477.

Sandbach, F. (1980) *Environment, Ideology and Policy*. Blackwell, Oxford, UK.
Sandell, K. (1998) The public access dilemma: the specialization of landscape and the challenge of sustainability in outdoor recreation. In: Sandberg, L.A. and Sörlin, S. (eds) *Sustainability: the Challenge: People, Power and the Environment*. Black Rose Books, Buffalo, New York, pp. 121–129.
Sanger, D.E. and Alvarez, L. (2001, June 29) Conservation-mindful Bush turns to energy research. *New York Times*, p. A18.
Satterfield, T. and Gregory, R. (1998) Reconciling environmental values and pragmatic choices. *Society and Natural Resources* 11, 629–647.
Schellenberg, J.A. (1982) *The Science of Conflict*. Oxford University Press, Oxford, UK.
Schilling, M. and Schulz, M. (1998, March) Improving the organisation of evironmental management: ecosystem management, external interdependencies, and agency structures. *Public Productivity and Management Review* 21(3), 293–308.
Schmid, A.A. (1978) *Property, Power, and Public Choice: an Inquiry into Law and Economics*. Praeger, New York.
Schneider, S.H. (1996, Summer) Engineering change in global climate. *Forum for Applied Research and Public Policy* 11(2), 92–97.
Schultz, P.W., Zelezny, L. and Dalrymple, N.J. (2000) A multinational perspective on the relation between Judeo-Christian religious beliefs and attitudes of environmental concern. *Environment and Behavior* 32(4), 576–591.
Scott, A. (1983) Property rights and property wrongs. *Canadian Journal of Economics* 16(4), 555–573.
Scott, A. and Johnson, J. (1983) *Property Rights: Developing the Characteristics of Interests in Natural Resources* (Resource Paper No. 88). University of British Columbia, Department of Economics, Vancouver, British Columbia, Canada.
Scott, D. (1993) Time scarcity and its implications for leisure behavior and leisure delivery. *Journal of Park and Recreation Administration* 11(3), 51–60.
Seib, G.F. (2001, 2 May) We want energy! We want green! We can't decide. *Wall Street Journal*, p. A34.
Shands, W.E. (1999) Leadership in a community. In: Aley, J., Burch, W.R., Conover, B. and Field, D.R. (eds) *Ecosystem Management: Adaptive Strategies for Natural Resources Organizations in the Twenty-first Century*. Taylor and Francis, Philadelphia, pp. 117–123.
Sherry, E. and Myers, H. (2002) Traditional environmental knowledge in practice. *Society and Natural Resources* 15, 345–358.
Shrader-Frechette, K. (1987) Four land ethics: an overview. *Environmental Professional* 9, 121–132.
Shurts, J. (2000) *Indian Reserved Water Rights*. University of Oklahoma Press, Norman, Oklahoma.
Siegel, J. and Taeuber, C. (1986) Demographic perspectives on the long-lived society. *Daedalus* 115, 77–177.
Silvern, S.E. (2000) Reclaiming the reservation: the geopolitics of Wisconsin Anishinaabe resource rights. *American Indian Culture and Research Journal* 24(3), 131.
Simmel, G. (1955) *Conflict*. Free Press, Glencoe, Illinois.
Simmons, I.G. (1991) *Earth, Air and Water: Resources and Environment in the Late 20th Century*. Edward Arnold, London.
Simmons, I.G. (1996) *Changing the Face of the Earth: Culture, Environment, History*, 2nd edn. Blackwell, Oxford, UK.
Simon, H.A. (1947) *Administrative Behavior*. MacMillan, New York.
Sirmon, J.M. (1991) Evolving concepts of leadership: towards a sustainable future. *Forest Perspectives* 1(2), 8–9.

Slocombe, D.S. (1998) Defining goals and criteria for ecosystem-based management. *Environmental Management* 22(4), 483–493.

Smith, L.G. (1993) *Impact Assessment and Sustainable Resource Management.* Longman Scientific and Technical–John Wiley & Sons, New York.

Spencer, G. (1989) *Projections of the Population of the United States, by Age, Sex, and Race: 1983 to 2080* (Current Population Reports-Series P-25–1018). US Bureau of the Census, US Government Printing Office, Washington, DC.

Spencer, R., Kelly, J. and Van Es, J. (1992) Residence and orientations toward solitude. *Leisure Sciences* 14(1), 69–78.

Spitler, G. (1988) Seeking common ground for environmental ethics. *Environmental Professional* 10, 1–7.

Squires, M.T. (1997, October) An investigation of the effectiveness of the multiple-use concept using Kananaskis Country, Alberta, as a case study. Unpublished honour's thesis, Acadia University, Acadia, Nova Scotia, Canada.

Stankey, G.H., Bormann, B.T., Ryan, C., Shindler, B., Sturtevant, V., Clark, R.N. *et al.* (2003) Adaptive management and the Northwest forest plan. *Journal of Forestry* 101(1), 40–46.

Stein, T.V., Anderson, D.H. and Kelly, T. (1999) Using stakeholders' values to apply ecosystem management in an upper Midwest landscape. *Environmental Management* 24(3), 399–413.

Stern, P.C., Dietz, T., Abel, T., Guagnano, G.A. and Kalof, L. (1999) A value–belief–norm theory of support for social movements: the case of environmentalism. *Human Ecology Review* 6(2), 81–97.

Struglia, R. and Winter, P.L. (2002) The role of population projections in environmental management. *Environmental Management* 30(1), 13–23.

Stull, D.D. (1990) Reservation economic development in the era of self-determination. *American Anthropologist* 92(1), 206–211.

Susskind, L., Levy, P. and Thomas-Larmer, J. (2000) *Negotiating Environmental Agreements*. MIT-Harvard Disputes Program, Island Press, Washington, DC.

Sussman, G., Daynes, B.W. and West, J.P. (2002) *American Politics and the Environment*. Addison Wesley Longman, New York.

Syme, S., Butterworth, J. and Namcarrow, B. (1994) *National Whole Catchment Management. A Review and Analysis of Process* (Occasional Paper Series No. 01/94) Land and Water Resources, Research and Development Corporation, Canberra, Australia.

Tarrant, M.A. and Cordell, H.K. (2002) Amenity values of public and private forests: examining the value–attitude relationship. *Environmental Management* 30(5), 692–703.

Teeple, G. (1995) *Globalization and the Decline of Social Reform*. Garamond, Toronto, Ontario, Canada.

Tillet, G. (1991) *Resolving Conflicts – a Practical Approach*. Sydney University Press, Sydney, Australia.

Trainer, T. (1998) *Towards a Sustainable Economy: the Need for Fundamental Change.* Jon Carpenter, Oxford, UK.

Trumbull, W. (1999) State and county programs. In: Rafson, H.J. and Rafson, R.N. (eds) *Brownfields: Redeveloping Environmentally Distressed Properties*. McGraw-Hill, New York, pp. 102–108.

Turner, M.G., Gardner, R.H. and O'Neill, R.V. (2001) *Landscape Ecology in Theory and Practice: Pattern and Process*. Springer-Verlag, New York.

Turner, R.K. (ed.) (1993) *Sustainable Environmental Economics and Management: Principles and Practice*. Belhaven Press, London.

Tyler, T.R. (1990) *Why People Obey the Law*. Yale University Press, New Haven, Connecticut.

Usher, P. (1984) Property rights: the basis of wildlife management. *National and Regional Interests in the North: Third Annual Workshop on People, Resources, and the Environment North of 60°*. Canadian Arctic Resources Committee, Ottawa, Ontario, Canada.

Van Hise, C.R. (1910) *The Conservation of Natural Resources in the United States*. MacMillan, New York.

Van Liere, K. and Dunlap, R.E. (1980) The social bases of environmental concern: a review of hypotheses, explanations, and empirical evidence. *Public Opinion Quarterly* 44(1), 181–197.

Van Maaren, A. (1984) Forests and forestry in national life. In: Hemmel, F.C. (ed.) *Forest Policy: a Contribution to Resource Development*. Martinus Nijhoff/Dr W. Junk Publishers, The Hague, The Netherlands, pp. 1–19.

Virden, R.J. and Walker, G.J. (1999) Ethnic/racial and gender variations among meanings given to, and preferences for, the natural environment. *Leisure Sciences* 21, 219–239.

Wali, A., Darlow, G., Fialkowski, C., Tudor, M., del Campo, H. and Stotz, D. (2003) New methodologies for interdisciplinary research and action in an urban ecosystem in Chicago. *Conservation Ecology* 7(3). Retrieved 12 December, 2003 from http://www.consecol.org/vol7/iss3/art2

Walters, C. (1986) *Adaptive Management and Renewable Resources*. MacMillan, New York.

Walther, P. (1987) Against idealistic beliefs in the problem-solving capacities of integrated resource management. *Environmental Management* 11(4), 439–446.

Warne, G. (1981, January/March) A history of energy resource regulation in Alberta. *Journal of Canadian Petroleum Technology* 20(1), 33–34.

Weale, A. (1992) *The Politics of Pollution*. Manchester University Press, Manchester, UK.

Wehr, P. (1979) *Conflict Regulation*. Westview Press, Boulder, Colorado.

Weir, J. (ed.) (1987) *Dartmoor National Park*. Webb and Bower, Exeter, UK.

Westman, W. (1993) How much are nature's services worth? *Science* 197, 960–964.

White, L., Jr (1967) The historic roots of our ecological crisis. *Science* 55, 1203–1207.

Wilbanks, T.J. (1994, December) Sustainable development in geographic perspective. *Annals of the Association of American Geographers* 541.

Wildavsky, A. (1987) Choosing preferences by constructing institutions: a cultural theory of preference formation. *American Political Science Review* 81(1), 3–21.

Williams, D.R. and Patterson, M.E. (1996) Environmental meaning and ecosystem management: perspectives from environmental psychology and human geography. *Society and Natural Resources* 9, 507–521.

Wilson, E.O. (2002, February) The bottleneck. *Scientific American* 82–91.

Withnall, I.W. (1990) Geology and mining in the Mitchell River watershed. In: *Proceedings of the Mitchell River Watershed Management Conference*. Kowanyama Land and Natural Resource Management Office, Kowanyama, Queensland, Australia, pp. 14–32.

Wolosoff, S.E. and Endreny, T.A. (2002) Scientist and policy-maker response types and times in suburban watersheds. *Environmental Management* 29(6), 729–735.

Wondolleck, J.M. (1988) *Public Lands Conflict and Resolution: Managing National Forest Disputes*. Plenum Press, New York.

Wondolleck, J.M. (1991, December) *Natural Resource Management in the 1990s and Beyond: Problems and Opportunities*. Prepared for the Pinchot Institute's Project on Leadership in Natural Resources, Milford, Pennsylvania.

Wondolleck, J.M. (1992) Resource management in the 1990s. *Forest Perspectives* 2(2), 19–21.

World Commission on Environment and Development (WCED) (1987) *Our Common Future*. Oxford University Press, Oxford, UK.

Yun, J.M. (2002, April) Offsetting behavioral effects of the corporate average fuel economy standards. *Economic Inquiry* 40(2), 260–270.

Zaczek, I. (1998) *Ancient Ireland*. Collins and Brown, London.

Index